Analog Circuits and Signal Processing

Series Editors:

Mohammed Ismail, Dublin, USA
Mohamad Sawan, Montreal, Canada

More information about this series at http://www.springer.com/series/7381

Anis Kharisovich Gil'mutdinov
Pyotr Arkhipovich Ushakov • Reyad El-Khazali

Fractal Elements and their Applications

Ed. by A.Kh. Gil'mutdinov

Springer

Anis Kharisovich Gil'mutdinov
Kazan National Research
Technical University-KAI
(KNRTU-KAI)
Kazan, Russia

Pyotr Arkhipovich Ushakov
Kalashnikov State Technical University
Izhevsk, Russia

Reyad El-Khazali
Khalifa Univ. of Science, Tech. & Research
Abu Dhabi, United Arab Emirates

ISSN 1872-082X ISSN 2197-1854 (electronic)
Analog Circuits and Signal Processing
ISBN 978-3-319-45248-7 ISBN 978-3-319-45249-4 (eBook)
DOI 10.1007/978-3-319-45249-4

Library of Congress Control Number: 2017930413

Printed on acid-free paper

This Springer imprint is published by Springer Nature
The registered company is Springer International Publishing AG
The registered company address is: Gewerbestrasse 11, 6330 Cham, Switzerland

We dedicate this book
To the blessed memory
Of Rashid Shakirovich Nigmatullin,
The founder of the Kazan scientific school
Of fractional operators

Foreword

Today we can say with certainty that both scientists and engineers have widely recognized the need to use the fractals theory along with the theory of fractional integral-differential operators and fractal treatment to solve a variety of problems that emerge in various fields of modern science and technology. Terms "fractals" and "fractal" reflect the modern view of the physical nature of real objects and processes; this view was firmly established after publication of the pioneering works of B. Mandelbrot about fractal geometry of the Nature.

The "fractional operators" reflect the modern approach to mathematically describe and identify the properties of fractal objects and processes that used to be described by integer order differential equations. Today we cannot but describe these objects and processes using the non-integer (fractional) order equations that fill the gaps between equations of the first, the second, and other integer orders.

Despite the fact that the concept of fractional derivatives was known as early as the end of the seventeenth century, systematic use of fractional calculus in science and technology actually can be attributed to the time of appearance of the pioneering works by Rashid Shakirovich Nigmatullin, the founder and the scientific director of the Kazan scientific school of investigation and application of electrochemical converters of information (ECCI), and his students.

R. Sh. Nigmatullin was *The First* to physically implement fractional integration and differentiation (FID) operations on the basis of real elements (in particular, electrochemical ones). He also was the first to develop methods for synthesis of ladder-type resistive-capacitive and resistive-inductive circuits that implement these operations. He proposed a number of certain useful applications of such elements, in particular, to increase the resolution of oscillopolarographic spectrum. It was further widely used to develop the corresponding devices both in Russia and abroad.

R. Sh. Nigmatullin has also offered the block diagram of a computer meant for solving equations of linear, spherical, and cylindrical diffusion. The device is based on operational amplifiers with special RC two poles (fractional-order integrators and differentiators). He showed the possibility to use the semi-infinite RC cable in order to create some special functions out of the trigonometric. All these results were obtained and published within the period from 1962 to 1968.

Appearance of the following works has ultimately established interest in fractional calculus: well-known works of Oldham K.B. (e.g., Oldham KB, Spanier J. The Fractional Calculus. New York: Academic Press, 1974. 234 p.), of B.B. Mandelbrot (Mandelbrot B.B. Les Objects Fractals: Forme, Hasard et Dimension. Paris: Flammarion, 1975. 187 p.; Mandelbrot B.B. Fractals: Forme, Chance and Dimension. San-Francisco: Freeman, 1977. 365 p.; Mandelbrot B.B. The Fractals Geometry of Nature. NY: Freeman, 1982. 468 p.), and the fundamental book of S.G. Samko, A.A. Kilbas and O.I. Marichev (S.G. Samko, A.A. Kilbas, O.I. Marichev Fractional Order Integrals and Derivatives and Some of Their Applications. Minsk: Nauka I Tekhnika, 1987. 688 p.).

It took fractional calculus mathematics a historically short period to find application in various fields of science, such as classical and quantum physics, field theory, electrodynamics, solid state physics, fluid dynamics, turbulence, general chemistry, biology and medicine, stochastic analysis, nonlinear control theory, image processing, seismology, geology, and social sciences. Numerous scientific publications and monographs approve this fact. Here are just three of them. The first is the monograph by A.A. Potapov (Fractals in Radiophysics and Radar: Sample Topology. M.: Universitetskaya kniga, 2005. 848 p.). The second is the monograph by V.V. Uchaikin (The Method of Fractional Derivatives. Ulyanovsk: "Artichoke" publishers, 2008. 512 p.). Each monograph contains more than 1000 references. And the third one is the following fundamental book: The Modern Image Processing Techniques (authors: A.A. Potapov, Y.V. Gulyaev, S.A. Nikitov, A.A. Pakhomov, V.A. German)/Ed. by A.A. Potapov. M.: FIZMATLIT, 2008. 496 p.

Another indicator of great interest in fractional analysis and its applications is the fact that a variety of international conferences on these issues are conducted annually. For example, representative conferences "Fractional differentiation and its applications" (FDA' 02, FDA' 04, FDA' 06, FDA' 08, FDA' 12, FDA' 14, FDA' 16 etc.) were organized by specialized scientific magazines, such as *Chaos, Solutions and Fractals*, *Nelineyny mir* (Nonlinear world), and *Journal of Fractional Calculus and Applied Analysis* (Bulgaria).

However, the Russian science, not to mention the industry, demonstrates absolutely insufficient use of these concepts and of emerging opportunities to understand the Nature and acquire new knowledge, to create new methods and measurement tools, and to create better models of technical equipment. One reason for this is that there is not enough scientific-technical and especially educational literature that would reflect both theoretical understanding of fractional differentiation and integration operations and their hardware implementation along with practical use. That is why scientists and engineers do not have the required knowledge on fractional analysis as well in the field of design of fractional-order elements ("fractal elements") that would make it possible to physically implement fractional operators and other devices for information and signal processing.

This book is the result of systematized outcomes of theoretical and experimental research works of the authors. The book partly fills the mentioned gap. The book can be used to develop general engineering and special education courses along with the corresponding teaching materials. The aim is to actively introduce the concepts of fractal geometry and fractional analysis into the minds of the future engineering professionals and scientists who would be able to work at the production industry and in research laboratories and would be able to embody these ideas into new instruments, devices, and systems.

Chapter 1 summarizes the fundamentals of the fractals theory, fractal dimension, and scaling. The concept of fractal signals and some methods for their processing are given.

Chapter 2 provides the essential information from the fractional analysis theory. This information will further be used to describe fractional-order systems and to perform frequency domain analysis of circuits containing fractal elements (FE). This chapter introduces few examples of electrical and electrochemical engineering that exhibit fractional-order dynamics.

Chapter 3 introduces the concept of fractional elements (FEs) and gives their mathematical description. Several versions are given of known devices and electrical circuits of frequency-dependent fractional-order input impedances. The multilayer RC structure is substantiated as the base for creating FEs.

Chapters 4, 5, and 6 describe design, schematic, and fundamental techniques to implement FEs based on multilayer resistive-capacitive medium. Powerful capabilities to obtain the required parameters and characteristics of FEs by means of static and dynamic heterogeneous medium are shown.

Chapter 7 discloses physical effects that are used to create controlled resistors and capacitors. We can assume that the application of these effects to resistive and dielectric materials in multilayer resistive-capacitive structures will make it possible to create parametric and nonlinear FEs; and the latter will significantly expand the capabilities of these structures.

Chapter 8 provides an overview of FE applications for modeling, signal processing, designing control systems, hybrid computers, etc. Breadth of applications demands also a wide range of FE characteristics and parameters that can be implemented based on the multilayer RC medium. Therefore, the authors proposed a universal structural framework suitable for implementing FEs in various application areas. This structural framework contains seven alternating layers of resistive, dielectric, and conductive materials. The whole of these layers constitute a generalized virtual element.

Chapter 9 describes the technique of forming a system of partial differential equations for potential distribution in the resistive layers of the proposed virtual element. An example of FEs classification by resistive layers potential distribution is shown for the particular case of the fractal element with “resistor-insulator-resistor” layers structure.

Chapters 10, 11, and 12 describe in detail how to calculate the y-parameters of two-terminal and (in the general case) multi-terminal elements formed on the basis of RC multilayer medium which contains static and dynamic heterogeneities. These chapters are of particular importance for the practical implementation of FEs as long as this kind of problems had no satisfactory solutions till now. The authors used their proposed method of finite distributed elements to show that the external parameters of FEs can be calculated regardless of the complexity of the structure, the heterogeneous nature, and distribution in the RC medium. Several types of algorithms for calculating the *y*-parameters for different design implementations of FEs are proposed to the readers.

Practical exercises and test questions are given at the end of every chapter with the aim to consolidate the given material and to provide self-studying.

The authors suppose that this book sufficiently fulfills educational and innovative objectives in applying the ideas of fractal geometry and fractional analysis aimed to create fractal radio-electronics devices, communication systems, and for system identification and control of distributed and fractional-order processes.

Director of Kotel'nikov Institute
of Radioengineering and Electronics
of Russian Academy of Sciences,
Doctor of Sciences in Physics
and Mathematics
Moscow, Russia

S.A. Nikitov

Contents

Abbreviations

The abbreviations are given in the order of appearance in the text.

RC	Resistive-capacitive
LC	Inductance-capacitive
FE	Fractal element
3D	Three-dimensional
2D	Two-dimensional
GBM	Generalized Brownian motion
RL	Riemann–Liouville
GL	Grunwald–Letnikov
L-Transform	Laplace transform
CPE	Constant phase element
ELP	Element with lumped parameters
FOE	Fractional-order element
ECCI	Electrochemical converters of information
CFR	Complex frequency response
LMFR	Logarithmic magnitude frequency response
MFR	Magnitude frequency response
PFR	Phase frequency response
FID	Fractional integration and differentiation
PID	Proportional-integral-differential
EDP	Element with distributed parameters
TEM	Transmission electronic microscopy
PE	Polyethylene
OD	One-dimensional
TD	Two-dimensional
ODHom	One-dimensional homogeneous
ODHet	One-dimensional heterogeneous
C*RC*-EDP	Complementary *RC*-EDP
PLD	Programmable logic device
PAD	Programmable analog device

TRC	Temperature resistance coefficient
TCC	Temperature capacitance coefficient
VP	Vapor phase
OEC	Organoelement compound
IC	Integrated circuit
HIC	Hybrid integrated circuit
SOS	Silicon-on-sapphire
FOO	Fractal-order oscillator
AEC	Alumina electrolytic capacitors
GA	Genetic algorithm
DME	Distributed measurement environment
TDHet	Two-dimensional heterogeneous
TDHom	Two-dimensional homogeneous
FDM	Finite differences method
FEM	Finite elements method
FinE	Finite element
FDEM	Finite distributed elements method
ODC	One-dimensional complementary
GFDEM	Generalized finite distributed elements method
GFDE	Generalized finite distributed element
ODSHet	One-dimensional structure-heterogeneous
SHet	Structure-heterogeneous

Chapter 1
Modeling of Fractal Elements and Processes

When we take a glance at the world around us and see forests, trees, grass, rivers with confluents, and clouds high in the sky and mountain peaks, we can hardly believe that all these natural objects are easily described by smooth functions. As for artificial man-made objects, such as cubic buildings, smooth roads, and cars shape, they can more likely be described by such functions.

Of course, this is not a proof but it is quite enough to understand that the process of developing physical theories or sketches of natural phenomena can be performed with another style of graphics without smooth lines. Moreover, if such kind of sketches allow to reflect the described world in more accurate way and provide easier understanding of it and of the observable facts, then such models will be able to compete with conventional sketches of nature description based on smooth lines.

Benoit Mandelbrot was the one who established this new geometry and thus allowed to produce more precise models of natural objects and processes. In the 1970s, he introduced the concept of ***fractal***. This concept has exerted a strong influence on the forthcoming development of all areas of knowledge.

1.1 Concept of Fractals, Self-Similarity, and Scaling

Fractals are geometrical objects (lines, surfaces, spatial bodies) with strong angularity of shape that perform self-similarity. The founder of the fractals theory, mathematician Benoit Mandelbrot, formed the term "*fractal*" out of the Latin participle "*fractus*." The corresponding verb "*frangere*" is translated as "*to break*," "*to wreck*," "*to fracture*," i.e., it means to create irregular shape fragments.

It is a surprising fact that fractal shapes are prevailing in the Nature (coastline, mountain relief, a river, a tree etc.). There are fractal structures of substances as well as fractal structures of sets and random processes.

Fractal showings can be found in the structure of the detected signals and fields in the course of physical experimental studies. Fractal nature often declares

A.Kh. Gil'mutdinov et al., *Fractal Elements and their Applications*, Analog Circuits and Signal Processing, DOI 10.1007/978-3-319-45249-4_1

itself in behavior of functions that describe time and space distribution of physical quantities.

So, what is fractal? Scholars still do not have a unified definition of this phenomenon. However, when you see an object of fractal shape once, you will always be able to recognize it whatever shape it has. We can say that the concept of fractal significantly depends on intuitive understanding. Nevertheless, there are definitions.

In the simplest case, a fractal (a noun) is a special type of geometrical figure and "fractal" (an adjective) is an attribute of a structure, a phenomenon or a process with properties of fractal.

Mandelbrot proposed the following definition of fractal: "Fractal is a structure comprised of parts that, in some manner, are similar to the whole of this structure." In other words, *self-similarity* is one of the attributes of the fractal.

Self-similarity as the basic feature of the fractal means that the fractal has a more or less uniform structure in wide range of scales. For example, when zooming in, small fragments of the fractal look more and more like large fragments. This involves scale invariance (scaling) of the basic geometric features of the fractal object and their permanence at scale variation.

Self-similarity principle is well known (obvious examples are the Russian matryoshka dolls, paintings on Japanese kimono wear). In spite of this fact, self-similar structures were not used until recent time when they have been kind of rediscovered and have caused real explosion of theoretical and experimental activity.

Consider the following scaling definition of homogeneous functions:

$$\Phi_\mu(l) = \frac{1}{\Gamma(\mu)} l_+^{\mu-1}, \quad l_+ = \begin{cases} t, & t > 0, \\ 0, & t < 0, \end{cases} \tag{1.1}$$

where $\Gamma(\mu)$ is the gamma function, which obeys the uniformity condition.

It is known that any single, or multiple, variable function is homogenous if the following condition is fulfilled; if all of the functions' variables are multiplied by one and the same factor m at the same time, then the value of this function is multiplied by some power of this factor. Otherwise, if there is a power function $\Phi_\mu(t)$ and if the equality $\Phi_\mu(mt) = m^{\mu-1}\Phi_\mu(t)$ holds, then such function is homogenous.

Scaling the variable t by a factor of m, i.e., from t to $t' = mt$, does not change the power function type. It just makes the function multiplied by the factor $m^{\mu-1}$. Similarly, the time shift $t' = m + t$ does not change the type of exponential function $e^{t'}$; it only multiplies it by a factor of e^m.

Homogenous functions possess numerous properties that make them preferable to perform approximate description of real processes and objects. Such homogenous functions play a very important role in describing phase transition thermodynamics, percolation statistic properties in turbulence, and in the modern renormalization group theory of critical phenomena, etc.

We can say that power functions with integer or fractional orders are the self-similarity generators. In mathematics, power functions are used as the basis of fractional calculus, of the concept of poles, of residue theory, of asymptotic power series theory, and of stable distributions.

Of course, a real natural fractal is specific for some length scale denoted by a minimum length, l_{min}, and a maximum one, l_{max}. These values determine the scaling range, $l_{min} \ll l \ll l_{max}$, outside which the fractal loses its basic property, i.e., self-similarity property.

1.2 Dimension Types

The concept of dimension is fundamental in physics and mathematics. In case of Euclidean geometric objects (point, line, plane, volume), the concept of topological dimension is used.

Topological Dimension D_T or *Dim* of topological space X is the minimal of the integer numbers n such that a closed contour with multiplicity not exceeding $n+1$ can be inscribed into any open contour of space X. Topological dimension is also known as a dimension.

This definition is very simple to illustrate. For example, $n=1$ means that a straight line can be represented as a series of linear segments that adjoin each other. Every linear segment is adjacent with no more than two other segments (see Fig. 1.1a). In the same manner, for $n=2$, the definition says that every two-dimensional "site" can be paved with stones that can be whatever small (closed sets) in such a way that every stone adjoins no more than three other stones (Fig. 1.1b). At the same time, this site cannot be paved with whatever small stones with every stone siding with only two other stones. When filling a three-dimensional volume with sufficiently small stones (bricklaying, for instance), every such brick already sides with four other bricks.

But what is the dimension of a highly broken line, a crumpled up piece of paper, or a piece of pumice, i.e., geometrical objects that cannot be described with

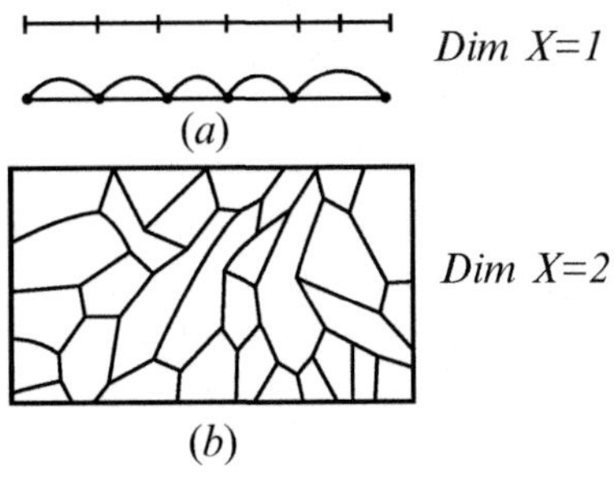

Fig. 1.1 Illustration of the method to determine the topological dimension: (**a**) for a straight line; (**b**) for a plane

conventional smooth functions? As it will be clear further, these objects can be described by the concept of fractal. For instance, fractal dimension is considered as the most important property of fractals.

Now we shall give the general definition of fractal dimension. Let d be conventional Euclidean spatial dimension where a fractal object is situated ($d=1$ for a line, $d=2$ for a plane, and $d=3$ for a conventional 3D-space). Now cover the whole of this object with l radius d-dimensional "balls." Assume that it took not less than $N(l)$ balls. Then, if l was small enough and let the relation between $N(l)$ and l be given by:

$$N(l) \approx 1/l^D, \tag{1.2}$$

then D is called **Hausdorff-Besicovitch** dimension or the fractal dimension of this object.

B. Mandelbrot used the concept of fractal dimension to give a more rigorous definition of the fractal as compared to the one given above. It says that the fractal is an object, the Hausdorff-Besicovitch dimension of which exceeds its topological dimension (0 for points, 1 for a curve, 2 for a surface, etc.).

Expression (1.2) can be written in another way:

$$D = -\lim_{l \to 0} \frac{\ln N(l)}{\ln l}, \tag{1.3}$$

which is used, as a rule, as the general definition of fractal dimension D. Hence, according to this definition, D is a local characteristic of the object.

It is clear that the result of estimation of the fractal dimension would be the same if the fractal was covered with cubes (or squares in case of 2D fractal object).

Sometimes covering is performed with basics of the fractal. In this case, expression (1.3) is used to determine a simplified fractal dimension. Suppose that at some stage of covering a fractal at least $N(l)$ elements of size l were used, and on another stage $N(l')$ elements of a typical size l'. Then the value of fractal dimension D can be calculated as follows:

$$D = -\frac{\ln \frac{N(l)}{N(l')}}{\ln\left(\frac{l}{l'}\right)}. \tag{1.4}$$

This expression can, obviously, be written as follows:

$$\frac{N(l)}{N(l')} = \left(\frac{l}{l'}\right)^D, \tag{1.5}$$

which follows from (1.2).

1.3 Regular Fractals

There are three examples of *regular fractals*: Koch curve, Cantor manifold, and Sierpinski gasket. Tracing a Koch curve begins with plotting a unit length $L(1)=1$ straight line segment (Fig. 1.2a). This initial segment is called a seed. In general case, such line segment can be replaced with a polygon, for example, a square. The seed is the zero generation of the Koch curve.

To produce the next generation it is necessary to replace every seed segment with a generating element. According to the definition performed by B. Mandelbrot, the n-th generation curve for any finite n is called prefractal.

In this particular case, to get the Koch curve of a first of prefractal, one-third (1/3) of the curve is cut in its middle section and replaced by a generating line as shown in Fig. 1.2b. The second generation of fractals are produced by means of replacing each of the four segments with a generating element reduced by 3 (see Fig. 1.2c) and so on. Figure 1.2d shows the third generation prefractal.

Now estimate the fractal dimension of Koch curve. Let the length of the initial line segment $l=1$. Then the amount of line segments of such length that cover the Koch curve at this (zero) stage (Fig. 1.2a) equals $N(l)=1$. At the next stage (Fig. 1.2b) the segment length becomes $l'=1/3$ and the amount of the segments becomes $N(l')=4$. That is why the fractal dimension of Koch curve (according to expression (1.4)) equals:

$$D=-\frac{\ln(1/4)}{\ln(3)}=\frac{\ln 4}{\ln 3}=1.2618. \tag{1.6}$$

This value exceeds one (the topological dimension of the straight line) but it is less than the Euclidean dimension of the plane $d=2$ where the curve is situated.

Another important property of Koch curve is its infinite length. Indeed, if the initial line segment length is one, then the first generation prefractal length is 4/3. The second-generation prefractal length is $4^2/3^2$. If we continue, then the n-th generation prefractal length will be $4^n/3^n$. Therefore, the length of the limit Koch curve will be infinite:

$$\lim_{n\to\infty}\frac{4^n}{3^n}=\infty. \tag{1.7}$$

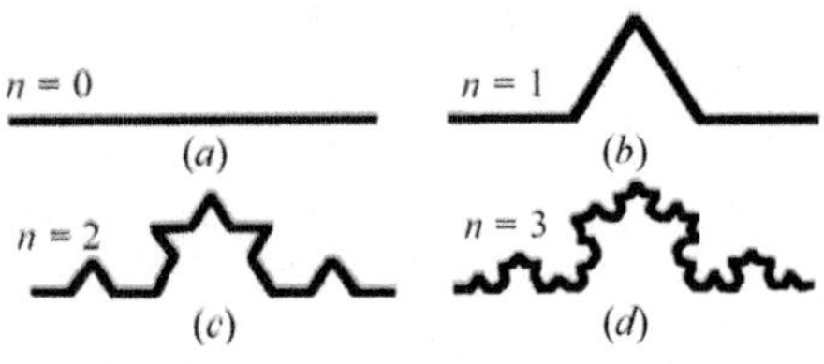

Fig. 1.2 Plotting the prefractals of the triad Koch curve: (**a**) the zero generation; (**b**) the first generation; (**c**) the second generation; (**d**) the third generation

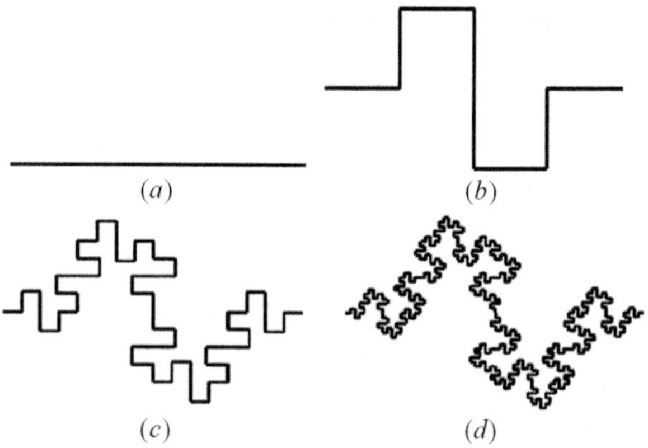

Fig. 1.3 Prefractals of the Koch curve with "pulse" generating element: (**a**) the zero generation; (**b**) the first generation; (**c**) the second generation; (**d**) the third generation

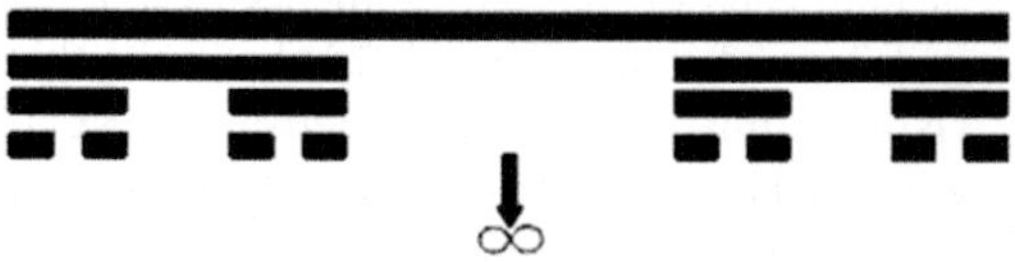

Fig. 1.4 Illustration of forming the homogenous Cantor manifold

It is obvious that using other types of generating elements will result in other dimension values. Hence, the proof is complete.

Let a unitary straight line segment be the seed. Let an open polygon comprised of $N = 8$ segments each $l' = 1/4$ in length be the generating element. The resulting Koch curve with its prefractals is shown in Fig. 1.3.

Expression (1.4) will be used to determine the fractal dimension of the produced Koch curve.

A very simple tracing proposed by Cantor allows to generate fractal manifolds with fractal dimension within the range $0 < D < 1$. In this case, a unitary line segment is also used as a seed. The generating element divides it into three equal (or, in general case, unequal) parts. Then, the middle part is removed leaving just two side parts. After that, the generating element is applied to each of the side parts and the process continues. Hence, the triad Cantor manifold appears. This process is shown in Fig. 1.4.

It is obvious that the total length (measure) of the resulting segments is zero in the limit because of exclusion of the unitary length. As a result, the appeared manifold is infinite number of separate points.

Now one can estimate the fractal dimension of this manifold using expression (1.3). It is apparent that at the n-th stage of tracing there are 2^n segments each of $1/3^n$ in length. That is why $N(l)$ can be equal to 2^n at this stage and l can be equal to $1/3^n$. Hence, as $l \to 0$ corresponds to limit $n \to \infty$, so the fractal dimension of Cantor manifold equals:

Fig. 1.5 Solid Cantor kernel

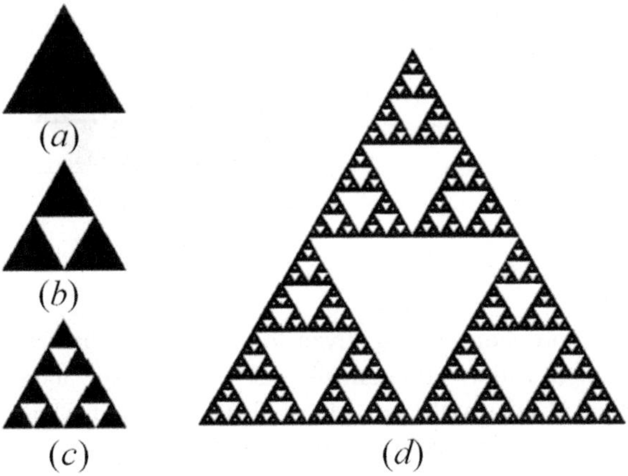

Fig. 1.6 Triangle Sierpinski gasket prefractals tracing: (**a**) the zero generation; (**b**) the first generation; (**c**) the second generation; (**d**) the fifth generation

$$D = -\lim_{n\to\infty} \frac{\ln 2^n}{\ln\left(\frac{1}{3^n}\right)} = \frac{\ln 2}{\ln 3} = 0.6309. \tag{1.8}$$

It turned out for D to be less than the Euclidean dimension of the space ($d = 1$), this manifold will be considered as points with zero dimension, which are the elements of the manifold. The dimension of Cantor manifold exceeds the topological dimension of the manifold elements.

An interesting variant of fractal geometry based on Cantor manifold is depicted in Fig. 1.5 (the fourth generation of prefractal). Unlike the Cantor manifolds considered before, the segments that appear in different generations are connected to each other thus forming kernels (Cantor kernels). The length of each segment is $1/r$ ($r < 2$) multiplied by the initial segment length. The fractal dimension of such manifold can be calculated with expression (1.3).

As for Sierpinski gasket, the seed here is an equilateral triangle and the generating element is the same kind of triangle but reduced with $r = 1/2$ factor. The generating element excludes the central triangle from the seed. Every time the generating element is applied, the initial triangle is replaced with $N = 3$ triangles (Fig. 1.6). In the limit, a holed geometric configuration consisting of infinite amount of separate points appears.

Fractal dimension of Sierpinski gasket can also be estimated from (1.4):

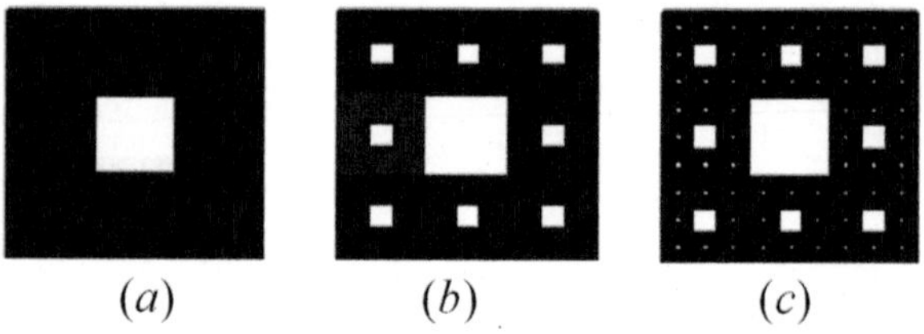

Fig. 1.7 Sierpinski carpet prefractals: (**a**) the first generation; (**b**) the second generation; (**c**) the third generation

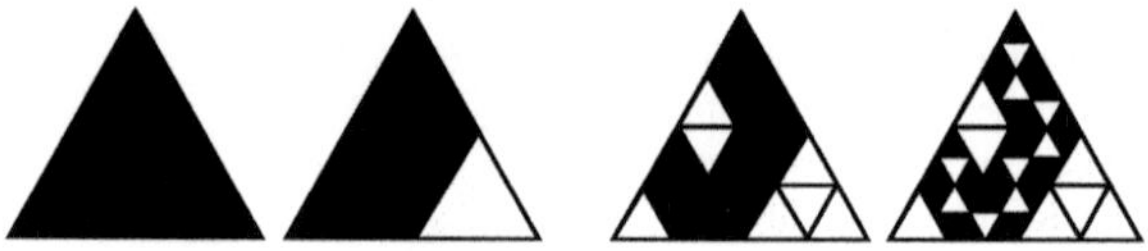

Fig. 1.8 Random irregular variant of Sierpinski gasket

$$D = \frac{\ln 3}{\ln 2} = 1.5849. \tag{1.9}$$

Expression (1.9) is given to account for the fact that there is one equilateral triangle with $l = 1$ length sides at the zero stage and three equilateral triangles with $l' = 1/2$ sides at the next stage. Hence, $N(l) = 1$ when $l = 1$ and $N(l') = 3$ at $l' = 1/2$. The gasket has zero square since it is easy to prove that the square equal to the square of the initial triangle was excluded in the process of tracing. This is proved also by the fractal dimension value $D < 2$, which is less than the dimension of the plane where this object located.

Now take a unitary four-square as the seed and a four-square with $r = 1/3$ similarity factor as the generating element. Then, the initial four-square is separated into 9 parts and the middle part is removed (i.e., 8 parts are left). The result is the first generation prefractal shown in Fig. 1.7a. The next generation prefractals are given in Fig. 1.7b,c.

The readers are suggested to calculate the fractal dimension of Sierpinski carpet themselves.

1.4 Irregular Random Fractals

The algorithms considered above that were used to trace regular fractals can be modified to trace irregular random fractals. For example, in case of regular Sierpinski gasket, the middle of the four triangle areas is removed (Fig. 1.6). In spite of this, we can randomly remove any of these triangles. The result is *a random irregular variant* of Sierpinski gasket. The sequence of its generation is given in Fig. 1.8.

Brownian motion path is a classic example of random fractal. Figure 1.9a depicts a microscope magnified typical motion path of a dust particle, which performs Brownian motion. The motion path depicted in Fig. 1.9a is a rough approximation.

Fig. 1.9 Particle's Brownian motion path, microscope zoomed: (**a**) initial image; (**b**) tenfold increase

If the resolution increases hundred wrinkles under the microscope that detects the particle's motion, the path interval between A and B (increase ten wrinkles) will appear as shown in Fig. 1.9b.

The straight line segment that links points A and B became a set of 100 line segments, and each of these segments has the same length (on the average) as the segments depicted in Fig. 1.9a (while they actually 10 times shorter as long as the motion path in Fig. 1.9b exhibits a tenfold increase).

The same situation is repeated with more precise detection of the particle's motion between points C and D as depicted in Fig. 1.9b. If the resolution is increased a 100-fold one more time, and then zoom in tenfold the detected motion paths, then the resulting picture will be statistically equal to the picture shown in Fig. 1.9b. That is exactly why in practice Brownian motion is called *statistically self-similar*.

Therefore, every real self-similar process should have the minimum and the maximum scale; the scale cannot be increased or decreased without limits. In case of Brownian motion, the range of scales within which self-similarity remains is very large, from the dimension of a vessel with liquid (0.1 m, for instance) to the free path length of molecules between their collisions, which can be as small as 10^{-9} m for small test particles.

1.5 Multifractals

Multifractals are heterogeneous fractal objects that, unlike regular fractals, cannot be completely described with just one parameter (fractal dimension D) and require a whole range of such dimensions, the amount of which is actually infinite. The reason is that such fractals possess not only pure geometrical characteristics determined by D but also some probabilistic properties.

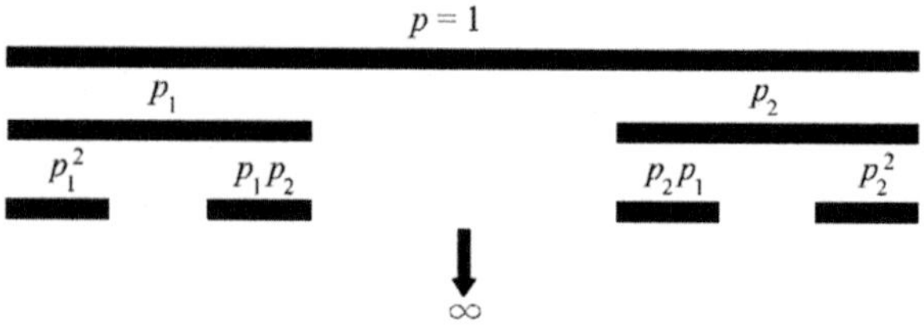

Fig. 1.10 The principle of creating homogeneous Cantor fractal manifold

Consider the so-called heterogeneous Cantor manifold as an example of heterogeneous fractal. Start with the familiar Cantor manifold of the excluded middle third parts (see Fig. 1.4). At the beginning of the procedure (the zero stage), let there be a unitary linear segment along which N points of the fractal manifold are distributed somehow. At the first step, there are already two 1/3 long segments at the sides of the initial unitary segment.

Let the initial N points be distributed over these segments as follows: the left segment is populated with p_1 probability and contains p_1N points, and the right segment is populated with $p_2 = 1 - p_1$ probability with a corresponding number of points p_2N. Then every segment is treated in the same manner. Again, at the second stage there are already four 1/9 long segments with population probabilities (from the left to the right): p_1^2, p_1p_2, p_2p_1, p_2^2 as depicted in Fig. 1.10.

At the n-th stage, the manifold is comprised of 2^n segments each $1/3^n$ in length. Their population probabilities are p^n, $p^{n-1}p_2$, $p^{n-2}p_2^2$, . . . , p^{2n} (this is not the order of their location!). Observe that the number of segments, with a probability of p^{n-m} p_m^m, equals the number of combinations C_m^n of n elements taken m at a time. As a result, at $n \to \infty$ and $p_1 \neq ½$ we finally acquire a heterogeneous fractal manifold.

If we apply the standard Hausdorf-Besikovitsch procedure to calculate the dimension, the result will be equal to the corresponding dimension of homogeneous Cantor manifold that is 0.6309. Despite this coincidence, there are evident differences in structures of homogeneous and heterogeneous manifolds. These differences can be found out by means of multifractal analysis.

1.6 Fractal Signals

Fractals can exist not just in space but also in time. In other words, there are not only fractal figures (objects) but also fractal processes. A classic example of such fractal process is Brownian motion of particles. Detection of time dependence of a Brownian particle position $X(t)$ results in a model of a fractal stochastic process (Fig. 1.11).

The values of a stochastic function $X(t)$ that describe variations of the signal are Gauss distributed, i.e., integral distribution $F_x(x)$ and probability density $f(x)$ are:

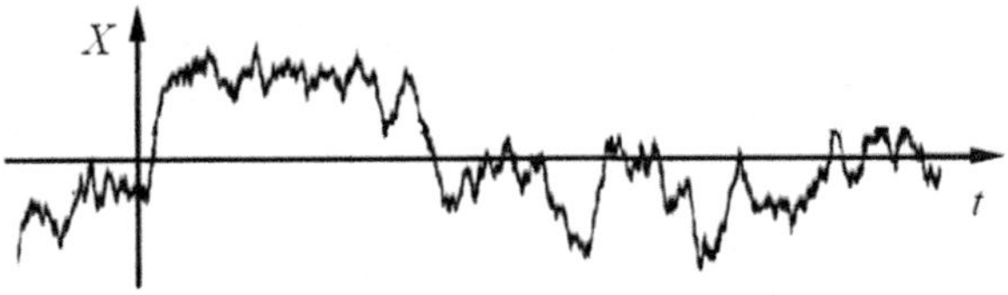

Fig. 1.11 Brownian signal time diagram

$$F_x(x) = \frac{1}{\sqrt{2\pi\sigma}} \int_{-\infty}^{x} \exp\left(-\frac{1}{2}\left(\frac{s - m_x}{\sigma}\right)^2\right) ds; \tag{1.10}$$

$$F(x) = \frac{1}{\sqrt{2\pi\sigma}} \exp\left(-\frac{1}{2}\left(\frac{x - m_x}{\sigma}\right)^2\right). \tag{1.11}$$

Here m_x defines the probabilistic mean, and σ represents the root-mean-square (rms) value of the standard deviation of random variable x.

The increments of the function $\Delta X = X(t_2) - X(t_1)$, $t_2 > t_1$ also have Gaussian distribution (with zero mean), so that the probability is determined as:

$$P(\Delta X < x) = \frac{1}{\sqrt{2\pi\sigma^2 (t_2 - t_1)}} \int_{-\infty}^{x} \exp\left(-\frac{1}{2}\left(\frac{u^2}{\sigma^2 (t_2 - t_1)}\right)\right) du. \tag{1.12}$$

It follows from the latter property that the expression of Brownian signal increments variance is

$$D[X(t_2) - X(t_1)] = \sigma^2 |t_2 - t_1| \tag{1.13}$$

for all t_1 and t_2 within the considered time period. With regard to the variance of increments, the term *delta-variance* is used.

It is indispensable to state that the value of delta-variance, $\sigma^2 |t_2 - t_1|$, only depends on the difference between t_1 and t_2 but not on their values. The expression of (1.13) is practically very important. If a signal processor discovers that (1.13) is true, it means that the structure of the signal belongs to the random fractals with $D = 1.5$ dimension.

The mean value of the increment (*the structure function of the signal*) is calculated using the following expression:

$$E[|X(t_2) - X(t_1)|] = \sqrt{\frac{2}{\pi}}\, \sigma \sqrt{|t_2 - t_1|}. \tag{1.14}$$

Any increment of a Brownian signal is statistically self-similar, i.e.:

$$X(t + \Delta t) - X(t) \triangleq \frac{1}{\sqrt{r}}(X(t + r\Delta t) - X(t)) \tag{1.15}$$

for any $r > 0$. The symbol $\triangleq$ means that the two random variables have the same distribution and, in particular, the same values of the mean and variance.

The fact that increment variations are self-similar at different intervals, it allows to apply the standard procedure of determining the fractal dimension to the Brownian signal diagram. This can be done as follows.

Let the time interval for which the signal is defined be one. Divide this interval into n equal subintervals each $\Delta t = 1/n$ in length. Divide the vertical axis in the same manner into subintervals each ΔX in length. The relation $|\Delta X|/|\Delta t|$ is used to estimate the number of squares or rectangles required to cover the part of the graph of $y = X(t)$ located over one subinterval. As long as the mean value of $|\Delta X|$ is proportional to $\sqrt{\Delta t}$, the number of rectangles required for one subinterval is also proportional to $\sqrt{\Delta t}$. Since there are $1/\Delta t$ such subintervals in all, the total amount of rectangles is proportional to:

$$N(\Delta t) \sim \Delta t^{-3/2} \tag{1.16}$$

or

$$D = -\lim_{\Delta t \to 0} \frac{\log N(\Delta t)}{\log \Delta t} = 1.5. \tag{1.17}$$

Hence, the fractal dimension of the Brownian signal equals 1.5.

In cases when there is a need to describe random signals with other values of D that possess some "memory," the *model of Generalized Brownian Motion* (GBM) is used.

It is assumed that the Gaussian signal $X(t)$ with standard deviation s obeys the GBM model if an increment $\Delta X = X(t_2) - X(t_1);\ t_2 > t_1$ has Gaussian distribution expressed with,

$$P(\Delta X < x) = \frac{1}{\sqrt{2\pi}\sigma(t_2 - t_1)^H} \int_{-\infty}^{x} \exp\left(-\frac{1}{2}\left(\frac{u}{\sigma(t_2 - t_1)^H}\right)^2\right) du. \tag{1.18}$$

It follows from (1.18) that the delta-variance for the GBM model equals to:

$$D[X(t_2) - X(t_1)] = \sigma^2 |t_2 - t_1|^{2H}. \tag{1.19}$$

The parameter H $(0 < H < 1)$ of the given expressions in (1.18) and (1.19) is called the *Hurst parameter*. When $H = 0.5$, the GBM model matches the classic model of Brownian motion.

The mean value of the signal increment (i.e., the first order structure function) is calculated as follows:

$$E[|X(t_2) - X(t_1)|] = \sqrt{\frac{2}{\pi}}\,\sigma(t_2 - t_1)^H. \tag{1.20}$$

The increments are statistically self-similar that is mathematically expressed as:

$$X(t + \Delta t) - X(t) \triangleq \frac{1}{r^H}(X(t + r\Delta t) - X(t)) \tag{1.21}$$

for any $r > 0$.

The fractal dimension of the generalized Brownian motion signal is calculated just like for the conventional one. The basic difference is that the expression (1.16) of estimation is replaced with a new one:

$$N(\Delta t) \approx \frac{\sigma}{\Delta t^{2-H}}, \tag{1.22}$$

which leads to the following relation:

$$D = -\lim_{\Delta t \to 0} \frac{\log N(\Delta t)}{\log \Delta t} = 2 - H. \tag{1.23}$$

It is convenient to simulate fractal signals for the full range of values of fractal dimension $1 < D < 2$ by means of Weierstrass function:

$$X(t) = \sqrt{2}\sigma \frac{\left(1 - b^{2D-4}\right)^{0.5} \sum_{n=0}^{N} b^{(D-2)n} \cos\left(2\pi s b^n t + \psi_n\right)}{\left[1 - b^{(2D-4)(N+1)}\right]}, \tag{1.24}$$

where σ is the standard deviation; b is the space-frequency scaling parameters; D is the fractal dimension that is related with the Hurst parameter like $D = 2 - H$; $N + 1$ is the number of harmonics (at $N \to \infty$, the Weierstrass function becomes an ideal mathematical fractal); Ψ_n is the randomly distributed phase over $[0, 2\pi]$ interval, and t represents the time.

1.7 Physical Meaning of Hurst Parameter

Let the functions that describe signals with certain values of H within the framework of the GBM model be designated by $X_H(t)$, which is more convenient to compare signals.

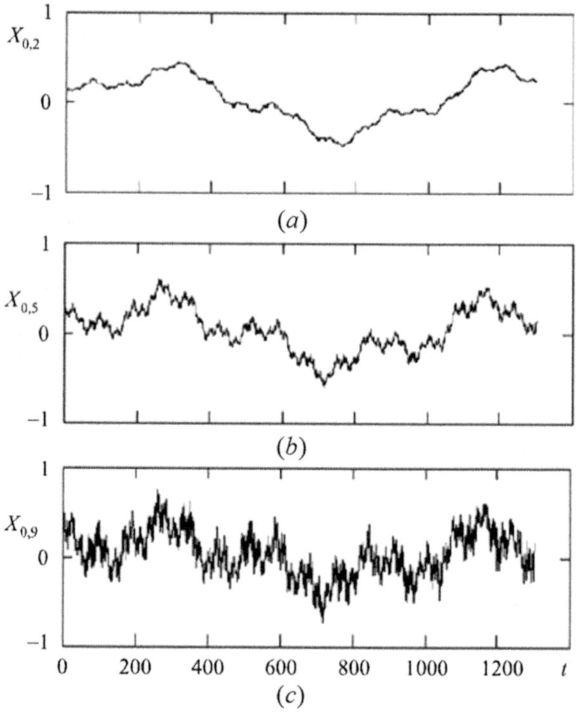

Fig. 1.12 Weierstrass function graphs for (**a**) $H = 0.2$; (**b**) $H = 0.5$; (**c**) $H = 0.9$

Figure 1.12 depicts X_H function graphs for different values of the Hurst parameter built according to (1.24). The picture makes it evident that H parameters specify the degree of angularity of the studied graph. At small values of H the graph is considerably flowing (although not smooth). At high values of H the graph is strongly angular.

In professional literature, Hurst parameter H is derived through range R of values of the studied signal during time period Δt and standard deviation S related to this time period as follows:

$$H = \ln(R/S)/\ln(\Delta t). \tag{1.25}$$

where the range R is calculated from:

$$R = \max X_H(t) - \min X_H(t). \tag{1.26}$$

In (1.26) the time moments, t, belong to Δt interval. If $H \neq 1/2$, then the signals the GBM model describes have infinite correlation time.

For convenience assume that $\langle X_H(t) \rangle = 0$. Now, the correlation of the future increments $X_H(t)$ with the past increments, $X_H(-t)$, normalized by the variance of $X_H(t)$, is equal to:

$$R(t) = \frac{\langle -X_H(-t)X_H(t) \rangle}{\langle X_H^2(t) \rangle} = 2^{2H-1} - 1. \quad (1.27)$$

When $H = 1/2$, the correlation $R(t)$ between the past and the future increments equals zero (i.e., there is no correlation) for all values of t; and this is just like supposed to be for a random process with independent increments. But when $H \neq 1/2$, $R(t) \neq 0$ regardless of t. This is the basic property of the GBM model called *persistency* (when the signal variation trend remains unchanged) or *antipersistency*. Assume that, for $H > 1/2$, the increments were positive during some past time interval, i.e., the signal was increasing. In the future, it is more likely that, on the average, the signal will keep rising. Hence, if a signal with $H > 1/2$ tends to rise in the past, it will tend to rise in the future, even for arbitrarily large t. Conversely, the tendency to decrease in the past involves, on the average, a tendency to decrease in the future. Observe that $H < 1/2$ signifies antipersistency. It means that increasing in the past involves decreasing in the future and, conversely, a tendency to decrease in the past implies a highly possible increase in the future. The graphs depicted in Fig. 1.12 proves this fact.

1.8 Relation Between Fractality and Spectrum Response

Spectrum response is an important characteristic of stochastic signals (that are often called noises). Spectral power density often obeys exponential laws with constant power index β:

$$S(f) \sim f^{\beta}, \quad (1.28)$$

here f is the frequency in Hz.

The power index of *White Noise* equals $\beta = 0$. There is *Pink Noise* with $\beta = 1$, which often occurs in real practice. Noise with $\beta = 2$ is called *brown*, and $\beta = 3$ noise is called *black*. Figure 1.13 shows representative time structures of the described noise signals. In case of fractal signals, the power index β is related to the Hurst parameter by:

$$\beta = 2H + 1. \quad (1.29)$$

This relation makes it evident that the values of β index of the fractal signals that are described with Brownian models ($0 < H < 1$) lie within the range $1 < \beta < 3$.

The signal that correspond to the classic Brownian model ($H = 0.5$) falls into the class of brown noises ($\beta = 2$). It follows from expression (1.28) and the signals depicted in Fig. 1.13 that when switching from white noise to, correspondingly, pink, brown, and black noises, the phenomenon of persistency comes more and more evident.

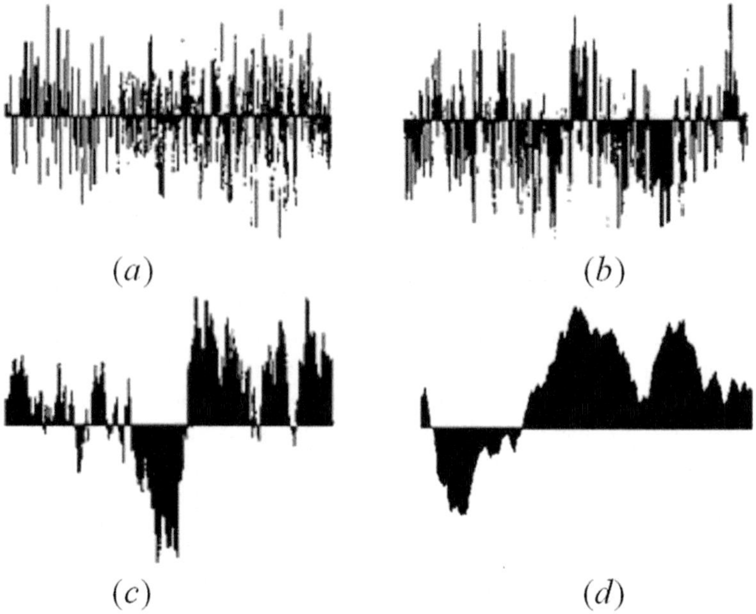

Fig. 1.13 Time diagrams of *white* (**a**), *pink* (**b**), *brown* (**c**), and *black* (**d**) noises

1.9 Examples of Signal Analysis

Fractal analysis of signals is performed when it is needed to determine the degree the fractal features of these signals appear. Here, the criterion of presence of these signals' fractal properties is trueness of expressions (1.19) and (1.20) that describe the behavior of the delta-variance and the structure function.

If log–log plots of the delta-variance of the signal and its structure function in the domain of time increments can be properly approximated with straight lines within large ranges of time increments (starting with the smallest increments), then we can say that this signal possesses fractal properties. The slope of the line makes it possible to estimate the value of $2H$ (when analyzing the behavior of delta-variance) of H (when the structure function plot is considered).

The calculated value of the Hurst parameter H allows to easily determine the fractal dimension D ($D = 2 - H$). The time interval within which delta-variance and structure function graphs can be properly approximated with straight lines determines the scaling range.

To illustrate that consider an example of simulated signals processing. It is practically convenient to process signals with software environment using indexed functions. That is why, instead of $X(t)$ functions, indexed functions X_k (the relation of time t and index $k = 1,2,3\ldots$ is $t = k\Delta t$, where Δt is the time interval between significant points of the signal) will be used. The diagrams of the simulated signals are depicted in Fig. 1.14.

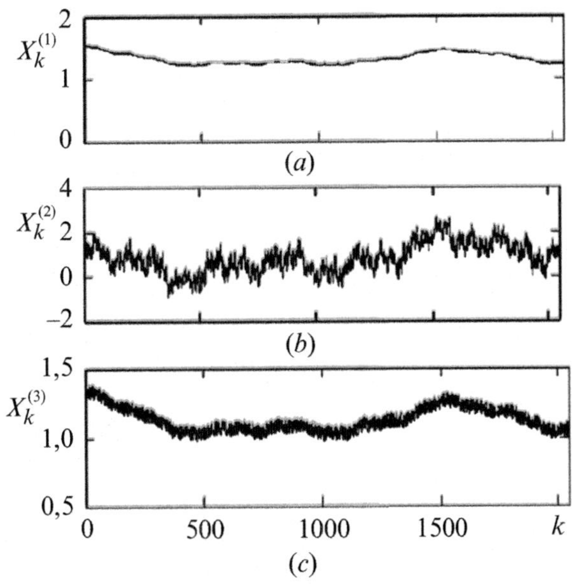

Fig. 1.14 The structure of the simulated signals: (**a**) signal $X_k^{(1)}$ with $D_1 = 1.1$; (**b**) signal $X_k^{(2)}$ with $D_2 = 1.8$; (**c**) signal $X_k^{(3)}$, $X_k^{(3)} = X_k^{(1)} + X_k^{(\text{Noise})}$

Weierstrass function (see expression (1.24)) was used to produce signals $X_k^{(1)}$ and $X_k^{(2)}$ at fractal dimensions $D_1 = 1.1$ and $D_2 = 1.8$ correspondingly. Signal $X_k^{(3)}$ is signal $X_k^{(1)}$superimposed with the noise. The expression of the noise is

$$X_k^{(\text{Noise})} = 0.1\frac{\text{rand}(k)}{k}.$$

Here rand(k) is a function generating a $k \times k$ array, the elements of which are the random numbers uniformly distributed over the [0,1] interval.

The log–log plots describing structure functions S_n of the signals $X_k^{(m)}$ ($m = 1,2,3$) are given in Fig. 1.15. The structure functions are determined with the following expression based on expression (1.20):

$$S_n^{(m)} = E\left[\left|X_{k+n}^{(m)} - X_k^{(m)}\right|\right] = \frac{1}{k-n}\sum_{k=1}^{k-n}\left|X_{k+n}^{(m)} - X_k^{(m)}\right|. \tag{1.30}$$

The expression of the Y-axis is $L_n^{(m)} = \log_2 S_n^{(m)}$. The expression of the X-axis is $\log_2 n$. The averaging in expression (1.30) is done for all values of k.

Figure 1.15a–c depicts functions $f_n^{(m)}$ that are linear approximations of $L_n^{(m)}$. It is clear that $L_n^{(1)}$ and $L_n^{(2)}$ relating to $X_k^{(1)}$ and $X_k^{(2)}$ can be properly approximated with linear functions within the wide range of n.

This result is proved by the fact that the signals described with Weierstrass functions are fractal. The Hurst parameters calculated through the slopes of $f_n^{(1)}$ and $f_n^{(2)}$ graphs equal $H = 0.86$ and $H = 0.21$. The corresponding fractal dimensions

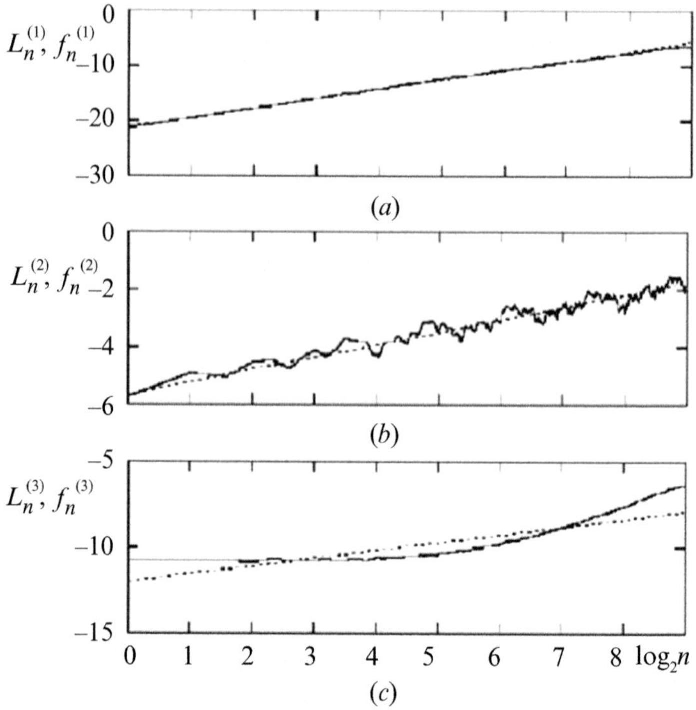

Fig. 1.15 Graphs of $L_n^{(m)}$ (*solid line*) and $f_n^{(m)}$ (*dashed line*): (**a**) $m = 1$; (**b**) $m = 2$; (**c**) $m = 3$

equal $D_1 = 1.14 \pm 0.03$ and $D_2 = 1.79 \pm 0.05$ (the errors of the fractal dimensions are calculated as a result of averaging the dimensions for certain variants of Weierstrass function (1.24) that are distinct from each other with random sets of phases Ψ_n). The average dimensions are much close to the initially set values ($D_1 = 1.1$ and $D_2 = 1.8$) when tracing graphs of $X_k^{(1)}$ and $X_k^{(2)}$.

Function $L_n^{(3)}$ graph (see Fig. 1.15c) deeply differs from graphs of $L_n^{(1)}$ and $L_n^{(2)}$ and can hardly be approximated with linear functions. Thereby we can say that signal $X_k^{(3)}$ has no fractal properties within the specified range of n.

Practice

1. Give an example of a function that would obey the homogeneity condition.
2. Build a regular fractal using a seed and a generating element that would differ from those considered in the chapter.
3. Calculate the fractal dimension of the Koch curve given in Fig. 1.3.
4. Calculate the fractal dimension of the Sierpinski carpet given in Fig. 1.7.
5. Draw a random irregular fractal of the Sierpinski carpet.
6. Prove that the Brownian motion fractal dimension equals 1.5.

7. Taking into account that the spectral density of a signal is related with its shape via the Fourier transform, depict the spectrums of the white, pink, brown, and black noises.

Test Questions

1. Formulate the definition of the self-similarity of fractals.
2. Formulate the definition of scaling.
3. Formulate the principle of calculation of the fractal dimension of regular fractals.
4. What is the difference between regular fractals and heterogeneous random fractals?
5. Formulate the definition of multifractals.
6. Formulate the basic properties of the generalized Brownian motion model.
7. Formulate the definition of the Hurst parameter and its physical meaning when estimating the manner of future change of a fractal signal.
8. Formulate the procedure of fractal analysis of random signals.

Chapter 2
Fractal Calculus Fundamentals

2.1 Preliminaries

Many real physical processes possess "memory," which comes as follows: time connection between the process cause, $f(t)$, and the process effect, $g(t)$, is not immediate, and the condition of $g(t)$ is specified with the condition of $f(t)$ not at the same moment but delayed. This property is called *hereditary*.

Hereditary property, in particular, shows itself in such phenomena and processes as metal fatigue, magnetic and electrical hysteresis, motion of bodies through viscous medium, propagation of sound waves, diffusion, etc.

The key part in the hereditary theory belongs to the influence (memory) function, which generally reflects the complexity of the system and the process. One of the important properties of the memory function is self-similarity that is determined based on some general considerations. As it was shown in Chap. 1, self-similarity is the "generic indicator" of power functions.

Exponential functions are well known to be the most widely used functions in electrical engineering, theoretical physics, and mathematics. These functions (or their superpositions) are used to solve many problems that result in ordinary differential equations with integer-order derivatives.

There is a need to mathematically describe physical processes and phenomena that possess hereditary effect, which obey power and even logarithmic laws. Thus, applications of nonconventional mathematics have emerged such as fractal dimensions, fractional-order integrals and derivatives, and nonstandard distribution functions with infinite moments (Levy distributions).

Although fractional derivatives and integrals were introduced by the famous mathematicians Abel and Liouville as early as in the 30s of the nineteenth century, the significance and popularity of the "new" mathematical concepts started growing rapidly just in recent decades. This quick growth of interest was undoubtedly stimulated by the introduction of fractal geometry.

A.Kh. Gil'mutdinov et al., *Fractal Elements and their Applications*, Analog Circuits and Signal Processing, DOI 10.1007/978-3-319-45249-4_2

This chapter describes the fundamental concepts of fractional-order integrals and derivatives, and some of their properties and transforms in several fields such as electrical engineering, electrical chemistry, and impedancemetry that exhibits fractional-order dynamics.

2.2 Properties of Fractional-Order Integrals and Derivatives

2.2.1 *Riemann-Liouville Fractional-Order Integral and Derivative*

The definition of fractional integral follows from the generalization of the integer-order Cauchy formula. Let the integral operator is denoted by I; then the integration of $f(x)$ is defined as

$$_0I_x^1 f(x) = \int_0^x f(t)dt.$$

Here, the subscripts specify the integration limits; the left one is the lower limit, while the right one is the upper limit. The superscript specifies the integration order.

The expression to calculate the integer-order n-fold multiple integral is defined by

$$_0I_x^n f(x) = \frac{1}{(n-1)!}\int_0^x f(t)(x-t)^{n-1}dt\,. \tag{2.1}$$

One may generalize (2.1) to a non-integer integral of order α, which is denoted as the Riemann-Liouville (RL) definition of fractional-order integral, as follows:

$$_0I_x^\alpha f(x) = \frac{1}{\Gamma(\alpha)}\int_0^x f(t)(x-t)^{\alpha-1}dt, \tag{2.2}$$

where $\Gamma(\alpha)$ is the well-known gamma function given by

$$\Gamma(x) = \int_0^\infty t^{x-t}e^{-t}dt,\ x\in\Re. \tag{2.3}$$

The simplest way to understand gamma function is to generalize the factorial of all real numbers. It may be shown, using integration by parts, that for integer values of $x=n$

$$\Gamma(n) = (n-1)\cdot(n-2)\cdot\ldots 2\cdot 1 = (n-1)!, \text{ and } \Gamma(n+1) = n\Gamma(n) = n!$$

Now, let D denotes a differential operator; then the relation between the fractional-order differential and integral operators is represented as $D^{\alpha}f(x) = I^{-\alpha}f(x)$. Consequently, the fractional-order derivative of order α can be deduced from (2.2) to yield

$$_0D_x^{\alpha}f(x) = \frac{1}{\Gamma(-\alpha)}\int\limits_0^x \frac{f(t)}{(x-t)^{\alpha+1}}dt. \tag{2.4}$$

It should be noted that the value of the lower limit of integration, which is zero here, could be arbitrary. In general, the integration (differentiation) limits are specified with subscripts. For example, a Riemann-Liouville derivative with non-zero lower limit will be defined as

$$_bD_x^{\alpha}f(x) = \frac{1}{\Gamma(-\alpha)}\int\limits_h^x \frac{f(t)}{(x-t)^{\alpha+1}}dt. \tag{2.5}$$

When the lower limit equals zero, for simplicity, one may replace $_0D_x^{a}f(x)$ by $D_x^{a}f(x)$. The RL derivatives and integrals are widely used in fractional calculus. It can be extended to the most general case when $n-1<\beta\le n$; for any integer number $n\ge 1$. The fractional-order derivative of order $n-1<\beta\le n$ can be expressed as

$$_aD_x^{\beta}f(x) = \frac{d^n}{dx^n}\left[{}_aD_x^{-(n-\beta)}f(x)\right] = \frac{1}{\Gamma(n-\beta)}\frac{d^n}{dx^n}\left[\int\limits_a^x \frac{f(t)}{(x-t)^{\beta-n+1}}dt\right]. \tag{2.6}$$

2.2.2 Grunwald-Letnikov Fractional-Order Derivative and Integral

Unlike the Riemann-Liouville approach, which is based on the concept of multiple integrals, the Grunwald-Letnikov definition (hereinafter referred to as GL definition) of fractional-order derivative follows from the classical definition of integer-order derivative.

Consider the following definition of the first-order derivative:

$$D^1 f(x) = \lim_{dx \to 0} \frac{f(x+dx) - f(x)}{dx}.$$

To acquire the second-order derivative, differentiate $D^1 f(x)$ once more to get

$$D^2 f(x) = \lim_{dx \to 0} \frac{D^1 f(x+dx) - D^1 f(x)}{dx}$$
$$= \lim_{dx_2 \to 0} \frac{\lim\limits_{dx_2 \to 0} \frac{f(x+dx_1+dx_2) - f(x+dx_1)}{dx_2} - \lim\limits_{dx_2 \to 0} \frac{f(x+dx_2) - f(x)}{dx_2}}{dx_1}.$$

Assume that the increments are of equal size, i.e., $dx_1 = dx_2 = dx$; then the expression of the second-order derivative can be simplified to

$$D^2 f(x) = \lim_{dx \to 0} \frac{f(x+2dx) - 2f(x+dx) + f(x)}{dx^2}$$

Similarly, the nth-order derivative of the function $f(x)$ can be obtained by running this procedure n times. Hence,

$$D^n f(x) = \lim_{dx \to 0} \frac{1}{h^n} \sum\nolimits_{m=0}^{n} (-1)^m \binom{n}{m} f(x - mh); \quad h \equiv dx, \tag{2.7}$$

where $\binom{n}{m} = \frac{n!}{m!(n-m)!}$ are the binomial coefficients.

This expression can be generalized for arbitrary real number, $\alpha \in \Re$, by replacing the standard factorials with gamma function. Furthermore, the upper limit of the summation, $(t-a)/h$, (not an integer number) tends to infinity as $h \to 0$ (where t and a are the upper and the lower limits of differentiation). Then the resulting GL fractional derivative of $f(x)$ will be described as

$${}_aD_x^\alpha f(x) = \lim_{h \to 0} \frac{1}{h^\alpha} \sum_{m=0}^{\left[\frac{t-a}{h}\right]} \frac{\Gamma(\alpha+1)}{m!\Gamma(\alpha-m+1)} f(x - mh); h \equiv dx, \tag{2.8}$$

where [.] is the flooring operator.

Just like in the case of the RL fractional integral transform into the RL fractional derivative, the GL fractional derivative can also be transformed into the GL fractional integral. The most natural way to do it is to determine the expression for binomial coefficients (2.8) at $n < 0$. It can be shown that

$$\binom{-n}{m} = \frac{-n(-n-1)(-n-2)(-n-3)\ldots(-n-m+1)}{m!} = (-1)^m \frac{(n+m-1)!}{(n-1)!m!}.$$

and

$$\binom{-\alpha}{m} = (-1)^m \frac{\Gamma(\alpha+m)!}{\Gamma(\alpha)m!}. \tag{2.9}$$

Hence, from (2.9), the GL fractional-order integral can be written as follows:

$$_aI_x^\alpha f(x) = \lim_{h\to 0} h^\alpha \sum_{m=0}^{\left[\frac{t-a}{h}\right]} \frac{\Gamma(\alpha+m)}{m!\Gamma(\alpha)} f(x-mh);\ h \equiv dx. \tag{2.10}$$

In general (2.8) can (2.10) can be combined in a single definition as follows:

$$_aD_x^{\pm\alpha} f(x) = \lim_{h\to 0} \frac{1}{h^{\mp\alpha}} \sum_{j=0}^{\left[\frac{t-a}{h}\right]} C_j^{\pm\alpha} f((x-j)h), \tag{2.11}$$

where

$$C_j^{\pm\alpha} \equiv (-1)^j \binom{\pm\alpha}{j} = \left(1 - \left(\frac{1\pm\alpha}{j}\right)\right) C_{j-1}^{\pm\alpha};\ C_0^{\pm\alpha} = 1;\ j = 1, \ldots, n. \tag{2.12}$$

Despite the difference between the RL and GL definitions of fractional-order integral and derivative, they are actually equivalent. In real practice, the RL definition is widely used to analytically calculate fractional-order integrals and derivatives of relatively simple functions (x^a, e^x, sin(x), ... etc.). The GL definition, on the other hand, can easily be used for numerical calculations, where its accuracy depends on the step size, h.

2.2.3 *Properties of Fractional-Order Derivatives*

Consider the following basic properties of fractional-order derivatives.

1. If $f(x)$ is an analytical function of x, then its fractional derivative $_aD_x^a f(x)$ is also an analytical function of x and α.
2. If $\alpha = n$ and n is an integer number, then $_aD_x^a f(x)$ operation converges to the same value of the conventional integer-order n differentiation.
3. If $\alpha = 0$, then $_aD_x^\alpha f(x) = {}_0D_x^0 f(x) = f(x)$.
4. Fractional-order differentiation, just like integer-order differentiation, is a linear operation:

$$_0D_x^\alpha [af(x) + bg(x)] = a\left[{}_0D_x^\alpha f(x)\right] + b\left[{}_0D_x^\alpha g(x)\right].$$

5. Fractional differentiation is commutative, i.e.,

$$ {}_0D_x^{\alpha}\left[{}_0D_x^{\beta}f(x)\right] = {}_0D_x^{\beta}\left[{}_0D_x^{\alpha}f(x)\right] = {}_0D_x^{\alpha+\beta}f(x). $$

6. Since $I^{\alpha}f(x) = D^{-\alpha}f(x)$, all previous five properties still apply for fractional-order integration.

2.3 Laplace Transform of Fractional-Order Operators

The Laplace transform is widely used to solve engineering problems, including problems in radio engineering. In this section, we introduce some fundamentals of the Laplace transform of integer-order calculus, which will then be generalized to fractional-order calculus.

2.3.1 *Fundamentals of Laplace Transform*

The Laplace transform of a function $f(t)$, denoted as $F(s)$, where $s = \sigma + j\omega$, is given by

$$ F(s) = \mathcal{L}\{f(t)\} = \int_0^{\infty} \mathrm{e}^{-st} f(t)dt. \tag{2.13} $$

Notice that (2.13) converges if $f(t)$ is both identically equal to zero for $t < 0$ and a single piecewise continuous function with finite number of the first kind discontinuities for $t > 0$. As $t \to \infty$, $f(t)$ should be bounded; that is, it should not grow faster than the exponential function $Me^{\sigma 0t}$ for some finite σ_0 and a positive number M. The functions in the Laplace domain are usually denoted with capital letters, while the original functions are denoted with lower case letters.

The original function $f(t)$ can be restored out from $F(s)$ by means of the inverse Laplace transform; that is,

$$ f(t) = \mathcal{L}^{-1}\{F(s)\} = \int_{\sigma - j\infty}^{\sigma + j\infty} \mathrm{e}^{st} F(s)ds, \quad \sigma = \mathrm{Re}(s) > \sigma_0. \tag{2.14} $$

where $\mathcal{L}^{-1}$ is the inverse Laplace transform operator or the inverse $\mathcal{L}$ transform; σ_0 is located in the right half of the absolute convergence plane of the Laplace integral (2.13).

The convolution of two functions, $f(t)$ and $g(t)$, denoted by $f(t)*g(t)$, is given by

$$f(t)*g(t) = \int_0^t f(t-\tau)g(\tau)d\tau = \int_0^t f(\tau)g(t-\tau)d\tau. \tag{2.15}$$

The Laplace transform of (2.15) yields the product of the Laplace transforms of the two corresponding functions, i.e.,

$$\mathcal{L}[f(t)*g(t)] = F(s)G(s) \tag{2.16}$$

provided that $F(s)$ and $G(s)$ both exist.

Another useful property is the Laplace transform of the nth-order derivative of $f(t)$:

$$\mathcal{L}\{f^n(t)\} = s^n F(s) - \sum_{k=0}^{n-1} s^{n-k-1} f^{(k)}(0) = s^n F(s) - \sum_{k=0}^{n-1} s^k f^{(n-k-1)}(0), \tag{2.17}$$

which can be obtained from (2.13) via integration by parts with the assumption that the corresponding integrals exist.

2.3.2 *Laplace Transform of Fractional-Order Integrals*

Consider the RL definition of fractional-order integral for $\alpha > 0$ given by (2.2). The Laplace transform of $f(t)$ can be obtained by taking the Laplace transform of the convolution of two functions, $g(t) = t^{\alpha-1}$, and $f(t)$ as follows:

$$\mathcal{L}\{{}_0D_t^{-\alpha} f(t)\} = \mathcal{L}\{{}_0I_t^{\alpha} f(t)\} = \frac{1}{\Gamma(s)} \int_0^t (t-\tau)^{\alpha-1} f(\tau)d\tau = t^{\alpha-1}*f(t). \tag{2.18}$$

Notice that the Laplace transform of $t^{\alpha-1}$ is equal to

$$G(s) = \mathcal{L}\{g(t) = t^{\alpha-1}\} = \Gamma(\alpha)s^{-\alpha}. \tag{2.19}$$

Using (2.18) and (2.19) implies

$$\mathcal{L}\{{}_0D_t^{-\alpha} f(t)\} \;=\; s^{-\alpha} F(s). \tag{2.20}$$

It should be noted that the Laplace transform of the GL fractional-order integral is just the same as that of the RL one.

2.3.3 *Laplace Transform of Fractional-Order Derivatives*

To determine the Laplace transform of the RL fractional-order derivative of $f(t)$, assume that

$$ {}_0D_t^{\alpha} f(t) = g^n(t), \tag{2.21} $$

or

$$ g(t) = {}_0D_t^{-(n-\alpha)} f(t) = \frac{1}{\Gamma(n-\alpha)} \int_0^t (t-\tau)^{n-\alpha-1} f(\tau)d\tau; \ (n-1 \le \alpha \le n). \tag{2.22} $$

Applying (2.17) onto (2.22) yields

$$ \mathcal{L}\{{}_oD_t^{\alpha} f(t)\} = s^n G(s) - \sum_{k=0}^{n-1} s^k g^{(n-k-1)}(0). \tag{2.22} $$

Moreover, from (2.19), the Laplace transform of $g(t)$ is then equal to

$$ G(s) = s^{-(n-\alpha)} F(s) \tag{2.23} $$

and using (2.2) yields

$$ g^{(n-k-1)}(t) = \frac{d^{n-k-1}}{dt^{n-k-1}} {}_0D_t^{-(n-\alpha)} f(t) = {}_0D_t^{\alpha-k-1} f(t). \tag{2.24} $$

Substituting from (2.23) and (2.24) into (2.22) yields the following Laplace transform of the RL fractional-order derivative for $\alpha > 0$:

$$ \mathcal{L}\{{}_0D_t^{\alpha} f(t)\} = s^n G(s) - \sum_{k=0}^{n-1} s^k \left[f^{(n-k-1)}(0)\right]_{t=0}, \tag{2.25} $$

where $n-1 \le \alpha < n$.

2.4 Fourier Transform of Fractional-Order Operators

2.4.1 *Fundamentals of Fourier Transform*

The Fourier transform of a continuous absolutely integrable function, $h(t)$, for $t \in (-\infty, \infty)$ and defined as follows:

$$H(\omega) = \mathcal{F}\{h(t)\} = \int_{-\infty}^{\infty} \mathrm{e}^{-j\omega t} h(t)dt. \quad (2.26)$$

The corresponding original function, *h*(*t*), can be restored from its Fourier transform *H*(*ω*) by means of the inverse Fourier transform:

$$h(t) = \frac{1}{2\pi} \int_{-\infty}^{\infty} H(\omega)\mathrm{e}^{+j\omega t} d\omega. \quad (2.27)$$

As in the case of Laplace transform, the original functions will be denoted with lower case letters, while their transforms with capital letters.

The Fourier transform of the convolution of two functions, *h*(*t*) and *g*(*t*), defined over $t \in (-\infty, \infty)$ equals to the product of their Fourier transforms, i.e.,

$$\mathcal{F}\{h(t)^* g(t)\} = H(\omega)G(\omega) \quad (2.28)$$

provided that *H*(*ω*) and *G*(*ω*) exist (here *G*(*ω*) is the Fourier transform of *g*(*t*) function). Property (2.28) will be used to determine the RL Fourier transform of both fractional-order integrals and derivatives.

Another useful property of the Fourier transform often used to solve applied problems is the Fourier transform of the derivatives of *h*(*t*). Namely if *h*(*t*), $h'(t), \ldots,$ $h^{(n-1)}(t)$ tend to zero as $t \to \pm\infty$, then the Fourier transform of the *n*th derivative of *h*(*t*) is equal to

$$\mathcal{F}\{h^n(t)\} = (-j\omega)^n H(\omega). \quad (2.29)$$

This expression is true provided that the function *h*(*t*) with all its derivatives up to and including the $(n-1)$th one tends to zero as $t \to \pm\infty$.

2.4.2 Fourier Transform of Fractional-Order Integrals

To find the Fourier transform of any function, first estimate the Fourier transform of the RL fractional-order integral. Let the lower limit be $a = -\infty$. Then,

$$_{-\infty}D_t^{-\alpha} g(t) = \frac{1}{\Gamma(\alpha)} \int_{-\infty}^{t} (t-\tau)^{\alpha-1} g(\tau)d\tau, \quad (2.30)$$

where $0 < \alpha < 1$.

Now consider the following subsidiary function:

$$h(t) = \frac{t^{\alpha-1}}{\Gamma(\alpha)}.$$

where, from (2.8), its Laplace transform can be calculated as

$$\mathcal{L}\{h(t)\} = \frac{1}{\Gamma(\alpha)} \int_0^\infty t^{\alpha-1} \mathrm{e}^{-st} dt = s^{-\alpha}. \tag{2.31}$$

Suppose $s = j\omega$, where ω is real. According to Dirichlet theorem, the integral (2.31) converges if $0 < \alpha < 1$. Then, the Fourier transform of the following causal signal, $h_+(t)$:

$$h_+(t) = \begin{cases} \frac{t^{\alpha-1}}{\Gamma(\alpha)} & (t > 0) \\ 0, & (t \le 0) \end{cases}$$

is given

$$\mathcal{F}\{h_+(t)\} = (j\omega)^{-\alpha}. \tag{2.32}$$

Hence, the Fourier transform of the RL fractional-order integral (2.30) can be found using (2.28) as the Fourier transform of the convolution of the functions $h_+(t)$ and $g(t)$ as follows:

$$h_+(t) * g(t) = {}_{-\infty}D_t^{-\alpha} f(t) \tag{2.33}$$

Consequently,

$$\mathcal{F}\{{}_{-\infty}D_t^{-\alpha}\, g(t)\} = (j\omega)^{-\alpha}\, G(\omega). \tag{2.34}$$

The expression (2.34) is also known as the Fourier transform of the GL fractional-order integral ${}_{-\infty}D_t^{-\alpha}\, g(t)$, which is the same as that of the RL definition of fractional-order integral.

2.4.3 Fourier Transform of Fractional-Order Derivatives

Let the lower limit in (2.4) be $a = -\infty$. Using integration by parts, (2.4) of the RL derivative yields

$$ {}_{-\infty}D_t^{\alpha}g(t) = \frac{1}{\Gamma(n-\alpha)} \int_{-\infty}^{t} \frac{g^{(n)}(\tau)}{(t-\tau)^{\alpha-n+1}} d\tau = {}_{-\infty}D_t^{\alpha-n}g^{(n)}(t). \tag{2.35} $$

From (2.34) and (2.29) we obtain the following expression of the Fourier transform of the RL derivative of $\mathcal{F}\ \{{}_0D_t^{\alpha}g(t)\}$, i.e.,

$$ \begin{aligned} \mathcal{F}\{{}_0D_t^{\alpha}g(t)\} &= (-j\omega)^{\alpha-n}\mathcal{F}\{g(t)\} \\ &= (-j\omega)^{\alpha-n}(-j\omega)^{n}G(\omega) = (-j\omega)^{\alpha}G(\omega). \end{aligned} \tag{2.36} $$

This expression is formally the same as the Laplace transform of the integer-order derivative by replacing s by $j\omega$.

2.5 Dynamics of Fractional-Order Transfer Functions

2.5.1 Fractional-Order Transfer Functions

An arbitrary dynamical system of fractional order can be described by the following fractional-order differential equation:

$$ \begin{aligned} & a_nD^{\alpha_n}y(t) + a_{n-1}D^{\alpha_{n-1}}y(t) + \ldots + a_0D^{\alpha_0}y(t) = \\ & = b_mD^{\beta_m}u(t) + b_{m-1}D^{\beta_{m-1}}u(t) + \ldots + b_0D^{\beta_0}u(t) \end{aligned} \tag{2.37} $$

One may assume, without loss of generality, that $0 \le \alpha_0 < \alpha_1 < \alpha_2 \ldots < \alpha_n$, and $0 \le \beta_0 < \beta_1 < \beta_2 \ldots < \beta_m$. It follows from (2.37) that the transfer function $Y(s)/U(s)$ is given by

$$ \frac{Y(s)}{U(s)} \equiv T(s) = \frac{b_ms^{\beta_m} + b_{m-1}s^{\beta_{m-1}} + \cdots + b_0s^{\beta_0}}{a_ns^{\alpha_n} + a_{n-1}s^{\alpha_{n-1}} + \cdots + a_0s^{\alpha_0}}. \tag{2.38} $$

Here $D^{\gamma} \equiv {}_0D_t^{\gamma}$; a_i $(i=0,\ldots, n)$, b_i $(i=0,\ldots, m)$ are constant coefficients, while α_i $(i=0,\ldots, n)$, β_i $(i=0,\ldots, m)$ are arbitrary real numbers.

2.5.2 Mittag-Leffler Function

The significance of Mittag-Leffler function lies in their importance of solving fractional-order differential equations, similar to the importance of the exponential functions for the case of integer-order differential equations. A one-parameter Mittag-Leffler function is defined as

$$E_{\alpha}(x) = \sum_{k=0}^{\infty} \frac{x^k}{\Gamma(\alpha k + 1)}, \ \alpha > 0, \tag{2.39}$$

where $\alpha \in \mathbb{C}$, $\mathrm{Re}(\alpha) > 0$, $x \in \mathbb{C}$.

When $\alpha = 1$

$$E_1(x) = \sum_{k=0}^{\infty} \frac{x^k}{\Gamma(k + 1)} = \sum_{k=0}^{\infty} \frac{x^k}{k!} = \mathrm{e}^x. \tag{2.40}$$

This is why the Mittag-Leffler functions are also known as the generalized exponential function. The two-parameter Mittag-Leffler functions, which are a generalization of (2.40), take the following form:

$$E_{\alpha,\beta}(x) = \sum_{k=0}^{\infty} \frac{x^k}{\Gamma(\alpha k + \beta)}, \tag{2.41}$$

where α, $\beta \in \mathbb{C}$, $\mathrm{Re}(\alpha) > 0$, $\mathrm{Re}(\beta) > 0$, $x \in \mathbb{C}$.

2.5.3 *Solving Fractional-Order Differential Equation (FoDEQ) Using Laplace Transform*

The response of a linear system to an arbitrary input signal can be found by convolving the input signal with the system's unit impulse response. The unit impulse response describes the system behavior due to a unit pulse, which corresponds to the homogeneous solution of the differential equation that describes the system dynamics.

In engineering practice, it is preferable to determine the unit impulse responses for linear dynamical systems, or their responses due to an arbitrary input signal, in the frequency domain by using Laplace transform and its inverse.

In the latter case, the Laplace transform of the differential equation (2.37) is the system's transfer function $T(S)$ given by (2.38). If $U(s)$, ($\mathcal{L}\{u(t)\}$, is known, then the response of the linear system in the frequency domain equals to $Y(s) = T(\mathrm{s})U(s)$.

If the dynamic system is linear and of an integer order, then the transfer function and the system's response can be considered as rational transfer functions of complex frequencies. Thus, the most common practice in determining the system's response of linear systems in time domain is by writing it as a sum of simple exponential functions of the form

$$y(t) = \sum\nolimits_{i=0}^{n} k_i \mathrm{e}^{s_i t},$$

where k_i are the residues of $Y(s)$, and s_i, $i = 1, 2, \ldots, n$ are the roots of the characteristic polynomial $Y(s) = 0$.

The same approach can be used in the case of linear systems that obey fractional-order dynamics. In this regard, there is a problem to expand the fractional-order system transfer function into simple fractions and to figure out the inverse Laplace transform for each of them since the characteristic polynomial exhibits infinite number of roots.

To clarify this point, let us consider some examples of solving fractional-order differential equations using Laplace transform.

Example 1 Consider a system with the following transfer function:

$$H(s) = \frac{1}{s^{\alpha}} \tag{2.42}$$

Here α can be both fractional and integer number.

From (2.31), the inverse Laplace transform of this function is the unit pulse response of the system; that is,

$$h(t) = \frac{t^{\alpha-1}}{\Gamma(\alpha)}, \ \alpha > 0. \tag{2.43}$$

Example 2 Consider a linear fractional-order system with the following transfer function:

$$H(s) = \frac{1}{s^{\alpha} - a}. \tag{2.44}$$

This function can be represented as a sum of the first q components of a geometric series as follows:

$$H(s) = \frac{1}{s^{\alpha} - a} = \frac{1}{s - a^{q}} \sum_{k=1}^{q} a^{k-1} s^{(1-k\alpha)}, \tag{2.45}$$

where $q = 1/\alpha$.

Taking the inverse Laplace transform of (2.45) yields

$$h(t) = \frac{1}{a} \sum_{k=1}^{q} a^{k} E_{1-k\alpha}(t, a^{q}). \tag{2.46}$$

Here $E_{1-k\alpha}(t, a^{q}) = \mathrm{e}^{a^{q}t} \sum_{k=0}^{\infty} \binom{1-k\alpha}{m} \frac{t^{-1+k\alpha+m}}{\Gamma(k\alpha+m)} a^{qm}$ is the Mittag-Leffler function.

Hence, the inverse Laplace transform of (2.44) using simple fractions expansion makes it possible to derive the time domain system's response as a sum of weighted Mittag-Leffler functions (i.e., generalized exponents).

2.6 Fractional-Order Electrical and Electronic Systems

2.6.1 *Semi-infinite Transmission Line*

Let us consider a homogeneous electric transmission line of length l. Let R, C, G, and L be the line parameters. The sending end of the line ($x = 0$) is supplied with electromotive force $e_0(t)$. The termination end of the line ($x = l$) is connected to a load Z. That described above is shown in Fig. 2.1.

Let $v(x, t)$ be the voltage difference between a point x on the line and the common rail. Let also $i(x, t)$ be the value of the current at point x and at the time moment t. The voltage drop across the unit length is a sum of the inductance voltage drop and the ohmic voltage drop:

$$\frac{\partial v(x,t)}{dx} = -L\frac{\partial i(x,t)}{dt} - Ri(x,t). \tag{2.47}$$

The current change in this area is a sum of the insulation leakage current and the capacitive leakage current:

$$\frac{\partial i(x,t)}{dx} = -C\frac{\partial v(x,t)}{dt} - Gv(x,t). \tag{2.48}$$

Taking the time Laplace transform of the (2.47) and (2.48) yields the following systems of equations:

$$\frac{dV(x,s)}{dx} = -(sL + R)I(x,s), \tag{2.49a}$$

$$\frac{dI(x,s)}{dx} = -(sC + G)V(x,s). \tag{2.49b}$$

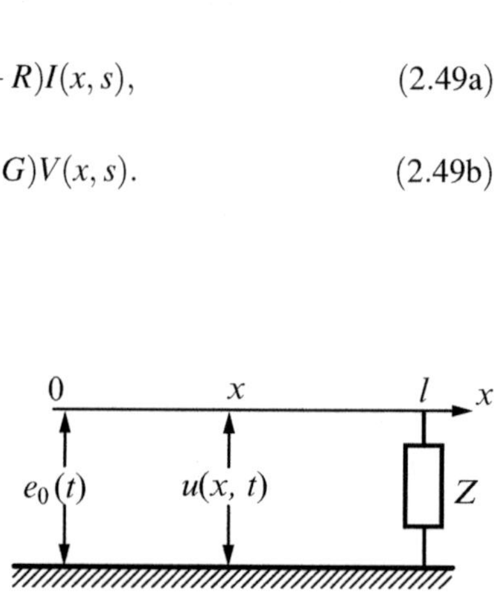

Fig. 2.1 Circuit diagram of a homogeneous electric transmission line

Let $Z_1(p) = sL + R$, and $Y_1(p) = pC + G$. The system of two first-order equations given by (2.49a and 2.49b) can be described as a single second-order equations for both the voltage and current as follows:

$$\frac{dV^2(x,s)}{dx^2} - \gamma^2 V(x,s) = 0; \tag{2.50a}$$

$$\frac{dI^2(x,s)}{dx^2} - \gamma^2 I(x,s) = 0, \tag{2.50b}$$

where $\gamma^2 = Z_1(s)Y_1(s)$.

The general solution of equation (2.50a) is given by

$$V(x,\ s) \ = \ Ae^{\gamma x} \ + \ Be^{-\gamma x} \tag{2.51}$$

and the general solution of equation (2.49a) is equal to

$$I(x,s) = \frac{\gamma\left(Be^{-\gamma x} - Ae^{\gamma x}\right.}{Z_1(s)}. \tag{2.52}$$

where the constant coefficients, A and B, are determined from the boundary conditions as follows:

$$V(0,\ s) \ = \ A \ + \ B; \tag{2.53a}$$

$$V(l,\ s) \ = \ Ae^{\gamma l} \ + \ Be^{-\gamma l}. \tag{2.53b}$$

It is obvious that $V(0,\ s) = E_0(s)$ as long as the load impedance equals $Z(s) = V(l,\ s)/I(l,\ s)$.

The second boundary condition can be expressed as follows:

$$Ae^{\gamma x} + Be^{-\gamma x} = Z(s)\frac{\gamma\left(Be^{-\gamma x} - Ae^{\gamma x}\right.}{Z_1(s)}. \tag{2.54}$$

where the impedance $Z_1(p)/\gamma$ is called the *wave impedance* or the *characteristic impedance* of the system, i.e.,

$$Z_c(s) = \sqrt{Z_1(s)/Y_1(s)} \tag{2.55}$$

Therefore, in order to calculate the constant coefficients, A and B, one has to solve the following system of equations:

$$A + B = E_0(s); \tag{2.56}$$

$$\left(\frac{z(s)}{z_c(s)}+1\right)Ae^{\gamma l}=\left(\frac{z(s)}{z_c(s)}-1\right)Be^{-\gamma l}. \tag{2.57}$$

If the line load impedance equals the line characteristic impedance (i.e., the case of matched transmission line without reflected waves), then it follows from (2.53) that

$$A = 0; B = E_0(s); V(s) = E_0(s)e^{-\gamma x}; I(s) = V(s) / Z_c(s).$$

If $L=0$ and $G=0$, we obtain a semi-infinite *RC* line, the characteristic impedance of which equals

$$Z_c(s)=\sqrt{\frac{R}{C}}S^{-1/2}, \tag{2.58}$$

and that implies

$$I(s)=\sqrt{\frac{C}{R}}s^{-1/2}V(s). \tag{2.59}$$

Then the time-domain fractional-order differential equation of the semi-infinite *RC* line will be

$$i(t)=\sqrt{\frac{C}{R}}D_t^{0.5}v(t). \tag{2.60}$$

According to (2.60), the current in such line is directly obtained from the half-order derivative of the voltage.

2.6.2 Electrochemistry

The principal objective of electrochemical analysis is to determine the concentration $\rho(x, t)$ of electrochemically active elements (for example, ions of some substance in the solution) on the electrode surface (at $x=0$). Direct measurement of $\rho(x, t)$ is quite difficult. But it is much easier to experimentally measure the surface current density $j_x(0, t)$, which relates with concentration $\rho(x, t)$ as follows:

$$j_x(0,t)=\left|-K\frac{\partial\rho(x,t)}{\partial x}\right|_{x=0}. \tag{2.61}$$

Using (2.61) requires solving the diffusion equation at the right half-plane (inside the electrode):

$$\frac{\partial \rho(x,t)}{\partial t} = K\frac{\partial^2 \rho(x,t)}{\partial x^2}.$$

The process of solving this equation can be simplified if the "square root" of the operators in both sides is found. Substituting $\frac{\partial \rho(x,t)}{dx}|_{x=0}$ from (2.61) into the above equation yields

$$_0D_t^{0.5}\rho(0,t) = K^{0.5}\frac{\partial \rho(x,t)}{\partial x}|_{x=0} = -K^{-0.5}j_x(0,t).$$

Hence, the concentration of chemically active elements on the electrode surface is calculated as

$$\rho(0,\ t) = -K^{-0.5}j_x(0,\ t).$$

The last expression is considered as the basis of creating modern chemical analysis and devices.

2.6.3 Rough Surface Impedance

The performance characteristics of electrochemical devices and their internal processes are largely specified with the properties of the surface of the metallic electrode that contacts the liquid or a solid electrolyte. The simplest model that affects the edges of the alternating current through the system is a serial connection of an external electrical capacitor and an electrolyte impedance resistor. In this case, the real part of the impedance of this model obviously does not depend on frequency, and the imaginary part is inversely proportional to frequency.

However, when the roughness of the electrode surface is quite significant, the system's behavior varies, at least within the limited frequency range. The frequency dependence impedance is also influenced by an additional power term of the form $A(j\omega)^{-\eta}$ where η is between zero and one ($0 < \eta < 1$).

The input impedance of the "rough electrode-electrolyte" interface is modeled with a special element in the equivalent circuit of its input impedance. Since the phase frequency response of such an element is constant, it is called a *constant phase element* (CPE). The index η of such a CPE depends on the surface roughness; that is, the smoother the surface the closer η to 1 and vice versa.

Studies of rough electrode surfaces by means of electronic microscopes revealed the fact that their images had no natural scale of length (just like in the case of fractal surfaces) and these images were the same at different rates of magnification.

To model the "rough electrode-electrolyte" system and to estimate its impedance, we will use a Cantor kernel regular fractal (see Fig. 1.5). Here all black parts symbolize an electrolyte contacting the metallic electrode (white color). Every

$$Z(j\omega) = R + \cfrac{1}{j\omega C + \cfrac{2}{aR + \cfrac{1}{j\omega C + \cfrac{2}{a^2R + \ldots}}}}.$$

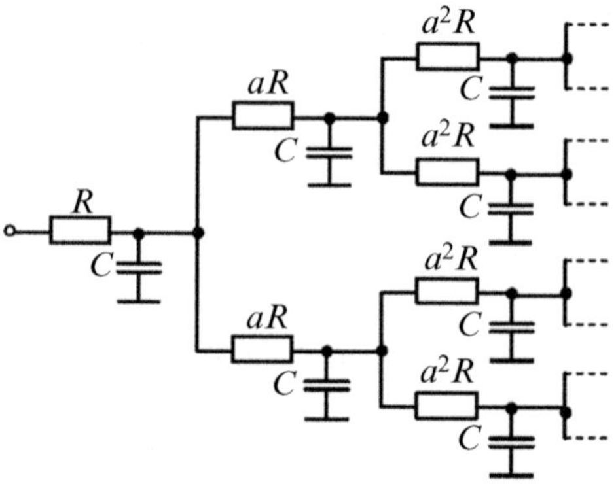

Fig. 2.2 Equivalent circuit of the "rough electrode-electrolyte" interface

stage of building the kernel for this system is as follows: the middle part of every segment is removed so that the length of each remaining segment equals $1/a$ ($a < 2$) of the initial segment length.

The current from the electrolyte towards the electrode encounters ohmic resistance of the electrolyte and the surface of electrical capacity at every area of the surface. The equivalent electrical circuit diagram of such system is shown in Fig. 2.2.

Every new stage of building of the Cantor manifold is reflected by the circuit branching. The impedance of every following branch increases a-fold as long as the thickness of the corresponding surface high point decreases a-fold. The number of capacity elements that model the surface capacities of the high points also increases at every following stage. All capacity elements have the same capacity value. The input impedance of the circuit depicted in Fig. 2.2 can be expressed with a continued fraction expansion (CFE) as follows:

The function $Z(\omega)$ can be written as the following scaling expression:

$$Z\left(\frac{\omega}{a}\right) = R + \frac{aZ(\omega)}{j\omega C \cdot Z(\omega) + 2}. \tag{2.62}$$

The expression of (2.62) is true when $Z(\omega) = A(j\omega)^{-\eta}$, where A is a constant coefficient, where $\eta = 1 - d$ and $d = \ln 2/\ln a$ represent the Cantor manifold fractal dimensions.

Hence, the circuit of Fig. 2.2 describes a constant phase element (CPE) as long as the Cantor manifold, d, satisfies $0 < d < 1$, and the parameter η is also limited between zero and one, $0 < \eta < 1$. Notice that $\eta = 3 - d_s$ for fractal dimension of the considered interface. When the surface is smooth, $d_s = 2$, which implies $\eta = 1$ and that coincides with the experimental results.

The power law dependence between impedance and frequency is specified with a reasonable set of resistive and capacitive current paths. Since capacitors block dc signals, a low-frequency signal propagates farther through the circuit before it "leaves" via the surface capacity, so the impedance is higher at lower frequencies. Real surfaces are self-similar only for scales of finite interval. This determines the frequency range within which the phase is constant.

The dependence between the effective complex value of the current and that of the voltage for rough surface (the generalized Ohm's law) is described as follows:

$$\dot{I}(j\omega) = \left[\dot{Z}(j\omega)\right]^{-1}\dot{U}(j\omega) = A^{-1}\cdot(j\omega)^{\eta}\dot{U}(j\omega).$$

The relation between the instantaneous current and voltage values will be

$$i(t) = A^{-1}{}_0D_t^{\eta}u(t).$$

The last expression shows that this dependence possesses fractional-order derivative in time domain.

Test questions

1. Explain why physical processes with "memory" are described in terms of fractional derivatives.
2. Give examples of physical and chemical processes that have hereditarity.
3. Formulate the concept of gamma-function.
4. Derive the Riemann-Liouville fractional integral expression via the integer-order multiple integral.
5. Derive the Grunwald-Letnikov fractional integral expression.
6. List the properties of fractional-order derivatives.
7. Develop the expression of the fractional-order integral (derivative) Laplace transform.
8. Demonstrate the relation between the fractional-order dynamic system differential equation and its transfer function.
9. Develop the Laplace transform method for solving the fractional-order differential equations.

Chapter 3
Fractal Elements

3.1 Fractal Impedances and Fractal Element

It is well known that state equations of electric circuits that contain passive *RLC* elements with lumped parameters (*RLC*-ELP) are ordinary differential equations, the order of which is the same as that of the number energy-storing elements. For example, the electric processes of an *LC* oscillator with losses due to the resistance *R* (see Fig. 3.1) are described by the following integer-order integral-differential equation:

$$C\frac{du(t)}{dt}+\frac{u(t)}{R}+\frac{1}{L}\int u(t)dt=i_{IN}(t).$$

This type of equation can be used to describe oscillations, not only in electric circuits, but also in other types of physical systems such as pendulum oscillations and vibration of buildings or platforms. That is why analog modeling is widely used in engineering applications where the behavior of some nonelectrical system is analyzed by their electric-equivalent circuit, which yields a reduction of time and cost of modeling.

Traditionally, the *RLC*-ELP-based equivalent circuits are described by integer-order differential equations since the energy-consuming *L* and *C* elements exhibit first-order dynamics; that is, the current-voltage relationship is described by first-order ordinary differential equations. In particular, the time-domain relation of the current, voltage, and electric charges in a capacitor, *C*, is given by

$$i_c(t)=\frac{dq_c(t)}{dt}=C\frac{du_c(t)}{dt}, \tag{3.1a}$$

where i_c, u_c, and q_c are the current, voltage, and electric charges of the capacitor, respectively. Taking the Laplace transform of (3.1a) gives

A.Kh. Gil'mutdinov et al., *Fractal Elements and their Applications*, Analog Circuits and Signal Processing, DOI 10.1007/978-3-319-45249-4_3

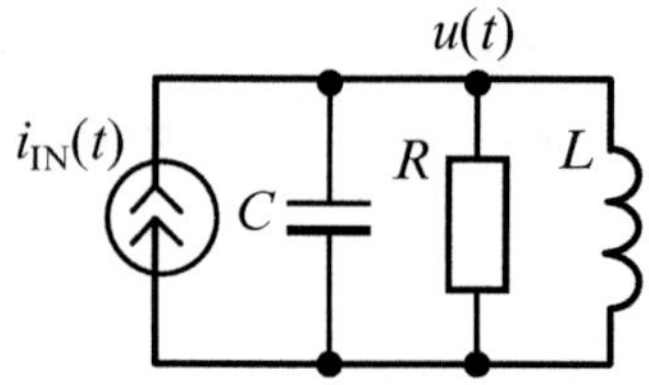

Fig. 3.1 Parallel LC oscillating circuit with losses

$$I_C(s) = sCU_C(s) = Y_C(s)U_C(s), \tag{3.1b}$$

where $s = \sigma + j\omega$ is the complex frequency.

Similarly, the differential equation that relates the voltage u, flux Ψ_L, and current i_L of an inductance L in time domain is described as

$$u_L(t) = -e_L(t) = \frac{d\Psi_L(t)}{dt} = L\frac{di_L(t)}{dt}$$

and its Laplace transform is:

$$U_L(s) = sLI_L(s).$$

However, when a dynamic system enjoys hereditary effect, it should be described by a fractional-order differential equation. The equivalent electric circuit of such systems should comprise at least one fractional order element (FOE), say a fractional-order capacitor of capacitance, C_{F_α}. Hence, from (3.1a), the time-domain current-voltage relation of a FOE (capacitor) can be expressed by a non-integer-order differential equation, of order α, as follows:

$$i_{F_\alpha}(t) = C_{F_\alpha}\frac{d^\alpha u_{F_\alpha}(t)}{dt^\alpha}, \quad 0 < \alpha < 1. \tag{3.1c}$$

Taking the Laplace transom of (3.1c) gives

$$I_{F_\alpha}(s) = C_{F_\alpha}s^\alpha U_{F_\alpha}(s) = Z_{F_\alpha}^{-1}(s)U_{F_\alpha}(s), \tag{3.1d}$$

where $Z_{F_\alpha}(s) = 1/C_{F_\alpha}s^\alpha$ is the input impedance of the fractional-order capacitor, C_{F_α}.

A comparison between (3.1c) and (3.1a) reveals the physical meaning of a fractional (fractal) capacitance. A similar comparison for a fractal inductor is deduced from (3.1c) and (3.1d).

The homogeneous semi-infinite RC line (see Sect. 2.6.1) can also be classified as a FOE element where the input current-voltage relation exhibits fractional-order dynamics of order $\alpha = 1/2$ and is given here for completeness:

$$i(t) = \sqrt{\frac{C}{R}}\frac{d^{1/2}u(t)}{dt^{1/2}}$$

Then,

$$I(s) = \sqrt{\frac{C}{R}}s^{1/2}U(s) \equiv \frac{\theta}{R}U(s), \quad \text{or} \quad I(s) = Z_{F_{0.5}}^{-1}U(s), \tag{3.1e}$$

where $Z_{IN}(s) = Z_{F_{0.5}}(s) = \left(\frac{R}{sC}\right)^{1/2}$ is the homogeneous semi-infinite *RC* line (hereinafter referred to as the *RC* line) input impedance. From (3.1e) it follows that $\theta = (SRC)^{\frac{1}{2}} = (s\tau)^{\frac{1}{2}}$, where $\tau = RC$ is the propagation time constant, *R* is the *RC* line overall resistance, and *C* is the *RC* line overall capacitance.

Since $Z_{IN}(s) = Z_{F_{0.5}}(s) = \left(\frac{R}{sC}\right)^{1/2}$ depends on the complex frequency of fractional power of order ½, it is usually called the half-order fractal impedance. Comparing (3.1e) and (3.1d) implies that with fractional differentiation of order $\alpha = 0.5$, the numerical value of the fractional (fractal) capacity is equal to $Z_{F_{0.5}} = \sqrt{C/R} = C\tau^{-0.5} = R^{-1}\tau^{0.5}$. In this case, the corresponding measurement units of fractal capacitance will be either Farad $s^{-0.5}$ or Ohm^{-1} $s^{0.5}$. This point is due to the fact that the overall electric capacitance of the *RC* line is not the same as that of the fractal capacitance of the line.

One way *to build a half-order capacitor is to use the homogeneous RC line* by constructing a lumped RC-ladder circuit (i.e., *R* and *C* elements with lumped parameters hereinafter referred to as an *RC*-ELP). It represents another way to implement *the half-order fractal capacitance*. In the early 1960s Professor R. Sh. Nigmatullin was *the first* to physically fabricate factional-order elements based on their electrochemical properties of transferring information. He used FOEs as the basic blocks of electric circuits and called them "*recond*" and "*reind*" (shortened combinations that correspond to resistor-capacitor (in Russian "condensator") and resistor-inductance, respectively).

One may generalize this particular case of fractal impedance; the half-order fractal (fractional) *RC* element, in terms of the homogeneous *RC*-line circuit as follows:

$$Z_{IN}(s) = Z_{F_{0.5}}(s) = \frac{R}{(sRC)^{1/2}} = \frac{R}{(s\tau)^{1/2}}.$$

As a result, an arbitrary fractional-order input impedance, $Z_{F_\alpha}(s)$, of any fractional order α, can be expressed as follows:

$$Z_{F_\alpha}(s) = \frac{|Z_c|}{(s\tau)^\alpha}, \tag{3.2}$$

where Z_c is the value of the input impedance of a fractal (fractional) *RC* element at frequency $\omega_c, \tau = 1/\omega_c$.

Just like in case of expression (3.1d), the fractal (fractional) element input impedance will hereinafter be called the fractal impedance, and the element will be called the fractal element.

It should be noted that, unlike the values of the integer-order capacitance and inductance *C* and *L*, the values of the fractal element are determined in terms of several parameters, $|Z_c|$, τ, and α, as defined in (3.2). Hence, the complex frequency response (CFR) of fractal impedance for real values of α can be expressed as

$$Z_{F_\alpha}(j\omega) = \frac{|Z_c|}{\tau^\alpha}(j\omega)^{-\alpha}. \tag{3.3a}$$

A normalized fractal impedance, $\bar{Z}_{F_\alpha}(j\omega)$, at $s = j\omega$ is defined as

$$\bar{Z}_{F_\alpha}(j\omega) = \frac{Z_{F_\alpha}(j\omega)}{|Z_c|}\tau^\alpha = (j\omega)^{-\alpha} = \omega^{-\alpha}e^{-\frac{j\alpha\pi}{2}}. \tag{3.3b}$$

Then the logarithmic magnitude frequency response (LMFR) and the phase frequency response (PFR) of the normalized CFR of the fractal impedance $\bar{Z}_{F_\alpha}(j\omega)$ will correspondingly be

$$\log \bar{Z}_{F_\alpha}(j\omega) = \alpha \log\omega, \tag{3.3c}$$

$$\varphi_z(\omega) = \alpha\frac{\pi}{2}. \tag{3.3d}$$

The LMFR and PFR graphs of a one-port normalized fractal impedance, $\bar{Z}_F$, described by (3.3c) and (3.3d) are shown in Fig. 3.2 for different values of α.

In extreme cases when $\alpha = 1$, $\alpha = -1$, or $\alpha = 0$, the one-port Z_F becomes a capacitive, inductive, or resistive element, respectively.

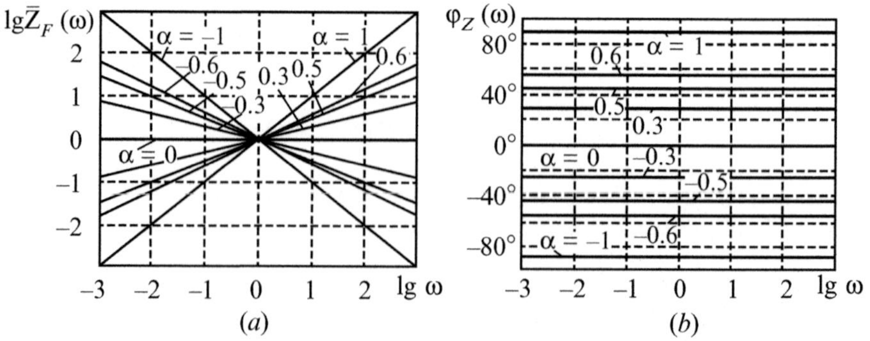

Fig. 3.2 (**a**) LMFR and (**b**) PFR of the one-port normalized fractal impedance $\bar{Z}_F$ (hereinafter α is omitted for simplicity)

When simulating objects and processes by means of *RLC*-ELP-based circuits, the response of the system is calculated through an input signal under certain boundary and initial conditions by solving and simulating integer-order differential equations, which describe the system, with a certain precision. In the same cases, simulation circuits of fractal elements can be used to solve the corresponding fractional-order differential equations.

It should be noted that conventional electronic elements (components), including capacitors and inductors, could demonstrate fractal impedance under specific conditions. For example, the inductor coil performs such properties within a frequency range of the presence of the conductor's skin effect, while for capacitors, their fractal properties appear at certain types of insulators with complex electrical permittivity. However, as a rule of thumb, from their impedance frequency response, the parameters do not significantly differ from one another since fractal properties are not dependent on such elements but most likely caused by manufacturing errors of these parameters.

It is then necessary to define basis of design and fabrication of fractal elements using basic performance parameters of certain values that would be provided at the stages of design development, manufacturing, and schematic application. In addition, the stability of these parameters and their performance should be maintained for specified operation conditions. The following section outlines some known implementation methods of fractal impedances.

3.2 Implementation of Fractal Impedances Using Electrochemical Converters

As it was stated in Chap. 2, as early as in 1926, research was performed on studying the electric properties of the interface between the metallic electrode and water electrolyte. It was discovered that at a limited frequency range the impedance frequency response contained the $(j\omega)^{-\alpha}$ term where $0 < \alpha < 1$. This indicated that a proper expression to model the input impedance of the "electrode-electrolyte" interface is the one that includes special element of constant phase frequency response (PFR). These elements are called *constant phase element* (CPE).

Thus, devices that operate on the basis of electrochemical processes that take place at electrodes submerged in electrolytes or polar liquids could be expected to perform the functions of fractal elements.

Thorough theoretical investigations of such elements (afterwards called the electrochemical converters of information (ECCI)) and their practical use to perform fractional (half)-order integration and differentiation were conducted first by Professor R. Sh. Nigmatullin in the early 1960s in Kazan Aviation Institute (today Kazan National Research Technical University named after A. N. Tupolev). This research then was continued and enhanced by his followers. R. Sh. Nigmatullin with his followers started using the ECCIs as discrete elements with constant

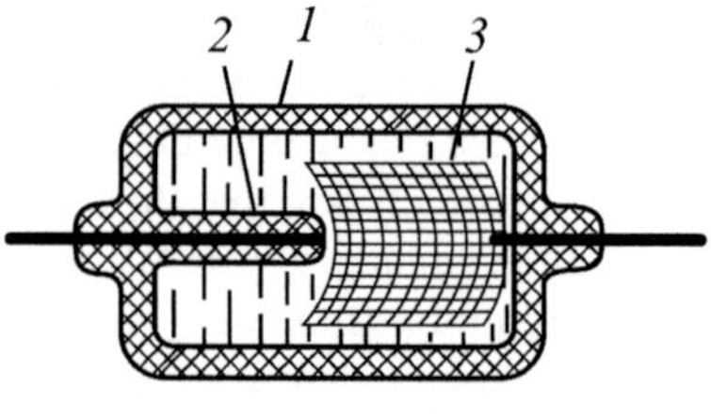

Fig. 3.3 Electrochemical diode structure: (*1*)—case; (*2*)—small electrode; (*3*)—large electrode

phase in polarography in the early 1960s. In 1963 A. I. Miroshnikov, a follower of R. Sh. Nigmatullin, was the first in the world to successfully pass his PhD thesis on physical implementation of ECCI-based CPE and its application.

3.2.1 Liquid Electrolyte Electrochemical Signal Converters

R. Sh. Nigmatullin with his followers proposed a structure of an electrochemical fractal element as shown in Fig. 3.3. It contains two platinum electrodes encased into a vacuum-sealed glass bulb *1* filled with water solution of potassium ferrocyanide $K_4[Fe(CN)_6]$ and potassium ferricyanide $K_3[Fe(CN)_6]$ at the same concentration. The macro-electrode *3* square surface is significantly larger (more than 40-fold) than the surface of the microelectrode *2*. Applying a voltage to the electrodes causes the following chemical reactions:

Reduction at the cathode:

$$\left[\mathrm{Fe(CN)_6}\right]^{3-} + e \rightarrow \left[\mathrm{Fe(CN)_6}\right]^{4-},$$

where *e* symbolizes an electron

Oxidation at the anode:

$$\left[\mathrm{Fe(CN)_6}\right]^{4-} + e \rightarrow \left[\mathrm{Fe(CN)_6}\right]^{3-}.$$

From the reactions above it follows that the given oxidation-reduction process does not change the concentration of the initial substances in the solution. Furthermore, if the applied voltage does not exceed the solvent decomposition potential (0.9 V for water), then such element is able to operate for an arbitrarily long period.

An equivalent circuit of the ECCI with hemispherical microelectrode is shown in Fig. 3.4. This circuit takes into consideration the double-electric layer electric capacity for small signals. Here, R_K represents the chemical resistance; $Z_1(\omega)$ and $Z_2(\omega)$ are the impedances of the transient diffusion of the oxidizer and the reductant, respectively (the Warburg impedance); R_1 and R_2 are the corresponding resistances of steady-state diffusion; R_0 is the volume resistance, while C_0 is the double-electric layer electric capacity.

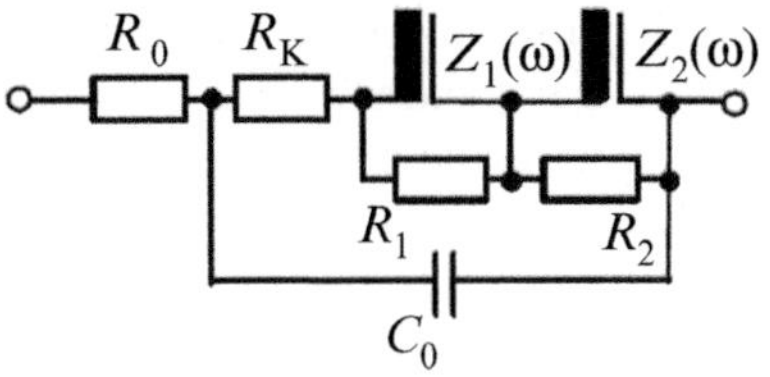

Fig. 3.4 Equivalent circuit of the electrochemical diode (ECCI)

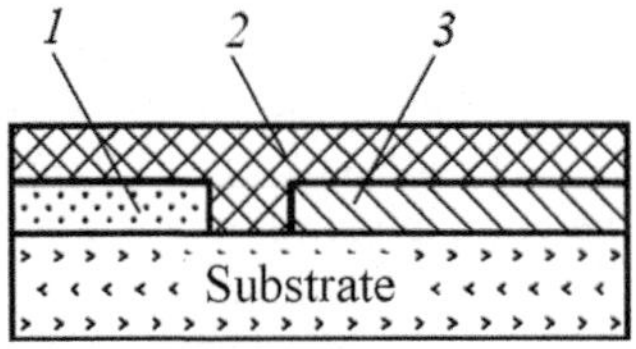

Fig. 3.5 Planar electrochemical converter: (*1*) electrode (Pt); (*2*) solid electrolyte; (*3*) electrode (Ag)

Assume that the following inequalities hold:

$$\begin{array}{lcl} R_1 >> Z_1(\omega); & & R_2 >> Z_2(\omega); \\ Z_1(\omega) + Z_2(\omega) & = & Z(\omega) >> R_K; \\ 1/(\omega C_0) >> Z(\omega); & & R_0 << Z(\omega). \end{array}$$

It means that the curvature of the microelectrode surface is insignificant, the electrochemical reaction velocity is infinitely high, and the double-electric layer capacity and the volume resistance can be omitted. Then the expression of the resistance of the considered one-port (the electrochemical diode) reduces to

$$Z(j\omega) = \frac{1}{A\sqrt{\omega}} e^{-j\frac{\pi}{4}},$$

where A is constant of a specific value that depends on the concentrations of the oxidizer and the reductant, the applied bias voltage, and the size of the microelectrode.

In the early 1960s, Professor R. Sh. Nigmatullin *first* recommended such electrochemical element (ECCI) (which is called today the fractal element) to perform mathematical operations of half-order fractional integration and differentiation (FID). However, the presence of liquid electrolyte and the need to vacuum and seal the electrochemical diode container made it difficult to use this device as compact components produced via the integrated circuit technology.

3.2.2 *Solid Electrolyte Electrochemical Signal Converters*

A planar or a solid electrochemical converter shown in Fig. 3.5 is easier to manufacture than the liquid one. It was proposed and developed by R. Sh. Nigmatullin with his follower professor F. A. Karamov. A rubidium silver iodide ($RbAg_4I_5$) solid electrolyte is used as an insulator.

The silver electrode is the only reversible substances like $RbAg_4I_5$ and α- and β-AgI that facilitate silver ion conductivity. Experiments proved that platinum electrodes could be used as polarizable electrodes.

The corresponding frequency dependence of the magnitude and phase of the impedance is described as follows:

$$|Z_{Ag}| = A \cdot \omega^{-\alpha};\ \varphi(\omega) = \arg Z_{Ag}(j\omega) = -\frac{\pi}{2}\alpha = \text{const.}$$

where A and α are constant coefficients.

The values of α for different samples of the converter vary from 0.66 to 0.68 and the corresponding phase angle $\varphi(\omega)$ is bounded between 60° and 62°.

It should be noted that the parameters of the input impedance of such ECCI are sufficiently sensitive to temperature variation. For example, when the temperature changes from −20 °C to +50°, the magnitude of the impedance increases hundred-fold, α changes from 0.64 to 0.82, and the corresponding phase angle $\varphi(\omega)$ varies from 57.5° to 74.5°.

3.2.3 Fractal Element Fractor™

In 2002, a new CPE design was proposed. This design was specific using fractal geometry properties of the electrode-electrolyte interface. It was called Fractor™ (G. Bohannan, USA). The structure and the general view of the device are depicted in Fig. 3.6.

The dimension of the device is 3.5 × 3.5 × 1.0 cm. The internal conducting plate possesses rough surfaces that were specially treated to provide the required surface structure that specifies the value of α. Gel with lithium ions is used as electrolyte.

Experimental graphs of the impedance MFR and PFR of one form of design of Fractor™ are shown in Fig. 3.7. The magnitude and phase angles lie within the following ranges: the phase angle varies from −30° to −60° with a ±5° fluctuation over five decades of frequency range. These parameters strongly depend on the temperature, electrolyte content, electrode material, degree of processing of the electrode surface, and electrode plate squeeze rate.

Since the operation of the considered fractal elements is diffusion and drift of the electrolyte ions, these elements operate properly over low- and infra-low-frequency ranges (from fractions of Hz to units of kHz).

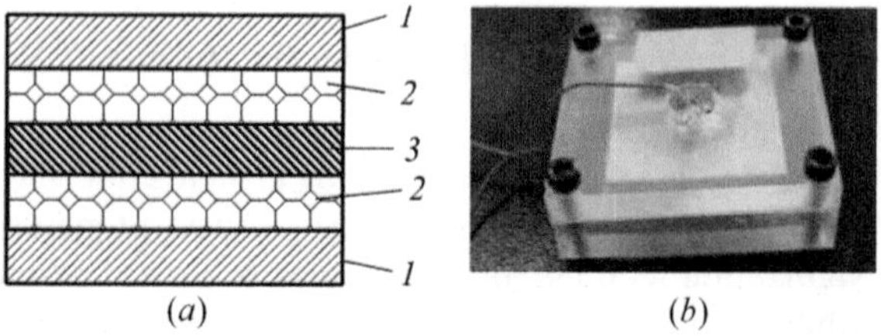

Fig. 3.6 Fractal element Fractor™: (**a**) section of the structure; (**b**) general view; (*1*) external electrode; (*2*) electrolyte; (*3*) internal conducting plate

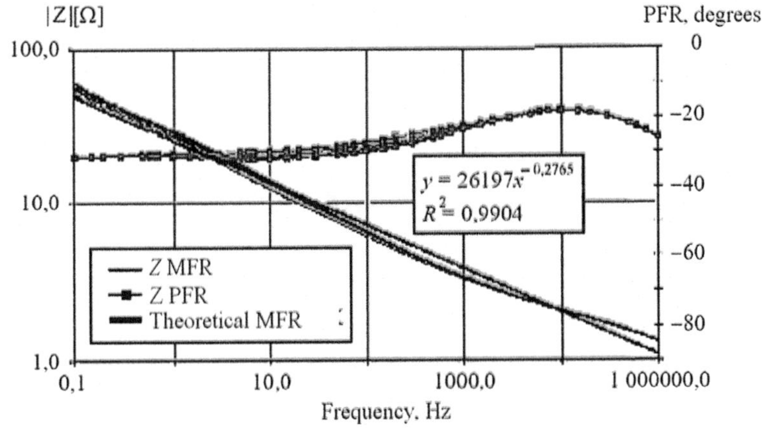

Fig. 3.7 Frequency response of experimental sample of Fractor™

3.3 Implementation of Fractal Impedances (Immittances) Using *RC* Circuits with Lumped Parameters

3.3.1 Preliminaries

The problem of synthesis of electric circuits by means of *RLC*-ELP using the flat phase response criterion of the input impedance (admittance) over a limited frequency range has already been successfully solved as early as in the late 1950s. At that time, such circuits were used as loads for transmission lines connected to one arm of the measurement bridge used to measure the biological tissue impedance at wide frequency ranges. Later, such circuits were widely considered as equivalent circuits of ECCI and of other devices, objects, and processes that exhibit frequency-dependent input impedances as described by (3.3a–3.3d). The CPEs were also commonly used to design fractional-order proportional-integral-differential (PID) regulators embedded into the control systems used to control fractional-order objects and processes.

Notice that regular structures like the Foster forms and the Cauer forms branching circuits are used to synthesize the described above circuits. By now, numerous methods of synthesis have been proposed. They differ by the approximation of their transfer functions. They are expanded using analytical methods to calculate the parameters of their equivalent circuits.[1]

[1]See, for instance, E.A. Guillemin. Synthesis of passive networks: theory and methods appropriate to the realization and approximation problems. John Wile and sons, New York. 1957.

Here we just consider the realization of fractal impedances (immittances) by means of *RC*-ELP since it is difficult to design fractal inductance suitable for frequencies lower than tens of MHz using integrated circuit (IC) technology.

3.3.2 Properties of Input RC-Transfer Functions

Let the input impedance of *RC* circuit be the sum of simple transfer functions as follows:

$$Z_{RC}(s) = Z_1(s) + Z_2(s) + \cdots + Z_n(s) = \sum_{i=1}^{n} Z_i(s). \tag{3.4}$$

Each impedance, $Z_i(s)$; $i = 1, 2, \ldots, n$, as given in (3.4), obviously represents the input transfer function of some elementary one-port *RC* circuits. In addition, $Z_{RC}(s)$ in (3.4) can also be implemented by serially connecting these elementary one-port-transfer functions $Z_i(s)$. The first summand represents a resistive element, the second one is a capacitive element, and the rest of them are parallel combination of resistive and capacitive elements. Hence, the input impedance function can be described as follows:

$$Z_{RC}(s) = R_1 + \frac{1}{sC_2} + \sum_{i=3}^{n} \frac{1/C_i}{s + 1/R_iC_i} \tag{3.5}$$

or

$$Z_{RC}(s) = k_\infty + \frac{k_0}{s} + \sum_{i=3}^{n} \frac{k_i}{s - q_i}, \tag{3.6}$$

where

$$k_\infty = Z_{RC}(\infty);\, k_0 = \operatorname{Res} Z_{RC}(s)|_{s=0} = sZ_{RC}(s)|_{s=0},$$
$$k_i = \operatorname{Res} Z_{RC}(s)|_{s=q_i} = (s - q_i)Z_{RC}(s)|_{s=q_i};\; i = 3, 4, \ldots, n.$$

Clearly, k_0 and k_i are real positive residues of the $Z_{RC}(s)$ transfer function.

$q_i = \sigma_i = -(1/R_iC_i) = -1/\tau_i$ is the pole of the *i*th-input impedance, $Z_i(s)$, with τ_i its corresponding time constant for $i = 3, 4, \ldots, n$.

Now we formulate some general properties of the *RC* one-port input impedance.

Property 1 All poles and zeroes of the $Z_{RC}(s)$ of the one-port *RC* form of the input transfer functions are located on the negative real axis of the open left-half *s*-plane.

Proof Find a common denominator for (3.6) and rewrite it as follows:

$$Z_{RC}(s) = \frac{N(s)}{s\prod_{i=3}^{n}(s-q_i)}. \tag{3.7}$$

In general, there is one pole at the origin and the other poles are located at $q_i = \sigma_i = -(1/R_iC_i) < 0$. Then, due to duality property of $Z(s)$ and $Y(s)$, we can state that the $Z_{RC}(s)$ defines an input immittance function with real poles located on the negative real axis of the open left-half s-plane. Since $Y_{RC}(s) = 1/Z_{RC}(s)$, i.e., the poles of $Y_{RC}(s)$ are the zeroes of $Z_{RC}(s)$, then Property 1 of Z_{RC} is verified.

Property 2 The one-port *RC* form of the $Z_{RC}(s)$ input impedance function cannot possess a pole at $s = \infty$.

Proof Since $\lim_{s\to\infty} Z_{RC}(s) = Z_{RC}(\infty) = k_\infty = R_1$ is a finite number, then $s = \infty$ cannot be a pole of $Z_{RC}(s)$.

Property 3 $Z_{RC}(\infty) < Z_{RC}(0)$.

Proof It is obvious that if $k_0 \neq 0$, then there is an elementary one-port consisting of a capacitive element C_1; that is, $Z_{RC}(s)$ possesses a pole at $s = 0$ and $Z_{RC}(0) = \infty$. If $k_0 = 0$, i.e., the capacitive element C_1 is absent, then

$$Z_{RC}(0) = k_\infty - \sum_{i=3}^{n}\frac{k_i}{q_i} = Z_{RC}(\infty) + \sum_{i=3}^{n} R_i > Z_{RC}(\infty).$$

Property 4 The input impedance function $Z_{RC}(s)$ monotonously decreases along the negative real axis of the open left-half s-plane, i.e., when $Z_{RC}(s)|_{s=\sigma} = Z_{RC}(\sigma)$ (excluding the poles of $Z_{RC}(s)$ when $s = q_i = \sigma_i = -(1/R_iC_i)$).

Proof Find the first derivative of (3.6) with respect to s as follows:

$$\frac{dZ_{RC}(s)}{ds}\bigg|_{s=\sigma} = \left\{-\frac{k_0}{s^2} - \sum_{i=3}^{n}\frac{k_i}{(s-q_i)^2}\right\}\bigg|_{s=\sigma} = -\frac{k_0}{\sigma^2} - \sum_{i=3}^{n}\frac{k_i}{(\sigma-q_i)^2}.$$

This expression is true for all $\sigma \neq q_i$ since $k_0 > 0$ and $k_i > 0$. This conclusion leads to the following two properties.

Property 5 The zeroes and poles of the one-port *RC* form of the input impedance transfer function, $Z_{RC}(s)$, are prime numbers and are located in alternating manner on the negative real axis of the s-plane. The nearest characteristic root to the origin is a pole of $Z_{RC}(s)$, and the characteristic point nearest to $s = \infty$ is a zero of $Z_{RC}(s)$.

Proof If the one-port *RC* representation of the input impedance function, $Z_{RC}(s)$, had two consecutive poles or zeros, then this function along the negative real axis of

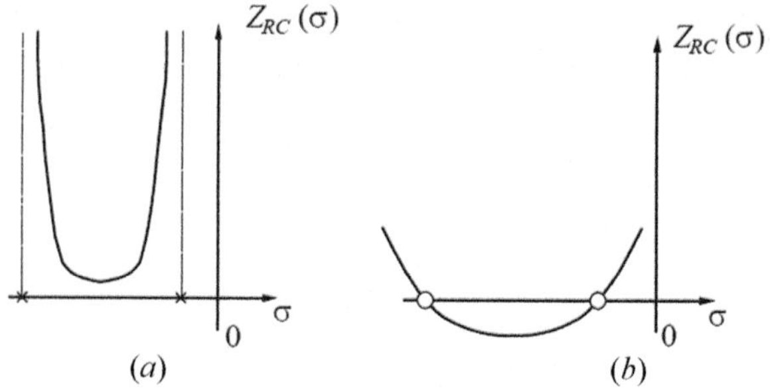

Fig. 3.8 Behavior of $Z_{RC}(\sigma)$ in case of having (**a**) two consecutive poles; (**b**) two consecutive zeroes

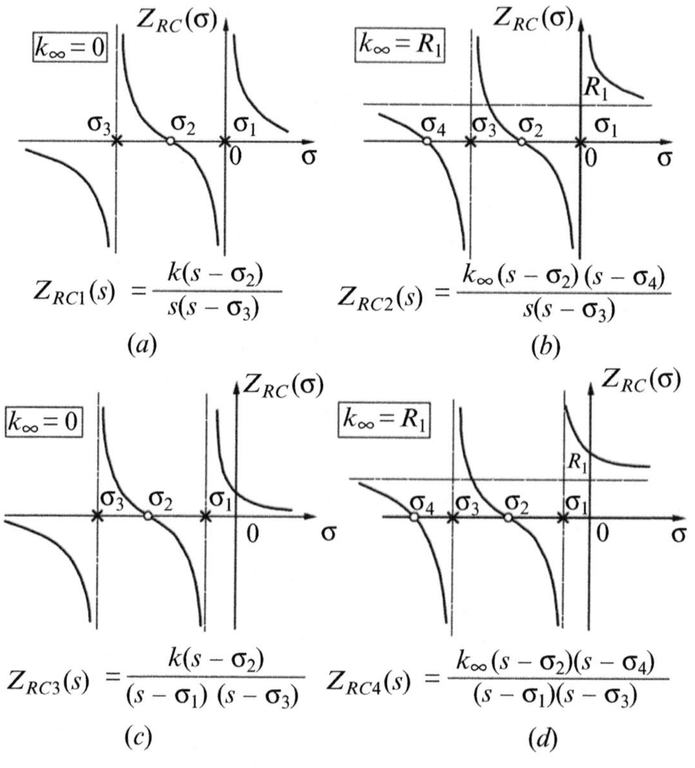

Fig. 3.9 Typical behavior of $Z_{RC}(\sigma)$ for (**a**) $k_\infty = 0$, $k_0 = 1/C_2$; (**b**) $k_\infty = R_1$, $k_0 = 1/C_2$; (**c**) $k_\infty = 0$, $k_0 = 0$; (**d**) $k_\infty = R_1$, $k_0 = 0$

the open s-plane (i.e., when $Z_{RC}(s)|_{s=\sigma} = Z_{RC}(\sigma)$) takes values as depicted in Fig. 3.8. However, such dependencies are impossible since this contradicts the monotonic property of $Z_{RC}(s)|_{s=\sigma} = Z_{RC}(\sigma)$ (Property 4 of Z_{RC}). That is why the poles and zeros of $Z_{RC}(s)$ alternate with each other.

Figure 3.9 depicts several cases of $Z_{RC}(s)$ and the location of their zeroes and poles at different values of its residues that depends on $Z_{RC}(\sigma)$.

Property 6 The order of the numerator $N(s)$ of $Z_{RC}(s)$ equals or less by one than the order of its denominator $M(s)$. In other words, if the input impedance transfer function $Z_{RC}(s)$ is described as

$$Z_{RC}(s) = \frac{N(s)}{M(s)} = \frac{a_m s^m + a_{m-1} s^{m-1} + \cdots + a_1 s + a_0}{b_n s^n + b_{n-1} s^{n-1} + \cdots + b_1 s + b_0}, \tag{3.8}$$

where a_i and b_i are real coefficients that depend on the parameters of the circuit elements, then$n - 1 \leq m \leq n$.

Proof If $k_\infty \neq 0$, then from (3.5) and (3.6) there is an elementary one-port consisting of a resistive element R_1. Then from Properties 4 and 5 of Z_{RC} (see Fig. 3.9b, d) the number of $Z_{RC}(s)$ zeroes will be equal to the number of its poles. If $k_\infty = 0$, then the resistive element R_1 is absent and $Z_{RC}(s)$ has a zero at $s = \infty$ (see Fig. 3.9a, c); that is, the number of finite zeroes of $Z_{RC}(s)$ will be less than the number of its poles by one.

In this section, the basic properties of the input impedance *RC* functions represented as sums of simple functions (3.6) have been considered.

3.3.3 Circuit Realization Using Foster Forms

Circuit realization using Foster forms has two possibilities depending on whether the input transfer function, $F(s)$, of the one-port *RC* circuits, represents an impedance function, $Z_{RC}(s)$, or an admittance function, $Y_{RC}(s)$.

The first Foster form utilizes the input impedance function $Z_{RC}(s)$. This form is a one-port *RC* circuit consisting of serially connected elementary one-port $Z_{RCi}(s)$, $i = 1, 2, \ldots, n$, as shown in Fig. 3.10.

The transfer function of the input impedance of such networks described by (3.8) can be rewritten as

$$Z_{RC}(s) = K\prod\nolimits_{j=1}^{m} (s - z_i) \Big/ \prod\nolimits_{i=1}^{n} (s - q_i), \tag{3.9}$$

where K is a scale multiplier, and q_i is the ith pole of the transfer function. For the one-port *RC* representation, $q_i = \sigma_i$, and z_j is jth zero of the numerator of (3.8) and it is equal to $z_j = \sigma_j$ for the one-port *RC* circuit.

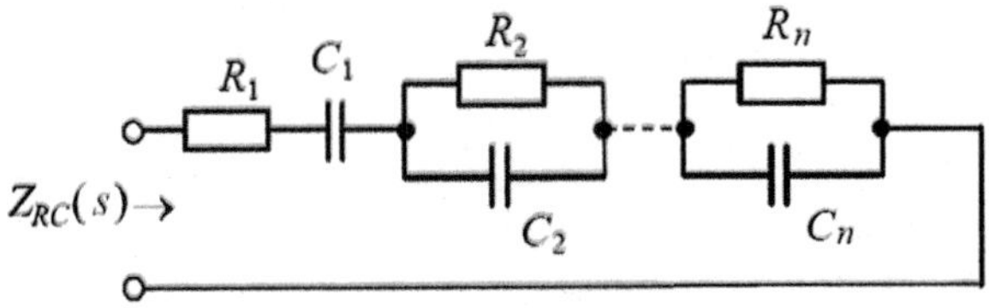

Fig. 3.10 *RC*-one-port realization by expanding $Z_{RC}(s)$ into simple fractions (the Foster-I form)

Fig. 3.11 Foster-II form of $Y_{RC}(s)$

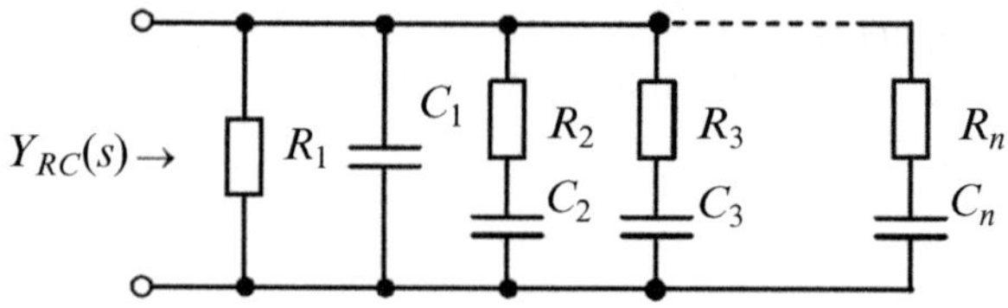

Let us now consider the procedure of calculating the parameters of the circuit shown in Fig. 3.10 using the first Foster form. If the *RC*-one-port input impedance function is represented as (3.8), then the first step is to determine the roots of the denominator, i.e., $M(s)=0$, which are denoted by $q_i=\sigma_i$. The second step is to expand $Z_{RC}(s)$ into a form similar to (3.6). Then, after putting together the summands of (3.5) and (3.6), we obtain

$$R_i = k_\infty;\ C_1 = 1/k_0;\ C_i = 1/k_i;\ R_i = k_i/\sigma_i;\ i = 2,\ \ldots,\ n. \tag{3.10}$$

The second Foster form utilizes the structure of the input admittance transfer function, $Y_{RC}(s)$. It represents one-port *RC*-ladder realization of one-port $Y_{RCi}(s)$ admittance, $i=1, 2,\ldots, n$, as shown in Fig. 3.11.

The first leg of the ladder is a resistance, the second one is a capacitance, while the rest are serial connections of resistive and capacitive elements.

Hence, the input immittance function can be written as follows:

$$Y_{RC}(s) = \frac{1}{R_1} + sC_2 + \sum_{i=3}^{n} \frac{s/R_i}{s + 1/R_iC_i} \tag{3.11}$$

or

$$Y_{RC}(s) = k_0 + k_\infty s + \sum_{i=3}^{n} \frac{sk_i}{s - q_i}. \tag{3.12}$$

Dividing both sides of (3.12) by s yields

$$Y^*_{RC}(s) = \frac{Y_{RC}(s)}{s} = k_\infty + \frac{k_0}{s} + \sum_{i=3}^{n} \frac{k_i}{s - q_i}. \tag{3.13}$$

Since (3.13) and (3.6) are identical we conclude the following:

$$k_\infty = Y^*_{RC}(\infty) = C_1;\ k_0 = \text{Res}Y^*_{RC}(s)|_{s=0} = sY^*_{RC}(s)|_{s=0} = 1/R_1,$$
$$k_i = \text{Res}Y^*_{RC}(s)|_{s=q_i} = (p - q_i)Y^*_{RC}(s)|_{s=q_i} = 1/R_i;\ i = 2,\ \ldots, n.$$

Here k_0 and k_i are the residues of $Y^*{}_{RC}(s)$; it is obvious that k_0 and k_i are real and positive, and $q_i=\sigma_i=-(1/R_iC_i)=-1/\tau_i$ represents the *i*th poles $Y_{RC}(s)$, where τ_i is the *i*th time constant of the one-port *RC* branch. Therefore, $C_1=k_\infty$; $R_1=1/k_0$; $R_i=1/k_i$; and $C_i=k_i/\sigma_i$, $i=2, 3,\ldots, n$.

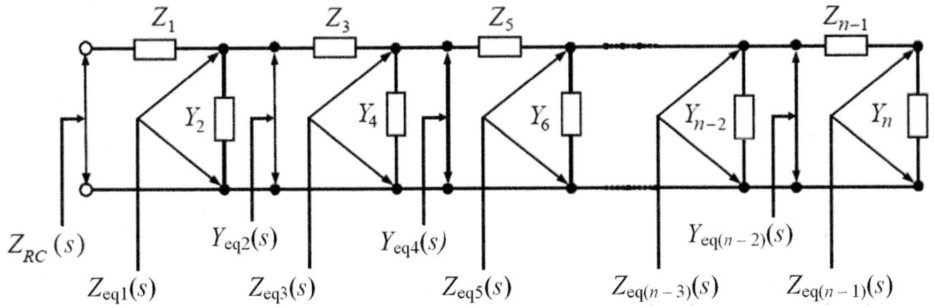

Fig. 3.12 Cauer-form realization

3.3.4 Circuit Realization Using Cauer Forms

The Cauer methods are used to realize input impedance transfer function, $Z_{RC}(s)$, or an input admittance function, $Y_{RC}(s)$, by means of a ladder circuit realization as depicted in Fig. 3.12. Notice that $Z_i(s)$ is the ith longitudinal impedance, and $Y_{i+1}(s)$ is the ith lateral admittance. For simplicity, the frequency-dependant elements of Z_i and Y_{i+1} are not shown in Fig. 3.12. The ith-equivalent impedance, $Z_{\mathrm{eq}_i}(s)$, and admittance, $Y_{\mathrm{eq}_{(i+1)}}(s)$, are defined to the right of their corresponding node for $i = 1, \ldots, n$.

The input impedance function $Z_{RC}(s)$ for this circuit can be sequentially rewritten as follows:

$$\begin{aligned} Z_{RC}(s) &= Z_1(s) + Z_{\mathrm{eq}_1}(s) = Z_1(s) + \frac{1}{Y_{\mathrm{eq}_1}(s)} \\ &= Z_1(s) + \frac{1}{Y_2(s) + Y_{\mathrm{eq}_2}(s)}, \end{aligned} \tag{3.14}$$

where

$$Y_{\mathrm{eq}_2}(s) = \frac{1}{Z_{\mathrm{eq}_2}(s)} = \frac{1}{Z_3(s) + Z_{\mathrm{eq}_3}(s)}$$

$$Z_{\mathrm{eq}_3}(s) = \frac{1}{Y_{\mathrm{eq}_3}(s)} = \frac{1}{Y_4(s) + Y_{\mathrm{eq}_4}(s)}$$

$$Y_{\mathrm{eq}_4}(s) = \frac{1}{Z_{\mathrm{eq}_4}(s)} = \frac{1}{Z_5(s) + Z_{\mathrm{eq}_5}(s)}$$

$$\vdots$$

$$Z_{\mathrm{eq}_{(n-3)}}(s) = \frac{1}{Y_{\mathrm{eq}_{(n-3)}}(s)} = \frac{1}{Y_{(n-2)}(s) + Y_{\mathrm{eq}_{(n-2)}}(s)}$$

$$Y_{\mathrm{eq}_{(n-2)}}(s) = \frac{1}{Z_{\mathrm{eq}_{(n-2)}}(s)} = \frac{1}{Z_{(n-1)}(s) + \frac{1}{Y_{\mathrm{eq}_n}(s)}}.$$

Then, inserting the expressions of $Z_{\mathrm{eq}_i}(s)$ and $Z_{\mathrm{eq}_{(i+1)}}(s)$ into (3.14) yields the ladder circuit of the input impedance, $Z_{RC}(s)$, as shown in Fig. 3.12. One may also rewrite $Z_{RC}(s)$ in a continued fraction expansion (CFE) form as follows:

$$Z_{RC}(s) = Z_1(s) + \cfrac{1}{Y_2(s) + \cfrac{1}{Z_3(s) + \cfrac{1}{Y_4(s) + \cdots \cfrac{\vdots}{Y_{n-2}(s) + \cfrac{1}{Z_{n-1}(s) + \cfrac{1}{Y_n(s)}}}}}} \tag{3.15}$$

There are two forms of Cauer circuits: (1) the first Cauer form with resistive longitudinal elements and capacitive lateral elements (see Fig. 3.13); (2) the second Cauer form with capacitance longitudinal elements and resistive lateral elements (see Fig. 3.14).

Inserting the values $Z_i(s) = R_i$ and $Y_{i+1}(s) = C_{i+1}$ into (3.15) results in the following first Cauer form of $Z_{RC}(s)$, i.e., Cauer-I form:

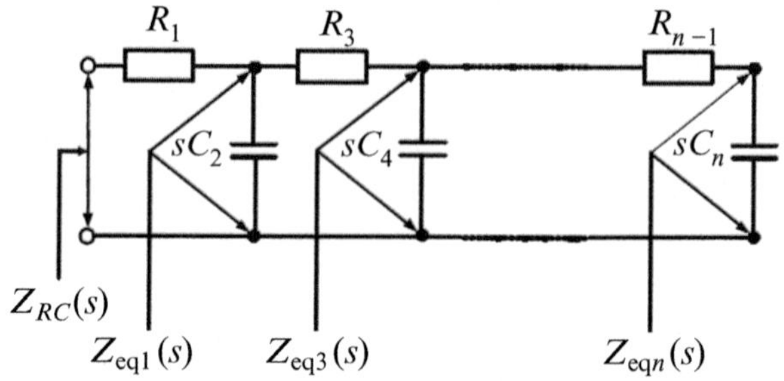

Fig. 3.13 Cauer-I form

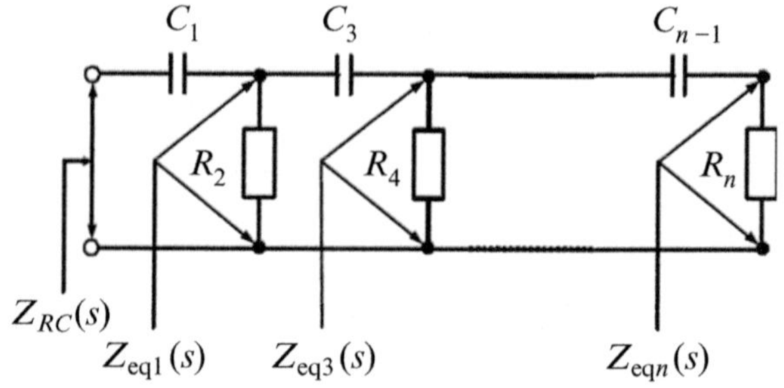

Fig. 3.14 Cauer-II form

$$Z_{RC}(s) = R_1 + \cfrac{1}{sC_2 + \cfrac{1}{R_3 + \cfrac{1}{sC_4 + \cdots \cfrac{\vdots}{sC_{n-2} + \cfrac{1}{R_{n-1} + \cfrac{1}{sC_n}}}}}} \tag{3.16}$$

Inserting the values of $Z_i(s) = 1/sC_i$ and $Y_{i+1}(s) = 1/R_{i+1}$ into (3.15) yields the following Cauer form-II realization of the input impedance $Z_{RC}(s)$:

$$Z_{RC_{II}}(s) = \frac{1}{sC_1} + \cfrac{1}{\cfrac{1}{R_2} + \cfrac{1}{\cfrac{1}{sC_3} + \cfrac{1}{\cfrac{1}{R_4} + \cdots \cfrac{\vdots}{\cfrac{1}{R_{n-2}} + \cfrac{1}{\cfrac{1}{sC_{n-1}} + \cfrac{1}{\cfrac{1}{R_n}}}}}}} \tag{3.17}$$

It is obvious that if $\lim_{s\to\infty} Z_{RC_I}(s) = Z_{RC_I}(\infty) = R_1 = 0$, then $s = \infty$ is a zero of $Z_{RC_I}(s)$ and in this case the input admittance transfer function should be used:

$$Y_{RC_I}(s) = \frac{1}{Z_{RC_I}(s)}\Big|_{R_1=0} = SC_2 + \cfrac{1}{R_3 + \cfrac{1}{sC_4 + \cfrac{1}{R_5 + \cdots \cfrac{\vdots}{R_{n-1} + \cfrac{1}{\cfrac{1}{SC_n}}}}}}$$

The following examples clarify the realization of input impedance using Cauer-I form-I:

Example 1 Realize the following input impedance transfer function using Cauer-I form:

$$Z(s) = \frac{s^2 + 4s + 3}{s^3 + 6s^2 + 8s} \tag{3.18}$$

Solution

1. *Calculate* R_1: Since $Z(s)|_{s=\infty} = 0$, it has a zero at $s = \infty$; then $R_1 = Z(\infty) = 0$. It follows from (3.14) that $Z(s) = Z_1(\infty) + Z_{\mathrm{eq}_1}(s) = 0 + 1/Y_{\mathrm{eq}_1}(s)$, which can be described by the following circuit diagram:

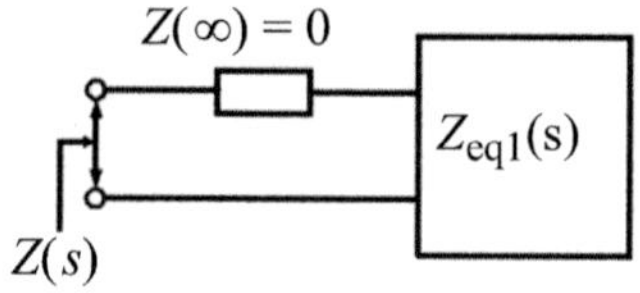

2. *Calculate* C_1: Now consider $Y_{eq_1}(s)$ as follows:

$$Y_{eq_1}(s) = \frac{1}{Z_{eq_1}(s)} = \frac{1}{Z(s)} = \frac{s(s+2)(s+4)}{(s+1)(s+3)}.$$

It is apparent that the admittance $Y_{eq_1}(s)$ has a pole at $s = \infty$ and can be represented in the following form:

$$Y_{eq_1}(s) = sC_2 + Y_{eq_2}(s).$$

From here, the value of C_2 is calculated as a residue at the pole at $s = \infty$,

$$C_2 = \text{Res}Y_{eq_1}(s)|_{s=\infty} = \left[\frac{1}{s}Y_{eq_1}(s)\right]|_{s=\infty} = \frac{s(s+2)(s+4)}{(s+1)(s+3)}|_{s=\infty} = 1.$$

Then the equivalent conductivity $Y_{eq1}(s)$ can be associated with the following circuit:

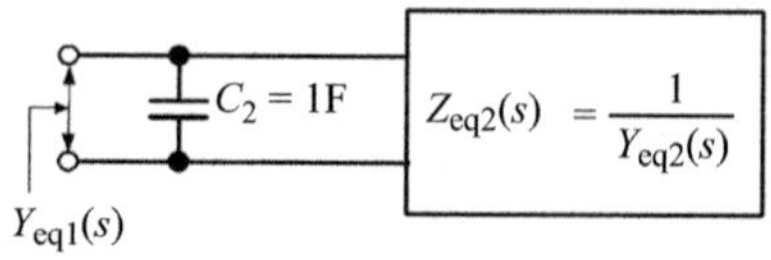

It is evident that

$$Y_{eq_1}(s) = sC_2 + \frac{1}{Z_{eq_2}(s)}.$$

Then,

$$Y_{eq_2}(s) = \frac{1}{Z_{eq_2}(s)} = \frac{s(s+2)(s+4)}{(s+1)(s+3)} - s.1 = \frac{2s^2+5s}{s^2+4s+3}.$$

3. *Calculate* R_3: Now define $Z_{eq_2}(s)$ as follows:

$$Z_{\text{eq}_2}(s) = \frac{1}{Y_{\text{eq}_2}(s)} = \frac{s^2+4s+3}{2s^2+5s}.$$

Represent $Z_{\text{eq}_2}(s)$ as

$$Z_{\text{eq}_2}(s) = Z_{\text{eq}_2}(\infty) + Z_{\text{eq}_3}(s),$$

where $Z_{\text{eq}_2}(\infty) = R_3 = \frac{s^2+4s+3}{2s^2+5s}|_{s=\infty} = \frac{1}{2}$.
Similarly, the following circuit diagram describes the equivalent impedance $Z_{\text{eq}_2}(s)$:

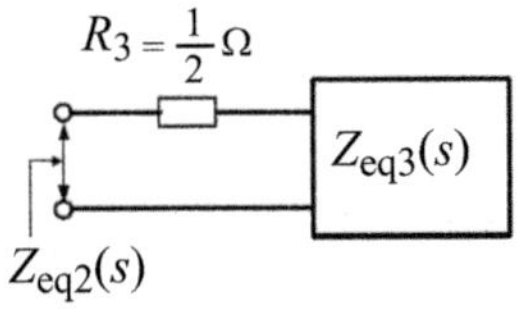

And

$$Z_{\text{eq}_3}(s) = Z_{\text{eq}_2}(s) - R_3 = Z_{\text{eq}_2}(s) - \frac{1}{2} = \frac{(s^2+4s+3)}{2s^2+5s} = \frac{\frac{3}{2}s+3}{2s^2+5s}.$$

Since $Z_{\text{eq}_2}(s) = \frac{1}{Y_{\text{eq}_2}(s)}$, then $Z_{\text{eq}_2}(s) = R_3 + 1/Y_{\text{eq}_2}(s)$.

4. *Calculate* C_4: Let $Y_3(s)$ be written as follows:

$$Y_3(s) = \frac{1}{Z_3(s)} = \frac{2s^2+5s}{\frac{3}{2}s+3.}$$

Since $Y_3(s)$ has a pole at $s = \infty$, then $Y_{\text{eq}_3}(s) = sC_4 + Y_{\text{eq}_4}(s)$, where

$$C_4 = \text{Res}\, Y_{\text{eq}_3}(s)|_{s=\infty} = \left[\frac{1}{s}Y_{\text{eq}_3}(s)\right]|_{s=\infty} = \frac{4}{3},$$

and

$$Y_{\text{eq}_3}(s) = \frac{4}{3}s + Y_{\text{eq}_4}(s).$$

The equivalent conductivity $Y_3(s)$ can be associated with the following circuit:

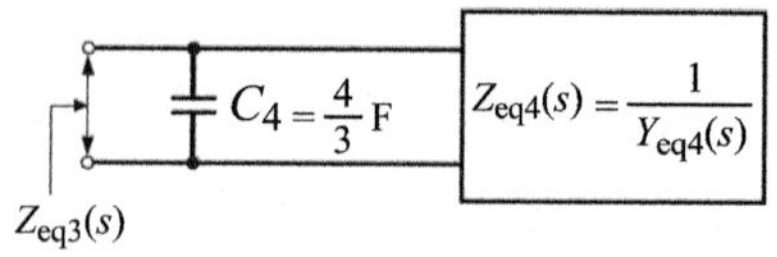

where

$$Y_{eq_4}(s) = Y_{eq_3}(s) - \frac{4}{3}s = \frac{2s^2+5s}{\frac{3}{2}s+3} - \frac{4}{3}s = \frac{2s}{\frac{3}{2}s+6.}$$

5. *Calculate* R_5: Now consider $Z_{eq_4}(s) = \frac{\frac{3}{2}s+6}{2s}$ and define the following:

$$R_5 = Z_{eq_4}(\infty) = \frac{3s+6}{2s}|_{s=\infty} = \frac{3}{2}.$$

Since $Z_{eq_4}(s) = R_5 + Z_{eq_5}(s)$, then

$$Z_{eq_5}(s) = Z_{eq_4}(s) - R_5 = \frac{3s+6}{2s} = \frac{3}{1.s}$$

and the equivalent impedance $Z_{eq_5}(s)$ will be

$$Z_{eq_5}(s) = R_5 + \frac{1}{Y_{eq_5}(s)},$$

where

$$Y_{eq_5}(s) = \frac{1}{Z_{eq_5}(s)} = \frac{1}{3}s.$$

Hence, the equivalent impedance $Z_5(s)$ is depicted by the following circuit:

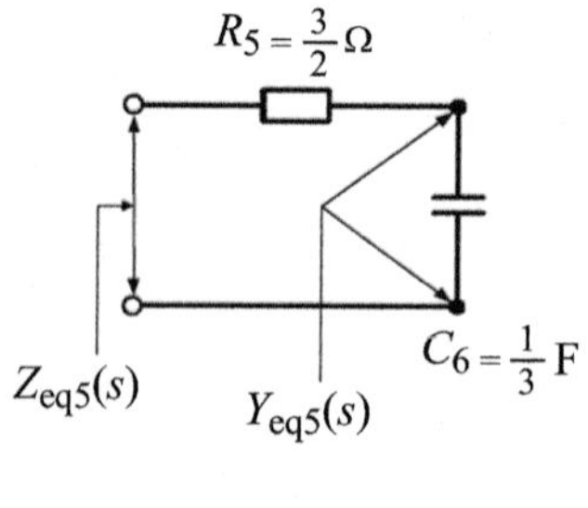

Finally, the complete RC realization of (3.18) using Cauer-I form is shown in Fig. 3.15.

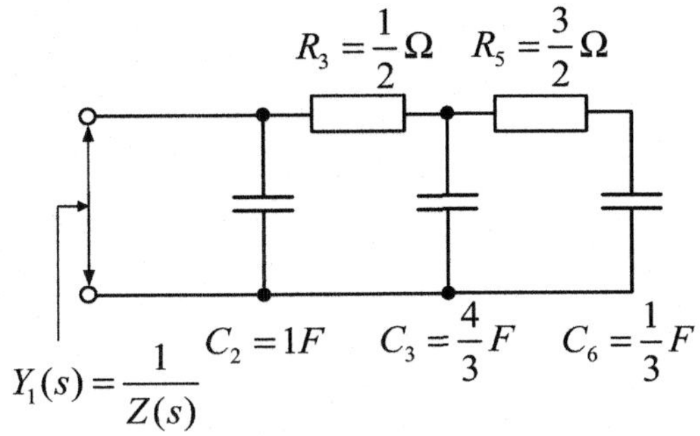

Fig. 3.15 RC realization of (3.18) using Cauer-I form

Since $Z_1(s) = Z(\infty) = 0$, one may expand $Z_{RC}(s)$ using the CFE method to sequentially identify $Z_{\text{eq}_i}(s)$, $Y_{\text{eq}_{(i+1)}}(s)$, where $Z_i(s) = R_i$, and $Y_{i+1}(s) = C_{i+1}$ as defined in (3.14). Applying the CFE on $Y_{\text{eq}_1}(s)$ yields

$$Y_{\text{eq}_1}(s) = 1.s + \cfrac{1}{\frac{1}{2} + \cfrac{1}{\frac{4}{3}s + \cfrac{1}{\frac{3}{2} + \cfrac{1}{\frac{1}{3}s}}}}. \tag{3.19}$$

Clearly, the CFE of (3.19) can be obtained by means of repeated division by inverting the point $s = \infty$. Hence, the following rules should be obeyed:

1. Place the polynomials $N(s)$ and $M(s)$ in (3.8) in decreasing or increasing order of powers of s (see items 2, 3, 4).
2. If $m = n$, starting from higher powers of s, apply the long division method on $Z(s)$.
3. If the order of the numerator of $Z(s)$ is less than the order of its denominator ($m < n$), then apply the CFE on the reciprocal transfer function $Y(s)$ of $Z(s)$.
4. If $m = n$ in $Y(s)$, start with the lower powers of s, and apply the long division method on $Y(s)$.
5. Divide one polynomial by another so that the resulting summands be positive and without s of power exceeding 1 and -1.
6. If the division results in the summands that are negative or has s of power exceeding 1 and -1, then we can change the sequence of the polynomial powers at the corresponding stage.

For example, let us perform repeated division of the input impedance function $Z(s)$ with inversion at $s = \infty$ using the CFE as in (3.8). Since $m < n$ then, from rule 3, the reciprocal of this function $Y(s)$ will be expanded using the CFE as follows:

$$Y(s) = \frac{N_\gamma(s)}{M_\gamma(s)} = \frac{b_n s^n + b_{n-1}s^{n-1} + \cdots + b_1 s + b_0}{a_m s^m + a_{m-1}s^{m-1} + \cdots + a_1 s + a_0}$$
$$= \frac{sC_1 M_\gamma(s) + O_1(s)}{M_\gamma(s)}, \tag{3.20a}$$

where $b_n s^n = sC_1 a_m s^m$, and $b_n = C_1 = a_m, n = m + 1$.

From (3.20a), we rewrite *Y*(*s*) as

$$Y(s) = sC_1 + \frac{O_1(s)}{M_\gamma(s)}$$
$$= sC_1 + \frac{1}{\frac{M_\gamma(s)}{O_1(s)}} = sC_1 + \frac{1}{sC_2 + \frac{O_2(s)}{O_1(s)}} = sC_1 + \frac{1}{sC_2 + \frac{1}{\frac{O_2(s)}{O_1(s)}}} = \ldots \tag{3.20b}$$

Example 2 Expand the input impedance function *Z*(*s*) of Example 1 using the CFE of its reciprocal *Y*(*s*).

Solution Since the order of the numerator of *Z*(*s*) is less than the order of its denominator $(m < n)$, then, the third rule, its reciprocal *Y*(*s*) will be expanded using the CFE according to (3.20a, 3.20b)

$$Y(s) = \frac{s^3 + 6s^2 + 8s}{s^2 + 4s + 3} = \frac{N_\gamma(s)}{M_\gamma(s)} = sC_1 + \frac{O_1(s)}{M_\gamma(s)}$$
$$= s.1 + \frac{2s^2 + 5s}{s^2 + 4s + 3} = sC_1 + \frac{1}{\frac{M_\gamma(s)}{O_1(s)}} = s + \frac{1}{\frac{s^2 + 4s + 3}{2s^2 + 5s}}$$
$$= s + \frac{1}{\frac{1}{2} + \frac{1}{\frac{2s^2 + 5s}{\frac{2}{3}s + 3}}} = s + \frac{1}{\frac{1}{2} + \frac{1}{\frac{4}{3}s + \frac{1}{\frac{\frac{3}{2}s + 3}{s}}}}$$
$$= s + \frac{1}{\frac{1}{2} + \frac{1}{\frac{4}{3}s + \frac{1}{\frac{3}{2} + \frac{1}{\frac{1}{3}s + 0}}}}$$

The process of repeated calculation of the ladder circuit elements can also be represented as follows:

$$
\begin{aligned}
Y_{\text{экв}1}(s) &= s + Y_{\text{экв}2}(s) = 1\cdot s + \frac{1}{Z_{\text{экв}2}(s)} = 1\cdot s + \frac{1}{\frac{1}{2} + Z_{\text{экв}3}(s)} \\
&= 1\cdot s + \frac{1}{\frac{1}{2} + \frac{1}{Y_{\text{экв}3}(s)}} == 1s + \frac{1}{\frac{1}{2} + \frac{1}{\frac{4}{3}s + Y_{\text{экв}4}(s)}} \\
&= 1s + \frac{1}{\frac{1}{2} + \frac{1}{\frac{4}{3}s + \frac{1}{Z_{\text{экв}4}(s)}}} == 1s + \frac{1}{\frac{1}{2} + \frac{1}{\frac{4}{3}s + \frac{1}{\frac{3}{2} + Z_{\text{экв}5}(s)}}} \\
&= 1s + \frac{1}{\frac{1}{2} + \frac{1}{\frac{4}{3}s + \frac{1}{\frac{3}{2} + \frac{1}{Y_{\text{экв}5}(s)}}}} == 1s + \frac{1}{\frac{1}{2} + \frac{1}{\frac{4}{3}s + \frac{1}{\frac{3}{2} + \frac{1}{\frac{1}{3}s}}}}\,.
\end{aligned}
$$

$$
\begin{array}{r|l}
N(s) \Leftarrow s^3 + 6s^2 + 8s & s^2 + 4s + 3 \Rightarrow M(s) \\
-\; s^3 + 4s^2 + 3s & s \Rightarrow Y_2(s)
\end{array}
$$

$$
\begin{array}{r|l}
M(s) \Leftarrow s^2 + 4s + 3 & 0 + 2s^2 + 5s \Rightarrow O_1(s) \\
-\; s^2 + \frac{5}{2}s + 0 & \frac{1}{2} \Rightarrow R_3
\end{array}
$$

$$
\begin{array}{r|l}
O_1(s) \Leftarrow 2s^2 + 5s + 0 & 0 + \frac{3}{2}s + 3 \Rightarrow O_2(s) \\
-\; 2s^2 + 4s + 0 & \frac{4}{3}s \Rightarrow sC_4
\end{array}
$$

$$
\begin{array}{r|l}
O_2(s) \Leftarrow \frac{3}{2}s + 3 & 0 + s + 0 \Rightarrow O_3(s) \\
-\frac{3}{2}s + 0 & \frac{3}{2} \Rightarrow R_5
\end{array}
$$

$$
\begin{array}{r|l}
O_3(s) \Leftarrow s & 0 + 3 \Rightarrow O_4(s) \\
-\; s & \frac{1}{3}s \Rightarrow sC_6
\end{array}
$$

$$0 \Rightarrow O_5(s)$$

The values of the elements calculated by expanding the input impedance function $Z(s)$ using CFE prove that the ladder circuit synthesized with this procedures is the same as the one presented in Example 1 (see Fig. 3.15).

3.3.5 Rational Approximation of Fractal-Order Impedances

To realize the fractal impedance by means of *RC* circuits discussed in Sect. 3.3, we need to represent the *irrational* input impedance (admittance) function (3.2) in terms of a rational transfer function (3.4). As it was stated in Sect. 3.3.2, the following conditions should be met to implement these functions by means of passive elements:

- The transfer function is real for real s.
- The transfer function possesses distinct features in the complex plane with negative real poles located in the open left-half s-plane.
- The input impedance $Z(s)$ pole or the input admittance $Y(s)$ zero is the closest to the origin of the coordinates.

It should be noted that if the number of the circuit's elements is limited, then the fractal impedance function implementation is possible only for a limited frequency range.

There are a number of ways to solve this problem. In this book, we will consider just few approximation methods for practical implementation of fractal impedances.

3.3.5.1 The Oustaloup Method of Rational Approximation[2]

Assume that it is required to approximate a fractal impedance of real-order α within the frequency range from ω_{start} to ω_{fin}. The normalized fractal impedance approximating function is written in the following form:

$$\bar{Z}(s) = \left(\frac{\omega_{\text{av}}}{\omega_{\text{start}}}\right)^{-\alpha} \left(\frac{1 + s/\omega_{\text{start}}}{1 + s/\omega_{\text{fin}}}\right)^{-\alpha}, \tag{3.21}$$

where $\omega_{\text{av}} = \sqrt{\omega_{\text{start}}\omega_{\text{fin}}}$.

From the properties of the input *RC* functions that were studied in Sect. 3.3.2, the approximating transfer function (3.21) is proposed to be implemented in the fractional-rational form as follows:

[2]See Oustaloup A. Systemes asservis lineaires d'ordre fractionnaire. Masson, Paris, 1983.

$$\bar{Z}(s) = \lim_{N\to\infty}\bar{Z}_N(s) = \lim_{N\to\infty}\left(\frac{\omega_{\text{start}}}{\omega_{\text{fin}}}\right)^{-\alpha}\prod\nolimits_{i=-N}^{N}\left(\frac{1+s/\omega_{zi}}{1+s/\omega_{pi}}\right). \tag{3.22}$$

The transfer function described by (3.22) can be realized by means of Foster or Cauer circuits. The order of the approximation is $2N+1$, where ω_{zi} and ω_{pi} are the zeroes and poles of the transfer function to be implemented, respectively.

Let $\mu = \omega_{\text{fin}}/\omega_{\text{start}}$ be a scale factor used to characterize the operating frequency range. Then the transfer function of the normalized input impedance of a one-port form, that is supposed to implement (3.21), can be written as follows:

$$\bar{Z}_N(s) = C\prod_{i=-N}^{N}(s+\omega_{zi})/(s+\omega_{pi}), \tag{3.23}$$

where $C = (\mu)^{-\frac{\alpha}{2}}\prod\nolimits_{i=-N}^{N}\frac{1+s/\omega_{zi}}{1+s/\omega_{pi}}$.

Figure 3.16 illustrates the principle of approximation of the transcendent function $\bar{Z}(s)$ with a fractional-rational function $\bar{Z}_N(s)$ (scale of frequency axis is logarithmic).

The value of N (i.e., the order of the transfer function) is calculated from

$$N = \frac{\lg(\omega_{pN}/\omega_{p0})}{\lg(\xi\eta)}, \tag{3.24}$$

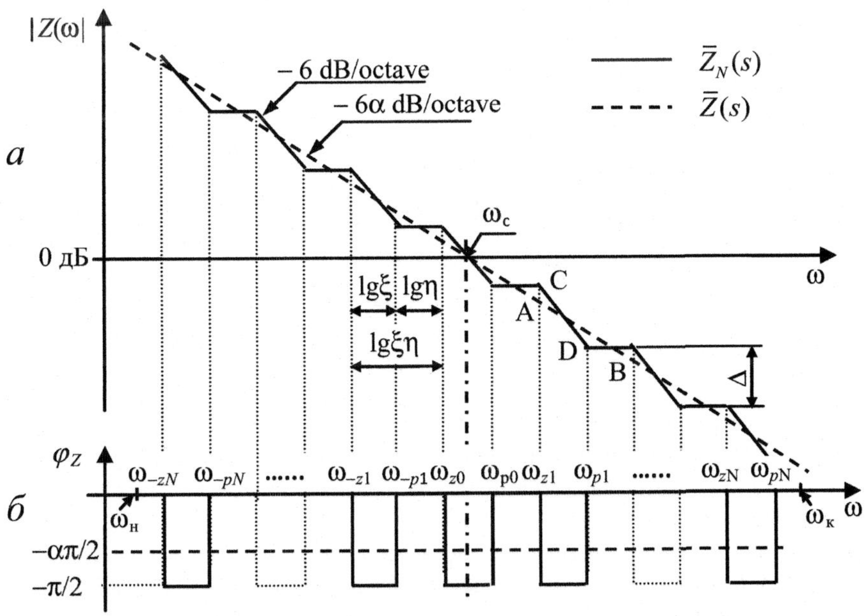

Fig. 3.16 Illustration of the Oustaloup approximation method: (**a**) logarithmic MFR; (**b**) PFR

where ω_{p0} is the first pole; ω_{pN} is the N-th pole; and ξ and η are the so-called *recursive factors* that define the location of zeroes and poles, which provide the needed fractal properties embedded in (3.22). The recursive factors are calculated as follows:

$$\xi = \frac{\omega_{pi}}{\omega_{zi}} = \text{const}, \eta = \frac{\omega_{z(i+1)}}{\omega_{pi}} = \text{const} \quad i = 1, 2, \ \ldots \tag{3.25}$$

Notice that these factors of $\bar{Z}_N(s)$ are constant for all i from $-N$ to $+N$.

The result of smoothing $\bar{Z}_N(s)$ that is being implemented (the solid line in Fig. 3.16) over an interval from ω_{start} to ω_{fin} can be represented with a straight line that matches the ideal logarithmic value of MFR of $\bar{Z}_N(s)$ (the dashed line).

The slope of approximation between points C and D, as depicted in the one-port logarithmic MFR shown in Fig. 3.16a, which corresponds to a transition from a zero of $\bar{Z}_N(s)$ to its next pole is calculated as follows:

$$\frac{\Delta[\text{dB}]}{\lg\xi} = -6\,[\text{dB/oct}].$$

Moreover, the slope of the smoothing line between points A and B in Fig. 3.16a can also be defined by

$$\frac{\Delta[\text{dB}]}{\lg\xi + \lg\eta} = -6\alpha\,[\text{dB/oct}].$$

Now the fractional order, α, can be expressed via the recursive factors ξ and η as follows:

$$\alpha = \frac{\lg\xi}{\lg(\xi\eta)}. \tag{3.26}$$

Fig. 3.16a shows that the required logarithmic fractal MFR intersects the middle points of lg η along horizontal segments within the interval from ω_{start} to ω_{fin}. Therefore, considering the logarithmic scale, the frequencies ω_{-zN} and ω_{pN} can be derived from ω_{start} and ω_{fin} in the following way:

$$\omega_{-zN} = \omega_{\text{start}}\sqrt{\eta}, \quad \omega_{pN} = \omega_{\text{fin}}/\sqrt{\eta}. \tag{3.27}$$

The phase frequency response of the one-port $\bar{Z}_N(s)$ fractal approximation is depicted in Fig. 3.16b with solid line in the form of a rectangular pulse. It is represented here with dashed line with a constant phase $\boldsymbol{\varphi}_{\mathbf{av}}$. The latter can also be expressed via the recursive factors ξ and η.

For example, if we imagine that the implemented one-port $\bar{Z}_N(s)$ phase frequency response (PFR) is a pulse sequence with lg($\xi\eta$) period, pulse width of lg(ξ), and magnitude of $\pi/2$, then the average of this sequence through one period can be expressed as follows:

$$\boldsymbol{\varphi}_{\mathbf{av}} = \frac{\frac{\pi}{2}\lg\xi}{\lg(\xi\eta)}. \tag{3.28}$$

From (3.26) and (3.28), it follows that

$$\varphi_{\text{av}} = \alpha\frac{\pi}{2}. \tag{3.29}$$

This expression completely matches the expression (3.3d).

Moreover, from (3.25), the product of the recursive factors ξ and η can be written as

$$\xi\eta = \frac{\omega_{z(i+1)}}{\omega_{zi}} = \frac{\omega_{p(i+1)}}{\omega_{pi}},$$

and the ratio of the border frequencies can also be expressed via the recursive factors ξ and η as follows:

$$\mu = \frac{\omega_{\text{fin}}}{\omega_{\text{start}}} = (\xi\eta)^{2N+1}. \tag{3.30}$$

Now the product of the recursive factors ξ and η can be written in the following form:

$$\xi\eta = \left(\frac{\omega_{\text{fin}}}{\omega_{\text{start}}}\right)^{\frac{1}{2N+1}} \tag{3.31}$$

Then from (3.26), the recursive factor can be written as

$$\xi = (\xi\eta)^{\alpha}. \tag{3.32}$$

Now substituting the value of ξ of (3.32) into the obvious relation $\eta = (\xi\eta)\xi^{-1}$ to get the value of the recursive factor η

$$\eta = (\xi\eta)^{1-\alpha}. \tag{3.33}$$

Substituting (3.31) into (3.32) and (3.33) yields the following recursive formulas of the factors ξ and η, respectively:

$$\xi = \left(\frac{\omega_{\text{fin}}}{\omega_{\text{start}}}\right)^{\frac{\alpha}{2N+1}} = \mu^{\frac{\alpha}{2N+1}}, \qquad \eta = \left(\frac{\omega_{\text{fin}}}{\omega_{\text{start}}}\right)^{\frac{1-\alpha}{2N+1}} = \mu^{\frac{1-\alpha}{2N+1}}. \tag{3.34}$$

Now it is evident that when fractional order α is specified along with the parameter μ that defines the range of the operating frequencies, the recursive factors ξ and η depend only on the order of approximation N.

From Fig. 3.16a, it follows that the frequency of a zero ω_{zi} can be written in the following form:

$$\lg\omega_{zi} = \lg\omega_{-zN} + (i+N)\lg(\xi\eta).$$

Then, $\omega_{zi} = \omega_{zN}(\xi\eta)^{(i+N)}$. From (3.27), the expression for the ith-frequency zero ω_{zi} is calculated from

$$\omega_{zi} = \sqrt{\eta}\omega_{\text{start}} \cdot (\xi\eta)^{(i+N)}. \tag{3.35}$$

Substitute (3.34) and (3.30) into (3.35) to get the approximating function of ith-zero frequency as follows:

$$\omega_{zi} = (\mu)^{\frac{(i+N+0,5(1-\alpha))}{2N+1}}\omega_{\text{start}}. \tag{3.36}$$

The frequency of the ith pole is found in the same manner as

$$\omega_{pi} = (\mu)^{\frac{(k+N+0,5(1-\alpha))}{2N+1}}\omega_{\text{start}}. \tag{3.37}$$

The frequencies of zeroes and poles calculated in this manner completely determine the normalized one-port input impedance function $\bar{Z}_N(s)$, which is supposed to be used to implement the fractal impedance as given by (3.21).

Now consider an example of synthesizing a fractal impedance one-port built using the given above method. Matlab® software will be used as an instrument.

Example 3 Synthesize a fifth-order one-port realization to approximate a fractional-order Laplacian operator, s^α, for $\alpha = 0.3$ within the frequency range from 0.001 rad/s to 1000 rad/s.

Solution The following ready-to-use Matlab function, named "**oust_fod**," is listed for completeness to find the rational approximation of fractional-order differential/integral operators. The calling command is G = ousta_fod(α, N, ω_b, ω_h), where α is the order of the fractional operator; $2N+1$ is the order of the transfer (circuit) function; $\omega_b = \omega_{\text{start}}$ and $\omega_h = \omega_{\text{fin}}$ are the start and the end frequency operating range.

The Matlab code to construct this realization is given as follows[3]:

N = 2; w$_1$ = 1e^{-3}; w$_2$ = 1e^3;
gl = ousta_fod(– 0.3, N, w1, w2);
function G = ousta_fod (r, N, w_L, w_H)

[3]See Dingyu Xue, Chen Yang Quan, Derek P. Atherton. Linear Feedback Control. Analysis and Design with MATLAB. London: SIAM, 2009. 354 p.

```
% r: fractional order α \ [-1,1]
% N: summation limits
% w_L: the lower limit of the working frequencies range
% w_H: the higher limit of the working frequencies range
% G: the type of the function that approximates the p^α fractal impedance
mu = w_H/w_L; k = -N:N;
w_kp = (mu).^((k + N + 0.5-0.5*r)/(2*N + 1) )*w_L;
w_k = (mu).^((k + N + 0.5 + 0.5*r)/(2*N + 1) )*w_L;
C = (mu)^(-r/2)*prod(w_k./w_kp);
G = tf(zpk(-w_kp', -w_k', C))
```

Using the above Matlab function, the following transfer function approximates a fractional-order integrator of order $\alpha = -0.3$:

$$s^{-0.3} = \bar{Z}_N(s) = \frac{0.1259s^5 + 51.09s^4 + 1230s^3 + 1862s^2 + 177.1s + 1}{s^5 + 177.1s^4 + 1862s^3 + 1230s^2 + 51.09s + 0.1259}. \quad (3.38)$$

The command **bode(zpk(-w_kp',-w_k', C))** generates the Bode diagram of the transfer function given by (3.38) as depicted in Fig. 3.17. It shows the frequency response of $\bar{Z}_N(s)$ for three values of N: $N = 2$, 3, and 4. It is evident that with $N = 2$, Oustaloup approximation yields an acceptable fifth-order transfer function of order ($2\ N + 1$) that equals to 5 (3.38) with a phase response precision of about ±4° over the specified frequency range (approximately four decades). Increasing N, however, reduces the PFR fluctuation, but that will narrow the bandwidth of approximation to only three decades.

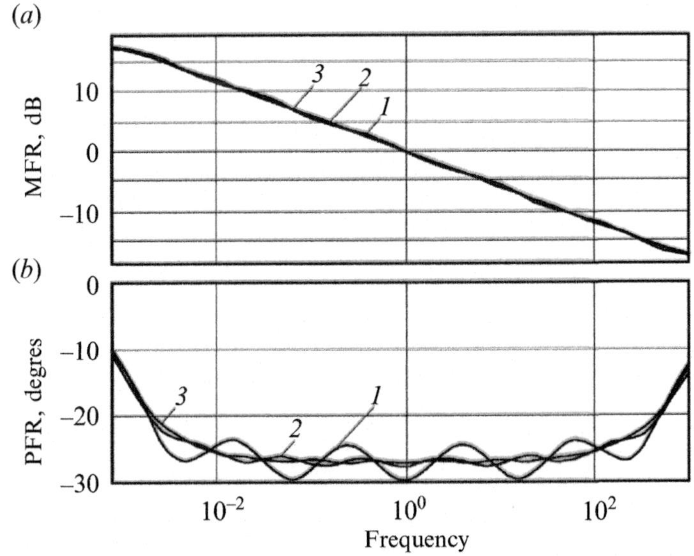

Fig. 3.17 Bode diagram of (3.38) using Oustaloup method with (*1*) $N = 2$; (2) $N = 3$; (*3*) $N = 4$

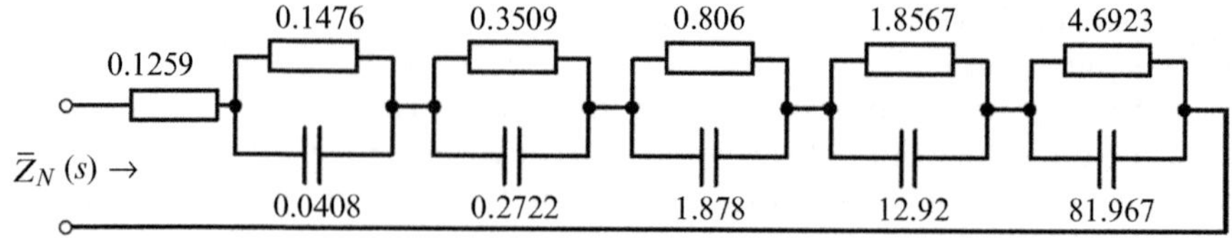

Fig. 3.18 A Foster-I RC-one-port realization of $\bar{Z}(s)$ given by (3.38)

In order to realize a one-port RC circuit for the fractional-order integrator given by (3.38), we use the following commands to develop a Foster-I form as discussed in Sect. 3.3.3:

[b,a] = zp2tf(-w_kp',-w_k', C)
[r,p,k] = residue(b,a)

This results in expanding the transfer function (3.38) into a sum of simple transfer functions of the form

$$\bar{Z}(s) = 0.1259 \quad + \frac{24.49}{s + 165.96} + \frac{3.6743}{s + 10.4713} + \frac{0.5325}{s + 0.6607} + \\ + \frac{0.0774}{s + 0.0417} + \frac{0.0122}{s + 0.0026}$$

Figure 3.18 shows Foster-I (see Sect. 3.3.3) one-port RC-circuit diagram of the fractional-order integrator approximated by (3.38)[4].

Looking carefully at the values of the circuit parameters implies that the ratio of the maximum and the minimum values of the resistors equals 37, while the same ratio for the capacitors exceeds 2000. This is the reason why this circuit cannot be produced using the IC technology without using materials with different specific values for resistors and insulators with the aim of minimizing the overall size of the device.

The frequency response of the one-port input impedance depicted in Fig. 3.18 was also verified using the electric circuit simulation software OrCAD. A screenshot of the simulation results of the MFR and the PFR is shown in Fig. 3.19.

The frequency response (3.19) of the one-port circuit diagram shown in Fig. 3.18 is the same as the one depicted in Fig. 3.17. This proves that the synthesized circuit diagram precisely describes the fractional-rational function acquired using Oustaloup method.

3.3.5.2 Shareff Method of Rational Approximation

The second approximation method of Shareff can be used to realize the normalized fractional-order integrator described by (3.3b) over a frequency range from ω_{start} to ω_{fin} using a set of first-order transfer functions of the form

[4]Charef A.“Analogue realization of fractional-order integrator, differentiator and fractional PID controller,” *IEE Proc. Control Theory Appl.* V. 153. №. 6.–pp. 714, 2006.

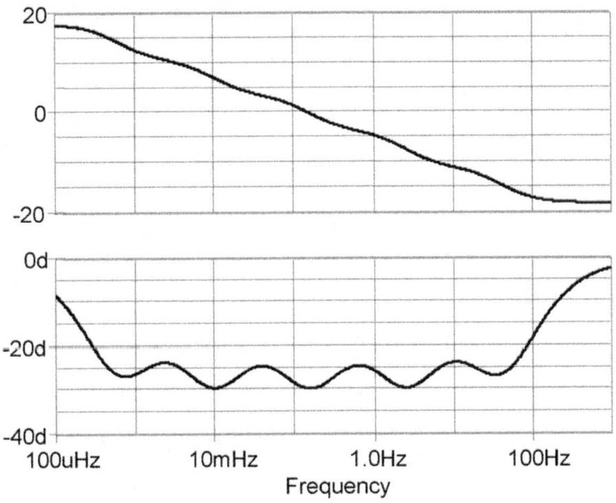

Fig. 3.19 A screenshot of OrCAD circuit simulation software of $\bar{Z}(s)$

$$\bar{Z}(s) = 1/(1 + s/\omega_{\text{av}}). \tag{3.39}$$

The approximating function (3.39) can be implemented by means of the following rational transfer function for the normalized input impedance $\bar{Z}_N(s)$:

$$\bar{Z}_N(s) = \frac{\prod_{i=0}^{N-1}\left(1 + \dfrac{s}{\omega_{zi}}\right)}{\prod_{i=0}^{N-1}\left(1 + \dfrac{s}{\omega_{pi}}\right)} = \prod_{i=0}^{N-1}\frac{\omega_{pi}}{\omega_{zi}}\frac{\prod_{i=0}^{N-1}(s + \omega_{zi})}{\prod_{i=0}^{N-1}(s + \omega_{zi})}\omega_{pN}. \tag{3.40}$$

Function (3.40) can be implemented by means of the Cauer or Foster circuit forms, where ω_{pi} and ω_{zi} are the ith pole and zero of the one-port input transfer function being implemented, respectively, and N is its order.

The poles and zeroes of the rational transfer function (3.40) are located on the negative real axis of the open-left half s-plane in an alternating manner ($p_0 < z_0 < p_1 < z_1 < \ldots < z_{N-1} < p_N$). Such position of the poles and zeroes of this function determines the piecewise linear form of its Bode diagram (i.e., MFR and PFR). It consists of linear segments with the slope equal to -20 dB/decade (-6 dB/octave) and 0 dB/decade as depicted in Figs. 3.16 and 3.20 (one section of this line).

Notice that the slope of the line plotted via the middle points of these linear segments (at certain location of poles and zeroes of the function) will be -20α dB/dec (-6α dB/oct).

Figure 3.20 illustrates the principle of forming the approximating function that would provide the maximum admitted deviation, ε dB, of the normalized input impedance, $\bar{Z}_N(s)$, of (3.40) using the Bode diagram of $Z_N(s)$ of (3.39) over a frequency range from ω_{start} to ω_{fin}. Hence, ε is considered the maximum error of implementation of the logarithmic MFR. This method implies the following calculation of zeroes and poles:

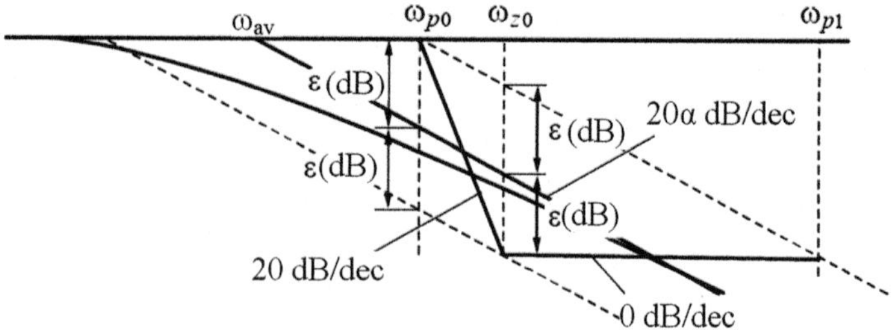

Fig. 3.20 Illustration of the principle of forming the approximating function of $\bar{Z}_N(s)$

$$\begin{aligned} \omega_{p_0} &= \omega_{\text{av}} 10^{\varepsilon/20\alpha},\ \omega_{z_0} = \omega_{p_0} 10^{\varepsilon/10(1-\alpha)}; \\ \omega_{p_1} &= \omega_{z_0} 10^{\frac{\varepsilon}{10\alpha}},\ \ \omega_{z_1} = \omega_{p_1} 10^{\frac{\varepsilon}{10(1-\alpha)}}; \\ &\vdots \\ \omega_{p_N} &= \omega_{z_{N-1}} 10^{\frac{\varepsilon}{10\alpha}},\ \omega_{z_{N-1}} = \omega_{p_{N-1}} 10^{\frac{\varepsilon}{10(1-\alpha)}}. \end{aligned} \tag{3.41}$$

Frequency ω_{av} can be found using the lower limit of the working frequency range, i.e., $\omega_{\text{av}} = \omega_{\text{start}} \sqrt{10^{\Delta/10\alpha} - 1}$ where Δ is the allowed error between ideal logarithmic MFRs (3.3b) and the approximated one of (3.39).

Define the following:

$$A = 10^{\varepsilon/10(1-\alpha)}; \quad B = 10^{\varepsilon/10\alpha}. \tag{3.42a}$$

Then

$$AB = 10^{\varepsilon/10\alpha(1-\alpha)} \tag{3.42b}$$

The following expressions uncover the physical meaning of these coefficients:

$$\frac{\omega_{z_0}}{\omega_{p_0}} = \frac{\omega_{z_1}}{\omega_{p_1}} = \cdots = \frac{\omega_{z_{N-1}}}{\omega_{p_{N-1}}} = A; \tag{3.43a}$$

$$\frac{\omega_{p_1}}{\omega_{z_0}} = \frac{\omega_{p_2}}{\omega_{z_1}} = \cdots = \frac{\omega_{p_N}}{\omega_{z_{N-1}}} = B; \tag{3.43b}$$

$$\frac{\omega_{z_1}}{\omega_{z_0}} = \frac{\omega_{z_2}}{\omega_{z_1}} = \cdots = \frac{\omega_{z_{N-1}}}{\omega_{z_{N-2}}} = AB; \tag{3.43c}$$

$$\frac{\omega_{p_1}}{\omega_{p_0}} = \frac{\omega_{p_2}}{\omega_{p_1}} = \cdots = \frac{\omega_{p_i}}{\omega_{p_{i-1}}} = \cdots = \frac{\omega_{p_N}}{\omega_{p_{N-1}}} = AB. \tag{3.43d}$$

From expression (3.43d) it follows that

$$\prod_{i=1}^{i}\left(\frac{\omega_{p_i}}{\omega_{p_{i-1}}}\right)=\frac{\omega_{p_i}}{\omega_{p_0}}=(AB)^i. \tag{3.43e}$$

Then from (3.41)–(3.43), the poles and the zeroes of the normalized input impedance fractional-rational transfer function $\bar{Z}_N(s)$ given by (3.40) can be expressed as follows:

$$\omega_{p0}=\omega_{\text{av}}\sqrt{B}:\ \omega_{pi}=\omega_{p0}(AB)^i;\omega_{zi}=\omega_{p0}A(AB)^i=A\omega_{pi}. \tag{3.44}$$

Take the logarithm of (3.43e) for $i=N-1$ (the maximum order of the zero). The result is a formula to calculate the order, N, of the normalized input impedance of $\bar{Z}_N(s)$ (3.40):

$$N=\frac{\lg\frac{\omega_{pN-1}}{\omega_{p_0}}}{\lg(AB)}+1.$$

One could simplify the order of approximation, N, by replacing A from (3.43a) and B from (3.43b). Assume that ω_{pN-1} is the upper limit of the frequency range of the approximating transfer function; $\omega_{pN-1}=\omega_{\max}$. Hence, now N may be expressed as follows:

$$N=\frac{10\alpha(1-\alpha)}{\varepsilon}\lg\frac{\omega_{\max}}{\omega_{p_0}}+1. \tag{3.45}$$

where $\omega_{\max}=L\cdot\omega_{\text{fin}}$, and L is a multiplying factor whose value can be chosen from 10 to 100.

Therefore, the wider the operating frequency range of the one-port configuration, the smaller the maximum allowable deviation ε of the normalized input impedance of the MFR. This definitely will increase the order, N, of the approximating transfer function, $\bar{Z}_N(s)$. Thus, a compromise has to be made between accuracy and complexity of the design.

Example 4 Use Shareff method to develop a rational transfer function for a normalized fractional-order integrator of order $\alpha=0.65$.

Calculate the order of approximation, N, as well as the values of poles and zeroes on the real frequency axis under the following initial conditions:

$\alpha=0.65$; $\varepsilon=1$ dB; $(\omega_{\text{start}},\omega_{\text{fin}})=(10^3$ rad/s, 10^5 rad/s) at $\Delta=10^{-5}$; $L=100$.

Solution

1. $\omega_{\text{av}}=100\sqrt{10^{\left(10^{-5}/10\times0.65\right)}-1}=1.8821\ \text{rad/s}.$
2. $\omega_{\max}=100\cdot\omega_{\text{fin}}=10^7\,\text{rad/s}.$
3. $A=10^{\varepsilon/10(1-\alpha)}=10^{1/10(1-0.65)}=1.9307;$
   $B=10^{\varepsilon/10\alpha}=10^{1/10\cdot0.65}=1.4251.$

4. $\omega_{p0} = \omega_{av}\sqrt{B} = 1.8821\sqrt{1.4251} = 2.2468\ \mathrm{rad/s}$.
5. $\omega_{z0} = A\omega_{p0} = 1.9307 \cdot 2.2468 = 4.3379\,\mathrm{rad/s}$.
6. $N = \frac{10\alpha(1-\alpha)}{\varepsilon}\log\frac{\omega_{\max}}{\omega_{p0}} + 1 = \frac{6.5(1-0.65)}{1}\log\frac{10^7}{2.2468} + 1 = 18.4 \approx 19$
7. $\omega_{pi} = \omega_{p0}(AB)^I = 2.2468 \times 2.7514^i\ \mathrm{rad/s};\ i = 1, 2,\ \ldots,\ 19.$
8. $\omega_{zi} = \omega_{z0}(AB)^I = 4.3379 \times 2.7514^i\,\mathrm{rad/s};\ i = 1, 2, \ldots, 18.$

The acquired fractional-rational function can be implemented by means of the Foster or Cauer method (see Sects. 3.3.3 and 3.3.4).

Remark 1 The previous two approximation algorithms of Oustaloup and Shareff are not suitable to design fractional-order differentiator since the order of the numerator of $\bar{Z}_N(s)$, m, is less than the order of the denominator, n, by one; i.e., $n = m + 1$.

3.3.5.3 El-Khazali Approximation of Fractional-Order Integro-Differential Operators[5]

El-Khazali method introduces an algorithm to approximate both fractional-order integrators and differentiators. It depends on using a special form, a biquadratic structure that approximates $s^{\pm\alpha}$, where s^{α} defines a differential operator, while $s^{-\alpha}$ defines a fractional-order integrator. Since equation (3.3b) represents a normalized integrator, then using El-Khazali method, it can be approximated as follows:

$$s^{\alpha} = \prod_{i=1}^{n} H_i(s) \equiv \prod_{i=1}^{n} \frac{N_i(s/\omega_i)}{D_i(s/\omega_i)}; \tag{3.46}$$

where ω_i, $i = 1, 2,\ \ldots,\ n$ is the center frequency of each biquadratic module, and where

$$H_i(s/\omega_i) = \frac{N_i(s/\omega_i)}{D_i(s/\omega_i)} = \frac{a_o\left(\frac{s}{\omega_i}\right)^2 + a_1\left(\frac{s}{\omega_i}\right) + a_2}{a_2\left(\frac{s}{\omega_i}\right)^2 + a_1\left(\frac{s}{\omega_i}\right) + a_0} \tag{3.47a}$$

and

$$\left.\begin{aligned} a_0 &= \alpha^{\alpha} + 3\alpha + 2 \\ a_2 &= \alpha^{\alpha} - 3\alpha + 2 \\ a_1 &= 6\alpha\ \tan\frac{(2-\alpha)\pi}{4} \end{aligned}\right\}. \tag{3.47b}$$

[5]See El-Khazali, R. "On the biquadratic approximation of fractional-order Laplacian operators", Analog Integrated Circuits and Signal Processing, Vol. 82 (3), pp. 503–517, 2015.

Substituting from (3.47b) into (3.47a) yields the following new form of $H_i(s/\omega_i)$ that only depends on the fractional order, α,

$$H_i(s/\omega_i) \approx \frac{(\alpha^\alpha + 3\alpha + 2)\left(\frac{s}{\omega_i}\right)^2 + \left(6\alpha \tan\frac{(2-\alpha)\pi}{4}\right)\left(\frac{s}{\omega_i}\right) + (\alpha^\alpha - 3\alpha + 2)}{(\alpha^\alpha - 3\alpha + 2)\left(\frac{s}{\omega_i}\right)^2 + \left(6\alpha \tan\frac{(2-\alpha)\pi}{4}\right)\left(\frac{s}{\omega_i}\right) + (\alpha^\alpha + 3\alpha + 2)}. \tag{3.48}$$

Observe that from (3.47b), $a_0 > a_2$ for $0 < \alpha \le 1$, and (3.48) yields $\arg\{H(j1)\} = \pi - 2\tan^{-1}\left(\frac{a_1}{a_0 - a_2}\right) > 0$. In this case, (3.47a) describes a fractional-order differentiator (FoD) around ω_i.

Observe that the biquadratic structure of (3.47a) allows one to use the reciprocal of (3.46) to define an integral action at ω_i. Thus, the following modular rational transfer function can also be used to approximate a fractional-order integrator (FoI):

$$s^{-\alpha} = \prod_{i=1}^{n} \hat{H}_i(s/\omega_i) \equiv \prod_{i=1}^{n} \frac{D_i(s/\omega_i)}{N_i(s/\omega_i)}; \tag{3.49}$$

For example when $\alpha = 0.5$ and $\omega_i = 1$ rad/s, equation (3.48) yields

$$s^{0.5} = H(s) \approx \frac{2.707s^2 + 4.828s + 0.707}{0.707s^2 + 4.828s + 2.707}$$

and its reciprocal equals to

$$s^{-0.5} = \hat{H}(s) \approx \frac{0.707s^2 + 4.828s + 2.707}{2.707s^2 + 4.828s + 0.707}.$$

Figure 3.21a, b displays the Bode diagram of both approximations. Clearly, they exhibit differential (integral) behavior around the center frequency, $\omega_i = 1$ rad/s. A simple test shows that the poles and zeros of $H(s)$ and obviously $\hat{H}(s)$ alternate. The zeros lead the poles when $a_0 > a_2$, while the poles lead the zeros when $a_0 < a_2$. In other words, these special forms of $s^{\pm\alpha}$ define new lead and lag operators.

Figure 3.22 shows a scenario of cascading three biquadratic forms centered at ω_1, ω_2, and ω_2, respectively. The spacing between each corner frequency must be carefully selected such that the phase value is as close as possible to $\frac{\alpha\pi}{2}$ over the entire frequency band. This can be accomplished by forcing the phase diagrams of every two consecutive biquadratic forms intersect at ω_{x_i}, the geometric mean of the corresponding center frequencies, where $\omega_{xi} = \sqrt{\omega_i \omega_{i+1}}$, $i = 1, 2, \ldots, N-1$. Hence, one should require

$$\arg\left\{H(s/\omega_i)\big|_{s=j\omega_{x_i}}\right\} = \frac{\alpha\pi}{4}; \quad i = 1, 2, \ldots, N-1. \tag{3.50}$$

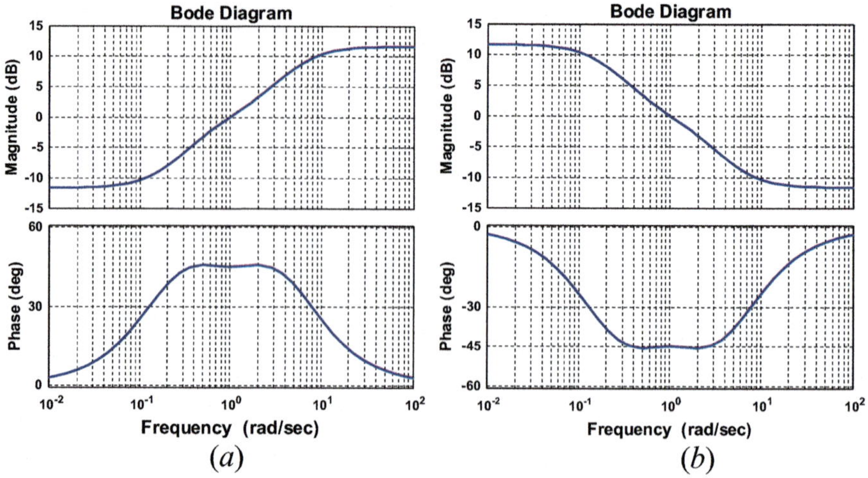

Fig. 3.21 Bode diagram of FoI and FoD when (**a**) $a_0 < a_2$, and (**b**) $a_0 > a_2$

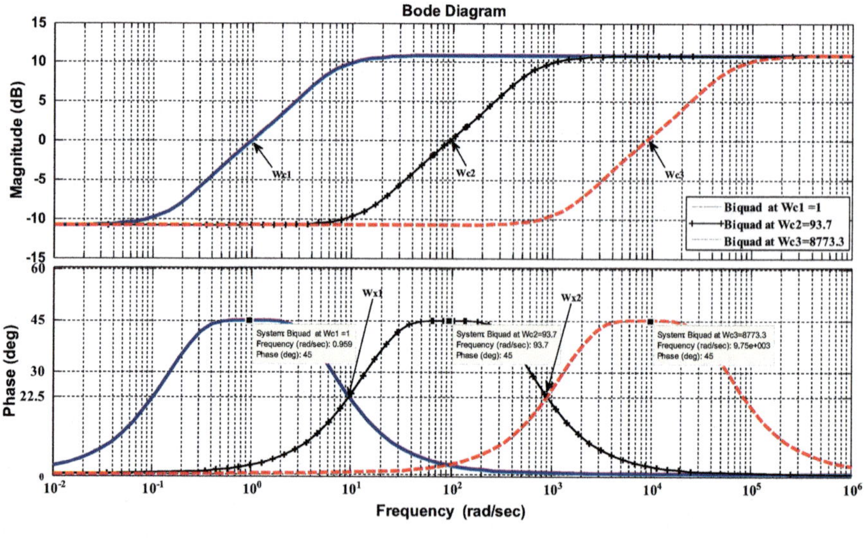

Fig. 3.22 Three biquadratic modules centered at $\omega_1 = 1$ rad/s, $\omega_2 = 95.35$ rad/s, and $\omega_3 = 9092.32$ rad/s for $\alpha = 0.5$

Theorem 1 Consider the cascaded system given by (3.50). Let (3.47) be the fundamental biquadratic approximation of $s^\alpha; 0 < \alpha \leq 1$. If $\arg(H(s/\omega_i))\big|_{\omega_{x_i}} = \frac{\alpha\pi}{4}$ for $i = 1, 2, \ldots, N-1$, then

$$a_0 a_2 \eta \Upsilon^4 + a_1(a_2 - a_0)\Upsilon^3 + \left(a_1^2 - a_2^2 - a_0^2\right)\Upsilon^2 + a_1(a_2 - a_0)\Upsilon + a_0 a_2 \eta = 0, \tag{3.51}$$

where $\Upsilon \equiv \frac{\omega_{x1}}{\omega_1}$, and $\eta \equiv \tan\left(\frac{\alpha\pi}{4}\right)$, and where

$$\omega_{x1} = \max\{\Upsilon\}, \omega_i = \omega_1 . \omega_{x1}^{2(i-1)}; \qquad i = 2, 3, \ \ldots, N. \tag{3.52}$$

Proof Consider the biquadratic transfer function given by (3.48) for $0 < \alpha \leq 1$, where $\arg\{H(\omega_1/\omega_1)\} = \frac{\alpha\pi}{2}$. Let $\Upsilon \equiv \frac{\omega_{x1}}{\omega_1}$, where ω_{x_1} be such that $\arg\{H(\omega_{x_1}/\omega_1)\} = \frac{\alpha\pi}{4}$. Without loss of generality, assume that $\omega_1 = 1$ rad/s; then $\Upsilon = \omega_{x_1}$. Hence, if $\arg(H(s/\omega_i))\big|_{\omega x_i} = \frac{\alpha\pi}{4}$, it implies

$$\tan^{-1}\left(\frac{a_1\Upsilon}{a_2 - a_0\Upsilon^2}\right) - \tan^{-1}\left(\frac{a_1\Upsilon}{a_2 - a_0\Upsilon^2}\right) = \frac{\alpha\pi}{4} \tag{3.53}$$

Solving (3.53) for Υ yields

$$a_0a_2\eta\Upsilon^4 + a_1(a_2 - a_0)\Upsilon^3 + (a_1^2 - a_2^2 - a_0^2)\eta\Upsilon^2 + a_1(a_2 - a_0)\Upsilon + a_0a_2\eta = 0, \tag{3.54}$$

where $\eta = \tan\left(\frac{\alpha\pi}{4}\right)$.

The solution of (3.54) for all $0 < \alpha \leq 1$ gives two complex roots that are discarded and two real ones. One of the real roots is very small, which is also neglected. Hence, the only solution of interest is the maximum real one. Since ω_{xi} ; $i = 1, 2, \ \ldots, N-1$ are located at the geometric mean of the corner frequencies, ω_i, and ω_{i+1}, $i = 1, 2, \ \ldots, N-1$, then ω_{x1} is placed at the geometric mean of the first two corner frequencies, ω_1, ω_2, i.e., $\omega_2 = \omega_{x_1}^2 . \omega_1$. Similarly, if one uses a third biquadratic module centered at ω_3, then $\omega_3 = \omega_{x_1}^2 . \omega_2 = \omega_{x_1}^4 . \omega_1$. It follows that $\omega_i = \omega_1 \cdot \omega_{x1}^{2(i-1)}$, $i = 2, 3, \ \ldots, N$. Hence, (3.53) is met.

El-Khazali algorithm can be extended to widen the bandwidth of the biquadratic approximation of the fractional-order operator by cascading N biquadratic modules centered at ω_i, $i = 2, 3, \ \ldots, N$, i.e., forming a 2Nth-order rational transfer function of the form

$$s^\alpha = \left(\frac{a_0\left(\frac{s}{\omega_1}\right)^2 + a_1\left(\frac{s}{\omega_1}\right) + a_2}{a_2\left(\frac{s}{\omega_1}\right)^2 + a_1\left(\frac{s}{\omega_1}\right) + a_0}\right)\left(\frac{a_0\left(\frac{s}{\omega_2}\right)^2 + a_1\left(\frac{s}{\omega_2}\right) + a_2}{a_2\left(\frac{s}{\omega_2}\right)^2 + a_1\left(\frac{s}{\omega_2}\right) + a_0}\right) \times \\ \ldots \times \left(\frac{a_0\left(\frac{s}{\omega_N}\right)^2 + a_1\left(\frac{s}{\omega_N}\right) + a_2}{a_2\left(\frac{s}{\omega_{cN}}\right)^2 + a_1\left(\frac{s}{\omega_N}\right) + a_0}\right) \tag{3.55}$$

where $\omega_{x1} = \max\{\Upsilon\}$; Υ is the solution of (3.54); and $\omega_i = \omega_1\omega_{x1}^{2(i-1)}$; $i = 2, 3, \ \ldots, N$.

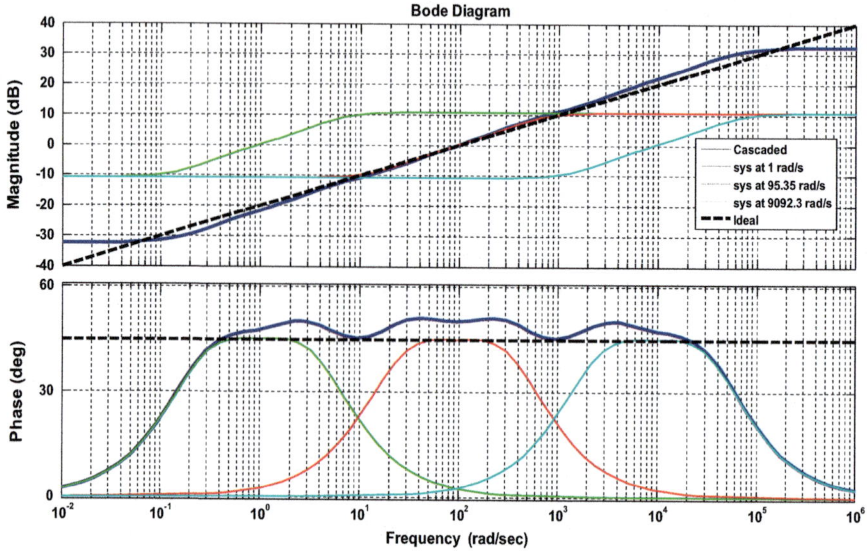

Fig. 3.23 Frequency response of an ideal differentiator, $s^{0.5}$, and its three cascaded approximation

Figure 3.22 shows the frequency response of three biquadratic approximations for $s^{0.5}$ cascaded at three consecutive frequencies. Without loss of generality, let $\omega_1 = 1$ rad/s; then equations (3.52), (3.54), and (3.55) yield $\omega_{x_1} = 9.78$ rad/s, $\omega_2 = 95.35$ rad/s, $\omega_{x_2} = 906.715\ \text{rad/s}$, and $\omega_3 = 9092.32$ rad/s. The frequency response of the corresponding three cascaded modules is also shown in Fig. 3.23. Even though each biquadratic module exhibits a flat and exact phase response within its own bandwidth, the three-cascaded modules, when compared with the ideal one, exhibit a phase error of around 7.5° as evident from Fig. 3.23. Such error can be minimized by tuning the approximation parameters in (3.53), or by tuning the fractional order.

3.3.6 *Realization of Fractional-Order Inductors (FoIs)*

The fractional-order inductors can be realized by using its voltage-current relationship, which yields a factional-order impedance of the form

$$\begin{aligned} Z_L(s) \equiv H_d(s) = Ls^\alpha &\approx L\frac{a_0 S^2 + a_1 S + a_2}{a_2 S^2 + a_1 S + a_0} \\ &= L\frac{a_0}{a_2}\frac{(s+z_1)(s+z_2)}{(s+p_1)(s+p_2)} = L\frac{a_0}{a_2}\frac{N(s)}{D(s)}. \end{aligned} \tag{3.56}$$

Since $a_0 > a_2 > 0$, (3.56) describes a differential operator with a net leading phase, i.e., $\underline{/Z_L(s)} = (\varphi_{z1} + \varphi_{z2}) - \left(\varphi_{p_1} + \varphi_{p_2}\right) > 0$, where φ_{z1}, φ_{z2} are the phase contribution of the zeros, z_1, and z_2 of $Z_L(s)$; $z_{1,2} = \frac{-a_1}{2a_0} \pm \sqrt{\left(\frac{a_1}{2a_0}\right)^2 - \left(\frac{a_2}{a_0}\right)}$, while φ_{p1}, and φ_{p2} are the phase contribution of the poles, p_1 and p_2 of $Z_L(s)$; $p_{1,2} = \frac{-a_1}{2a_2} \pm \sqrt{\left(\frac{a_1}{2a_2}\right)^2 - \left(\frac{a_0}{a_2}\right)}$.

The zeros and poles of $H_d(s)$ alternate where $p_2 < z_2 < p_1 < z_1 < 0$. One may use the *Foster-II* design method (Budak, 1974) in order to build a physically realizable FoI. To obtain nonnegative passive components, $Z_L(\hat{s})/\hat{s}$ is expanded using partial fraction expansion as follows:

$$\begin{aligned}\frac{Z_L(\hat{s})}{\hat{s}} &= L\left(\frac{a_0}{a_2}\right)\frac{(\hat{s}+z_1)(\hat{s}+z_2)}{\hat{s}\,(\hat{s}+p_1)(\hat{s}+p_2)}\\ &= L\left(\frac{a_0}{a_2}\right)\left\{\frac{k_0}{\hat{s}}+\frac{k_1}{(\hat{s}+p_1)}+\frac{k_2}{(\hat{s}+p_2)}\right\},\end{aligned} \tag{3.57}$$

where $k_0 = \frac{z_1 z_2}{p_1 p_2}$, $k_1 = \frac{(z_1-p_1)(z_2-p_2)}{-p_1(p_2-p_1)}$, and $k_2 = \frac{(z_1-p_2)(z_2-p_2)}{-p_2(p_1-p_2)}$, and where $\hat{s} \equiv s/\omega_c$ is a normalized Laplacian operator at the center frequency, ω_c.

Hence,

$$Z_L(\hat{s}) = L\frac{a_0}{a_2}k_0 + \left\{\frac{\hat{s}\,a_0k_1L/a_2}{(\hat{s}+p_1)} + \frac{\hat{s}\,a_0k_2L/a_2}{(\hat{s}+p_2)}\right\} \equiv Z_1(\hat{s}) + Z_2(\hat{s}) + Z_3(\hat{s}),$$

which is designed by the following single-stage one-port RL circuit as depicted in Fig. 3.24.

The *RL* circuit shown in Fig. 3.24 represents a single-module realization of a FoI centered at any frequency using two coils and three resistors. For example, to synthesize $Z_L(s) = s^{0.5}L$; $L = 1\,\text{H}$; then, for $\alpha = 0.5$, equation (3.47b) yields $a_0 = 4.207$, $a_1 = 7.2426$, and $a_2 = 1.2071$. The circuit diagram of $Z_L(s) = s^{0.5}$ and its frequency response are shown in Fig. 3.25a, b, respectively, for $\omega_c = 2\pi(0.159) = 1\,\text{rad/s}$.. The circuit diagram is depicted in Fig. 3.25a, while

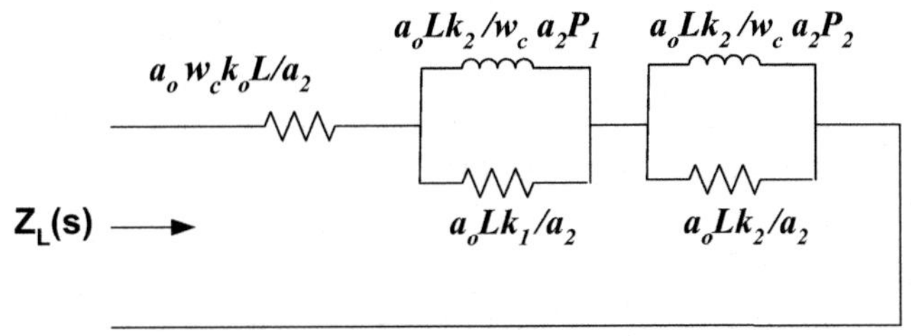

Fig. 3.24 Single-stage one-port *RL* realization of FoI

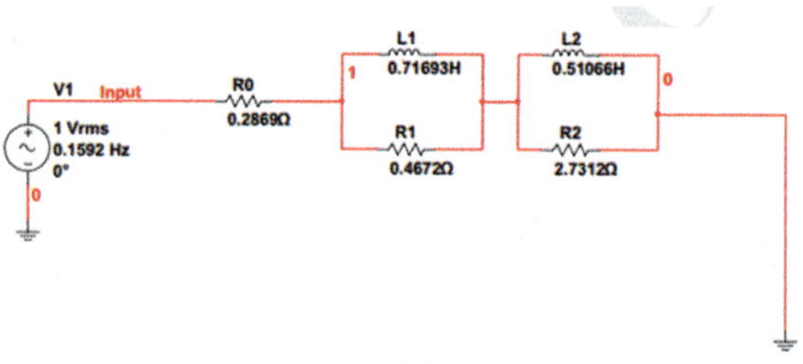

(a)

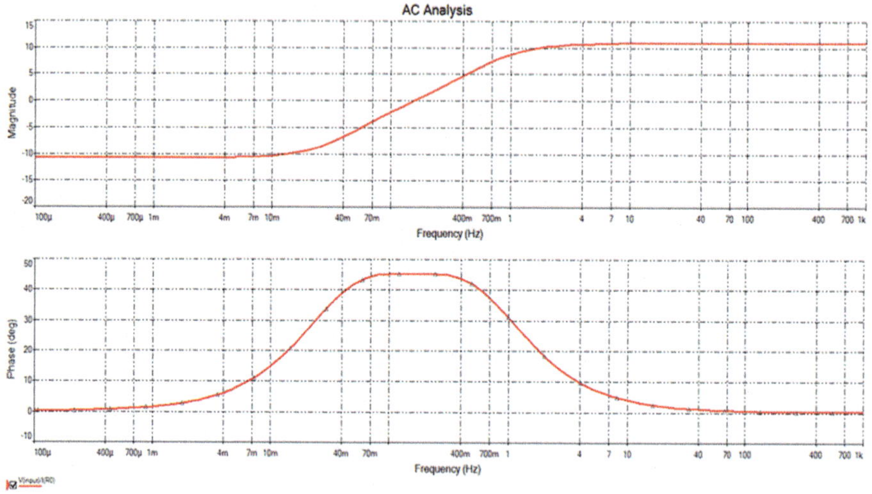

(b)

Fig. 3.25 (**a**) Single-module realization of $Z_L(s) = s^{0.5}L; L = 1$ H and (**b**) frequency response of $Z_L(s) = s^{0.5}$

the performance of the circuit is simulated using Multisim11® software. As expected, the simulated *RL* circuit given in Fig. 3.25b yields an exact phase value of $\pi/4$ and a gain of slope 10 dB/dec over a frequency spectrum centered at $\omega_c = 2\pi(0.159) = 1$ rad/s.

For *N*-cascaded modules, however, the partial fraction expansion (PFE) of (3.55) yields

$$\frac{Z_L(s)}{s} = L\left\{\frac{k_0}{s} + \frac{k_1}{(s+p_1)} + \frac{k_2}{(s+p_2)} + \ldots + \frac{k_N}{(s+p_N)}\right\};\ i = 1, 2, \ \ldots, N; \quad (3.58)$$

where the values of the *RL* circuit of the fractional-order differentiator are computed from

$$R_0 = Lk_0, \ R_i = \frac{1}{Lk_i}, \ L_i = \frac{Lk_i}{p_i}; \quad i = 1, 2, \ \ldots, N. \quad (3.59)$$

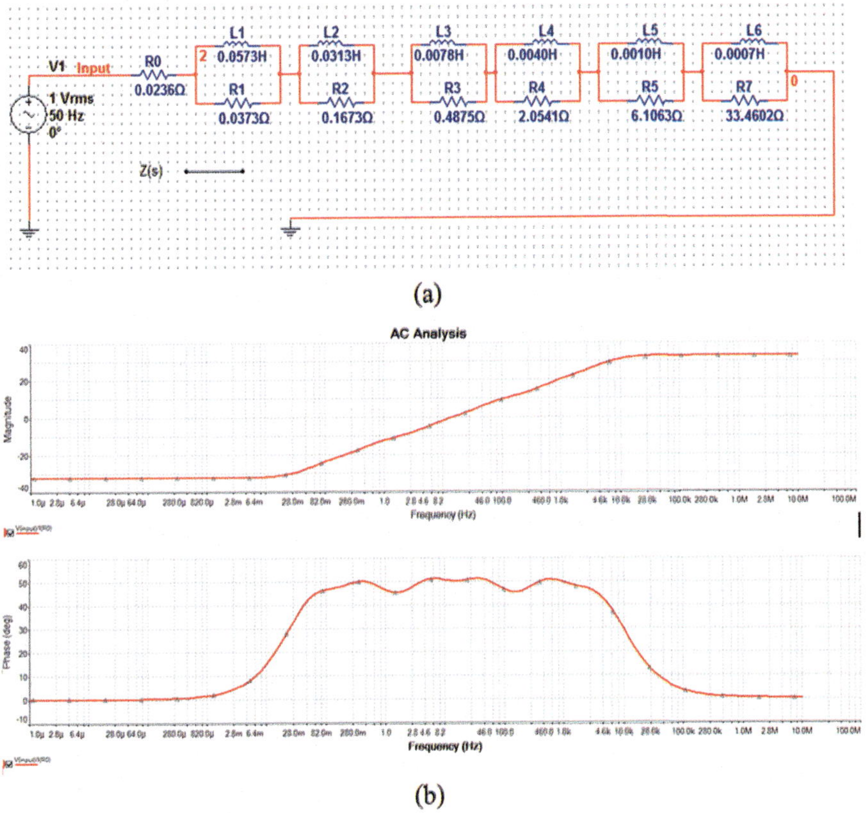

Fig. 3.26 (**a**) Three-cascaded biquadratic modules that approximate $s^{0.5}$, and (**b**) frequency response $Z_L(s) = s^{0.5}$

From (3.59), one may cascade several modules to synthesize FoIs over a wider frequency spectrum. Figure 3.26a, b shows the circuit diagram and the frequency response of three biquadratic modules centered at $\omega_{c_1} = 1$ rad/s, $\omega_{c_2} = 95.3537$ rad/s, and $\omega_{c_3} = 9092.3$ rad/s using Multisim11®. It is evident from the AC analysis of the designed *RL*-circuit diagram shown in Fig. 3.26b that the three stages are centered at their geometric mean, which is 95.35 rad/s, or at $f = 15.176\,\text{Hz}$.

Remark 2 In order to obtain a proper realization using positive passive elements to synthesize FoIs using multistage biquadratic modules, one must properly allocate each module at its corresponding center frequency using (3.52). Therefore, one should not use random values for the corner frequencies since the PFE would yield unacceptable negative values for the resistors or the inductors.

3.3.7 Realization of Fractional-Order Capacitors (FoC)

The FoC can also be synthesized in a similar fashion by considering the reciprocal of the Laplacian operator $\frac{1}{s^{\alpha}}$. From (3.3a) and (3.49), the basic building circuit that approximates anormalized FoC defined over a narrow band frequency can be described by $Z_c(s) = \frac{1}{s^{\alpha}C}$, where C is constant and represents the value of an integer-order capacitor, i.e.,

$$Z_c(s) = \frac{1}{s^{\alpha}C} \approx \frac{1}{C}\frac{a_2}{a_0}\frac{(s+p_1)(s+p_2)}{(s+z_1)(s+z_2)} = \frac{1}{C}\frac{a_2}{a_0}\frac{D(s)}{N(s)}. \tag{3.60}$$

In a similar argument, since $0 < a_2 < a_0$, (3.60) exhibits a lagging phase and the fractional expansion exists with positive residues, which yields positive passive elements, i.e.,

$$\begin{aligned} Z_c(s) &= \frac{1 a_2}{C a_0}\left\{1 + \frac{\lambda_1}{(s+z_1)} + \frac{\lambda_2}{(s+z_2)}\right\} \\ \text{where } \lambda_1 &= \frac{(p_1 - z_1)(p_2 - z_1)}{(z_2 - p_1)}, \ \lambda_2 = \frac{(p_1 - z_2)(p_2 - z_2)}{(z_1 - p_2)}. \end{aligned} \tag{3.61}$$

Figure 3.27 shows the *RC* realization of $1/Cs^{\alpha}$ using a single-stage biquadratic transfer function with integer-order capacitors.

Similar to (3.58), an FoC using *N*-cascaded modules, as depicted in Fig. 3.28a, yields

$$Z_c(s) = \frac{1}{C}\left\{k_0 + \frac{k_1}{(s+z_1)} + \frac{k_2}{(s+z_2)}, \ \ldots, \frac{k_N}{(s+z_N)}\right\} i = 1, 2, \ \ldots, N, \tag{3.62a}$$

$$R_0 = \left(\frac{k_0}{c}\right), R_i = \frac{k_i}{c z_i}, \ \ C_i = \frac{c}{k_i} \ \ i = 1, 2, \ \ldots N. \tag{3.62b}$$

Figure 3.28, on the other hand, shows a Bode plot diagram of a triple-stage approximation of $\frac{1}{s^{0.5}}$; $C = 1$ F, for the same parameters used in designing the FoIs in Sect. 3.3.6.

Fig. 3.27 Single-stage approximation of $1/Cs^{\alpha}$

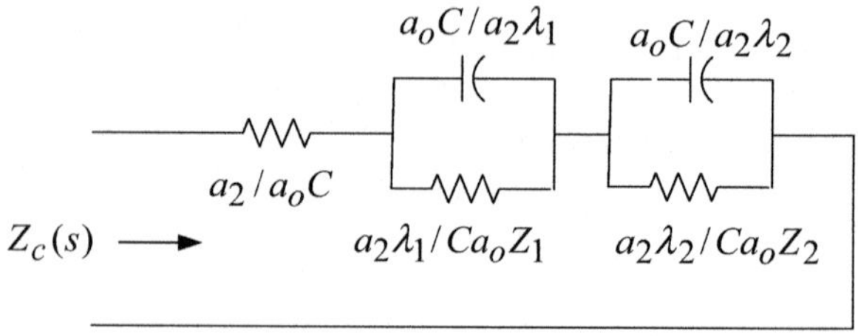

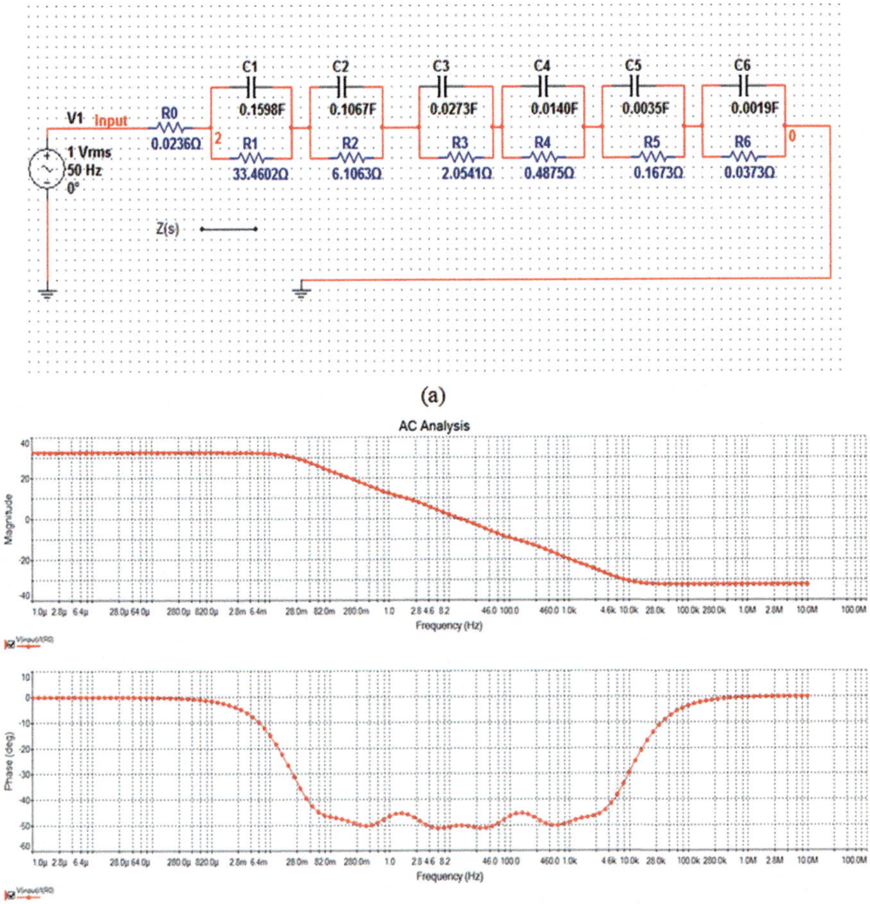

Fig. 3.28 (**a**) Realization of $1/Cs^{0.5}$ using three biquadratic modules, and (**b**) frequency response of $1/Cs^{0.5}$

3.4 Realization of Fractal Impedances Using *RC* Circuits with Distributed Parameters

The current-voltage relation at the input of an *RC* cable (a semi-infinite *RC* line) that complies with equation (2.60) is an example of an analog implementation of half-order integration and differentiation operations up to the middle of the twentieth century before the appearance of microelectronics. The ladder circuit depicted in Fig. 3.29a was used as a physical model of the *RC* cable. This circuit simulates the features of a segment of the *RC* line (see Fig. 3.29b), where r is the *RC*-line resistance per unit length, c is the *RC*-line capacitance per unit length, and Δx represents the *RC*-line unit length segment.

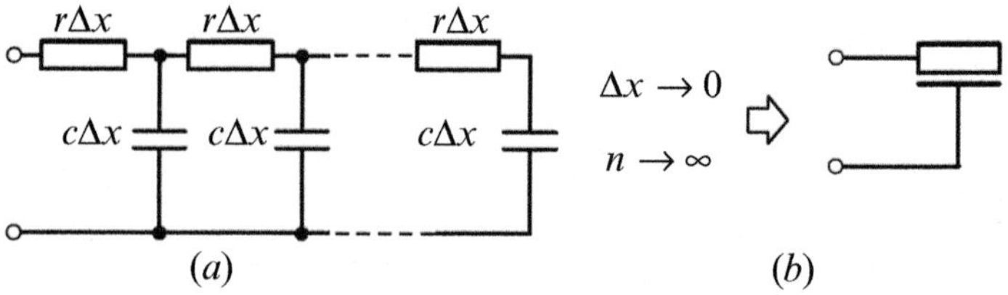

Fig. 3.29 Ladder circuit (**a**) that simulates *RC*-line segment (**b**): lumped representation

Fig. 3.30 Three-layer thin-film *RC*-EDP structure: (*1*) wafer; (*2*) plate; (*3*) dielectric film; (*4*) plate terminal; (*5*) resistive layer terminals; (*6*) resistive layer

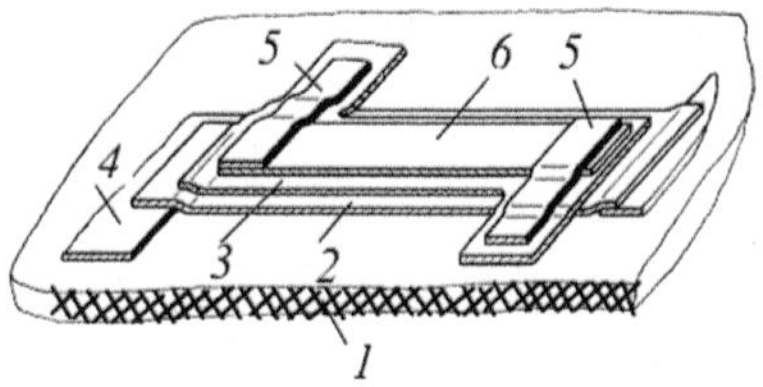

Fig. 3.31 A bipolar transistor structure-based *RC*-EDP (**a**) and its schematic diagram (**b**)

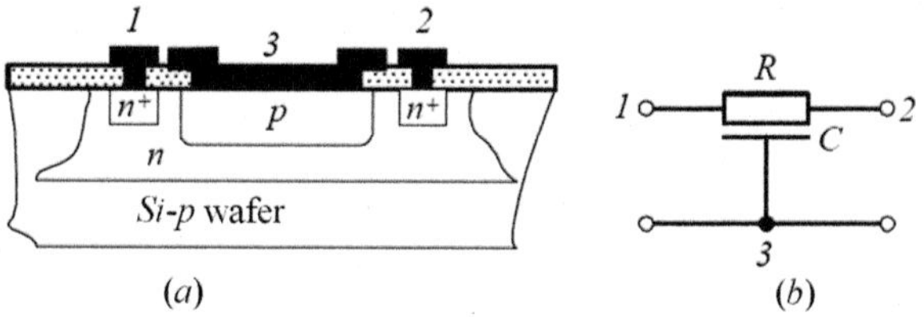

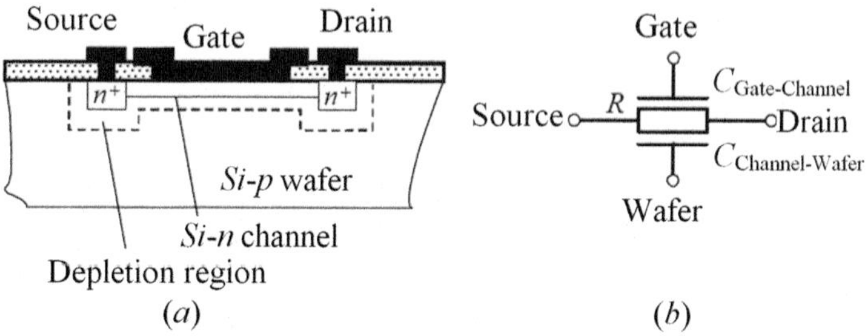

Fig. 3.32 An MOS structure-based *RC*-EDP (**a**) and its schematic diagram (**b**)

With the development of microelectronics, it became possible to fabricate ladder *RC* circuits of Fig. 3.29 by means of film and semiconductor structures as depicted in Figs. 3.30, 3.31, and 3.32. Since these structures could be considered as separate elements of resistive and dielectric layers (ribbons), they were defined as resistive-capacitive elements with distributed parameters (*RC*-EDP).

They performed the same functions as the ladder circuits but their size was significantly smaller, their precision of modeling the features of an ideal *RC* cable was higher, they were more thermally stable, and they possessed other advantages, which the *IC*-circuit technology provides.

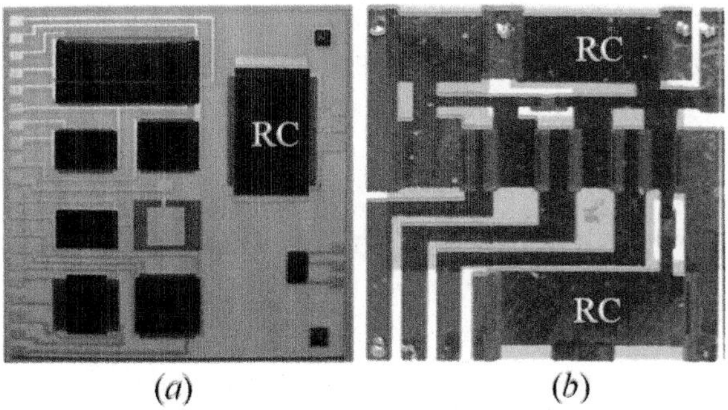

Fig. 3.33 Photo images of (**a**) thick-film wafers and (**b**) thin-film *RC*-EDP

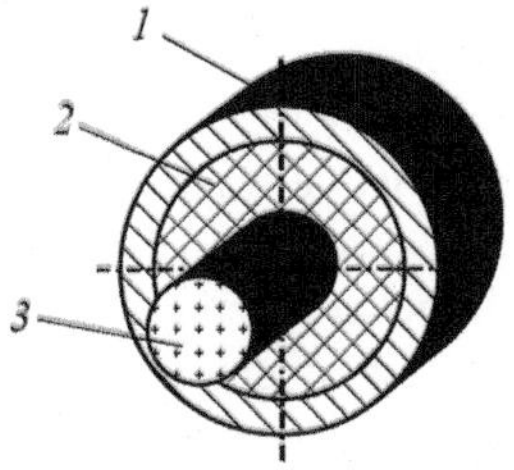

Fig. 3.34 Microwire-based *RC*-EDP: (*1*) conductive layer; (*2*) insulator; (*3*) resistive microwire

Photo images of some real thin-film and thick-film *RC* elements with distributed parameters are given in Fig. 3.33 (they are marked as *RC*).

As a rule of thumb, limitations of the square and the extreme parameters of dielectric layers of these elements did not allow to build low- and infra-low-frequency *RC*-EDP. Hence, in addition to the *RC*-EDP film, the *RC*-EDP were also created in the form of *RC* cable (see Fig. 3.34). The central core is a microwire made of material with high specific resistance. This wire is protected with an insulating coating. The insulating coating is covered with high-conductivity metallic layer. Such wire is wounded on a ceramic body, and the *RC*-EDP is then performed as an ordinary discrete resistor. The microwire-based *RC*-EDP provides large values of τ_{RC} (units of seconds). However, there are difficulties to provide heterogeneous specific parameters along the length of the cable. These difficulties limit the corresponding implemented fractional-order impedance with the value of 1/2. Therefore, the microwire *RC*-EDP is naturally a one-dimensional element, and it reduces its schematic potential.

Integral circuit production methods used to manufacture *RC*-EDP allow to make microelements with distributed parameters. The fractional-order impedance can be adjusted with different designs, schematics, and technology processes. The design-and-technology methods include the following: creation of film layers with different laws of specific parameter variation, introduction of local heterogeneities into the layers (cutouts, for instance), increase of the number of resistive, conducting, and dielectric layers, and increase of the amount of terminals.

The schematic methods include variation of interlayer connections and variations of connection layouts. Furthermore, the parameters (including the fractal order) of film elements and semiconductor *RC*-EDP can be adjusted using physical magnetic fields on the corresponding sensitive materials of resistive and dielectric layers.

3.5 Fabrication of Fractal Impedances Using Nanostructured Materials

It is a common knowledge that the fractional power, α, of an input impedance depends on frequency and that appears evident when investigating the conductivity of its composite materials. It also depends on the structure of the percolating cluster.[6]

The observed dependencies can be explained by the fact that there are numerous alternative ways for the electric current paths in composite materials. The current can exist through the high conducting channels consisting of a series of metallic grains contacting each other. It can also flow through the conducting paths that consist of grains of semiconductor impurities. The current can also appear via tunneling or hopping conduction through dielectric layers.

In this case, the conductivity is complex and its absolute value and phase will depend on frequency in nearly similar manner as the conductivity-frequency dependence of the "rough electrode-electrolyte" system.

Thus, the features of the composite materials will lead us to create fractal elements. The conducting medium of these materials is introduced into the polymeric matrix (polymeric base). The variation of the matrix essence, the filling medium material, its concentration, and particle size make it possible to change the characteristics of the formed percolating cluster, consequently determining the required parameters of the acquired fractal element.

Figure 3.35 depicts TEM images[7] of nanostructured polymeric composition materials. The images were obtained using JEM-1011 electronic microscope produced by JEOL Company. These composites are comprised of metal-containing nanoparticles distributed over the volume of the polyethylene (PE) matrix.

The designed composite materials based on high-pressure PE that contains nanoparticles with different content and structure are a heterogeneous disordered dispersion system. In this system, the polymeric matrix is the dispersion medium and the nanoparticles are the dispersed phase. Such materials exhibit evident concentration (mass content of the filling material), which depends on their physical properties (density, dielectric permittivity, electric conductance, etc.).

The samples with molybdenum nanoparticles are produced by thermal degradation of molybdenum hexacarbonyl in the PE matrix (its nanoparticle content:

[6]A *cluster* in physics is often a system of linked atom or molecules, *percolation*—the process of trickling through.

[7]TEM—Transmission electronic microscopy.

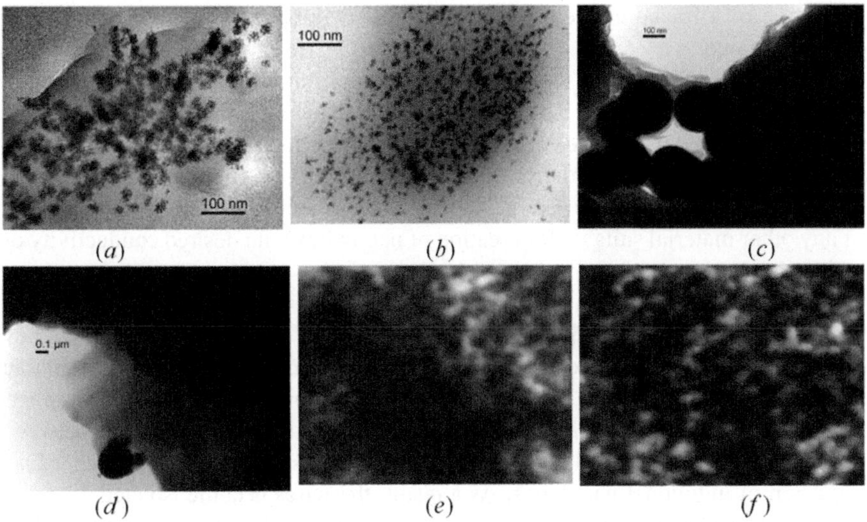

Fig. 3.35 TEM images of samples of PE-based nanostructured polymeric composition materials: (**a**) Mo (10 % mass); (**b**) Mo (20 % mass); (**c**) Bi (10 % mass); (**d**) Bi (20 % mass); (**e**) Fe (10 % mass); (**f**) Fe (20 % mass)

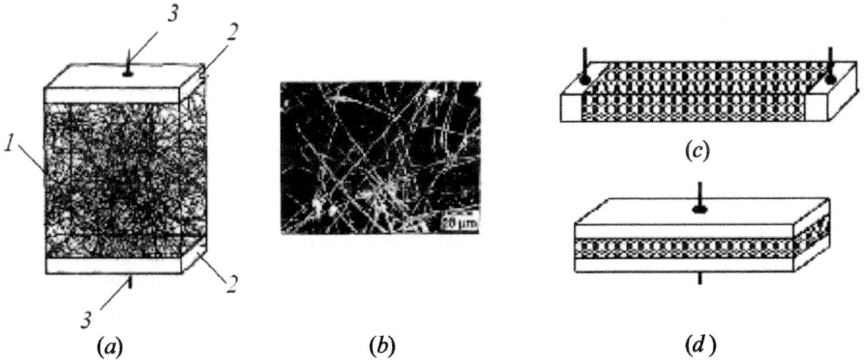

Fig. 3.36 Nanostructured material-based fractal element: (**a**) principle of formation; (**b**) fractal medium microstructure via electronic microscope; (**c**, **d**) some design structures: (*1*) nonconducting base (matrix) with nanowires; (*2*) conducting plates; (*3*) element terminals

MoO_2, metallic molybdenum). The samples with bismuth nanoparticles are produced by thermal degradation of bismuth acetate in the PE matrix (its nanoparticle content: Bi_2O_3, metallic bismuth). The samples with iron particles are produced by thermal degradation of iron pentacarbonyl in the PE matrix (its nanoparticle content: Fe_2O_3, metallic iron).

An example of a structure design fabrication of nanowire fractal elements is described in the US patent 2006/0267595 A1. Their operation is based on the electric current that flows through the matrix. Figure 3.36 depicts the principle of formation of a fractal element based on nanostructured materials and some of its design variants.

Material *1* located between two contact plates *2* comprises a three-dimensional complex of nanowires (Fig. 3.36a). Figure 3.36b depicts the fractal medium microstructure acquired using a scanning electronic microscope. The nanowires here are electrically conducting and their resistance value is either insignificant or specified. For example, the nanowires can be made of metal–metal polymeric components or of any other material suitable for creation of nanowires with desired conductivity or resistance.

Polymer components can be made of one-dimensional structure using partial oxidation of metallic complexes (aggregated metal atoms). For example, they consist of partially oxidized platinum complexes. Photooxidation is one possible method of partial oxidation. Some nanowires of the complex can touch each other. But in general, the complexes are produced to reduce interrelations between individual wires. This can be obtained using excess of charges, which cause repulsion of individual nanowires. As a result, the wires become isolated.

The nanowires are encapsulated into the base (matrix). The material of the matrix may possess some conductivity as well as it can be nonconducting. Such material can be polymer, copolymer (a polymeric molecule comprised of at least two monomers), or a combination of them. The material can be in any physical state: solid, liquid, gel, or sol-gel (transient state between solid and gel) that would preserve certain spatial distribution of wires to maintain the desired fractal impedance.

The nanowires can have random or preferred orientation. They can be uniformly distributed over the base and can have homogeneous measurements within it. On the other hand, they can possess different sizes and location distribution laws. The material of the base can be either homogeneous or heterogeneous. In the latter case, the nanowire dimensions and distribution over the base may vary. Figure 3.36c shows a schematic diagram of a fractal element where the conducting base is a three-dimensional solid body with terminals at each end. The conducting base shown in Fig. 3.36d is in the form of a thin-film element where both surfaces contact the conducting plates.

Figure 3.37 shows the frequency response (the MFR and PFR) of the experimental fractal element, which is comprised of $[NH_2BU_2]_x[Pt(Ox)_2]$ base of nanowires in PVA/Pani polymer (here is real numbers where $1 \leq x \leq 2$). It is evident that the impedance phase fluctuates from $-13°$ to $-18°$ over a frequency range of more than four decades, i.e., the corresponding PFR flatness $\Delta\varphi_Z < 5$.

Another way to create fractal elements, especially for the microwave frequency range, is to use the micro-relief of treated surfaces. Fractal features of such materials and medium show themselves simply by a change of the skin effect and the impedance.

The roughness of the desired structure can be provided by using the intensively developed methods of surface treatment with concentrated energy flows (laser, plasma, and electric-discharge treatment) using nanotechnology processes such as chemical assembly, sol-gel processes, vapor-phase deposition of metals, and atomic layered epitaxy.

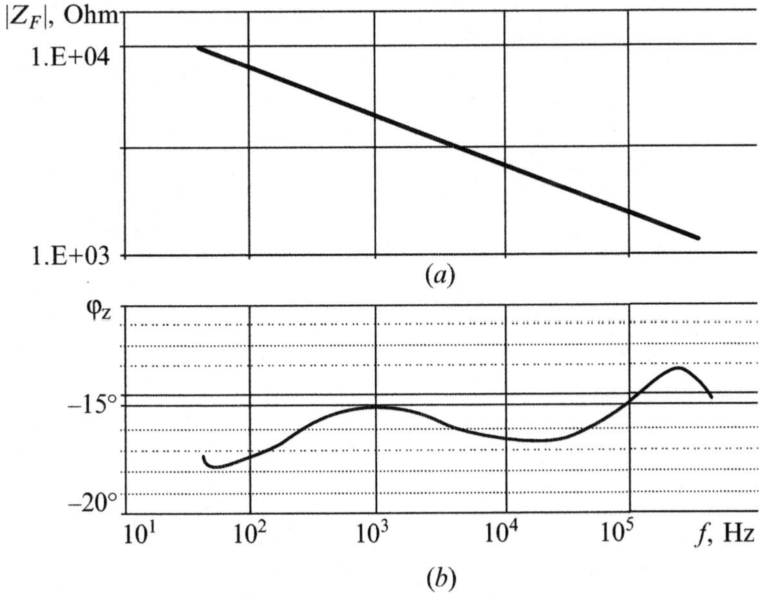

Fig. 3.37 (**a**) Logarithmic MFR and (**b**) PFR of the fractal element (US patent 2006/0267595 A1)

3.6 Comparison of Fractal Element Characteristics Using Different Fabrication Technology

The previous different methods of making fractal impedance provide opportunities to produce fractal elements for a wide range of frequencies. However, to make such hardware components as widely used as the conventional passive components are, it is necessary to make the design and the technology of fractal elements comply with the following requirements:

- Compatibility with the manufacturing technology of semiconductors or thin-film integrated circuits
- Feasibility of the desired frequency response for a wide frequency spectrum
- Feasibility of fractal impedance dependence on the fractional-order index α for a maximum range of allowed values ($0 < |\alpha| < 1$)
- Feasibility of precise adjustment of the fractal impedance parameters and characteristics
- Capability of parameter dynamical adjustment

Table 3.1 provides comparative evaluation of the above-given requirements. It provides different fulfillment levels of the considered design-and-technology variants of the *RC*-EDP. The maximum fulfillment of a requirement is marked with three asterisks.

Table 3.1 Comparison of fractal elements using different design-and-technology methods

Criteria	Fractal elements on the basis of			
	Electrochemical converters	*RC*-ELP	Film *RC*-EDP	Nanotechnology
Compatibility with technological processes of manufacturing of semiconductor or film elements of integrated circuits	*	**	***	*
Feasibility of precise adjustment of the fractal impedance parameters	*	***	***	?
Fractal impedance parameter dynamical adjustment capability	**	*	***	?
Feasibility of the fractal impedance for a wide range of frequencies	*	*	***	***
Overall measurements	*	*	***	***
Feasibility of simulation of real objects with fractal dimension	*	**	***	*
Existence of methods and software for development	*	***	***	?
Summary indicator	8	13	21	8

Fractal elements based on electrochemical converters with liquid electrolyte are hardly compatible with the modern integrated circuit technologies. Reproduction of the fractal impedance parameters of these elements strongly depends on the reproducibility of the "electrode-electrolyte" interface parameters. The latter reproducibility is specified with a large amount of uncontrolled factors. The obvious advantage of such fractal elements is their capability of working at ultralow frequencies.

Fractal elements based on the semiconductor technology also have good potential, especially due to small size and capability to control their parameters with electric field. Nevertheless, there is no defined method to fabricate fractal impedances with a desired and accurate value of fractional order, α. This problem is caused by the difficulties associated with recording the surface and volume effects in the semiconductor crystal and the interaction of the integrated circuit elements.

The most serious disadvantages are those of *RC*-ELP fractal circuits due to their significant size, the difficulty of precisely adjusting the fractal impedance parameters, and the impossibility of dynamically varying their parameters.

There are new principles of making fractal impedances based on conductivity of their basis (complexes) of nanowires or nanoparticles of conductors encapsulated into polymeric matrixes as well as the fractal impedances appearing due to interaction of electromagnetic waves with conducting fractal structure surfaces. However, these principles are still at the research stage.

It is evident from the table above that the best summary indicator for the whole of the criteria belongs to the *RC*-EDP fractal film elements. Hence, the authors suppose that the *RC*-EDP film has a great potential to become the design-and-technology base of the new elements of fractal radio-electronics.

Practice

1. Plot the logarithmic MRF and PFR of the fractal impedance for fractional orders $\alpha = 0.2; 0.8; 0.4 + j0.3; j0.5$.
2. Prove the identity of the properties of the input impedance *RC*-transfer functions expanded into a continued fraction expansion (CFE) form.
3. Prove the identity of the properties of the input admittance *RC*-transfer functions expanded into CFE form.
4. Using Example 1 as a template, apply the Foster and the Cauer methods to synthesize the electrical circuit of the following two input impedance transfer functions:

$$Z(s) = \frac{s^3 + 4s + 1}{s^4 + 6s^2 + 8}; \quad Z(s) = \frac{s^2 + 12s + 3}{s^4 + 6s^2 + 8s + 2}.$$

5. Synthesize the circuit with the following input impedance function:

$$Z(s) = \frac{s^5 + 55s^4 + 330s^3 + 462s^2 + 165s + 1}{11s^5 + 165s^4 + 462s^3 + 330s^2 + 55s + 1}$$

   Determine the order of the fractional operator of this function approximates.
6. Develop a Matlab program that would figure out the implemented one-port normalized input impedance function $\bar{Z}_N(s)$ using *Shareff* method. Use the solution results of Example 4 to generate the Bode diagram to sketch the MFR and PFR of the given transfer functions to ensure that the fractal impedance parameters are achieved.
7. Design two cascaded biquadratic structures using *El-Khazali* method to approximate the same fractional-order operator found in Problem 6. Determine and sketch a proper circuit of passive elements to realize the same operator. Verify your design by sketching the Bode diagram and the AC circuit analysis of the transfer function you just found.

Test Questions

1. Explain why the state equations of the circuits with lumped parameters are the integer-order differential equations.
2. Write down the fractal impedance expression and clarify the physical meaning of its variables.
3. Recite the advantages and disadvantages of the electrochemical cell-based fractal elements.
4. Represent the Foster circuit method of *RC*-ELP-based synthesis of fractal impedance using the specified function.
5. Represent the Cauer circuit method of *RC*-ELP-based synthesis of fractal impedance using the specified function.
6. Represent the graphical interpretation of the Oustaloup method.

7. Show the correspondence between the regions of the bipolar and the MOS structures and the elements of the *RC*-EDP graphical symbol (see Figs. 3.30 and 3.31).
8. Recite the basic advantages and disadvantages of the film *RC*-EDP-based fractal elements.
9. Represent the physical interpretation of the possibility to create the fractal elements on the basis of composition materials and nanowires encapsulated into polymeric matrixes.
10. List the basic advantages and disadvantages of the nanostructured material-based fractal elements.
11. Represent the comparative analysis of the fractal impedance implementation methods studied in Chap. 3.
12. Design and sketch a fourth-order *RL* circuit to design a fractional-order differentiator of order 0.77.

Chapter 4
Design and Implementation of Thin *RC*-EDP Films

It was noted in Chap. 3 that the resistive-capacitive elements with distributed parameters (*RC*-EDP) circuits appeared more than half a century ago. A great variety of designs of these circuits with different functional capabilities and performance parameters have been studied and described in various sources of information. To solve the design and fabrication problem of the *RC*-EDP fractal elements, the design fundamentals and procedures should be standardized, and the *RC*-EDP design parameters should be easily adjusted to achieve the desired performance. This chapter is devoted to highlight the basis of design fundamentals and implementation of *RC*-EDP films.

4.1 Classification of Static Heterogeneous Characteristics of *RC*-EDP Films

The *RC*-EDP film design given in Fig. 3.30 defines a fractal input impedance with a fractional order, $\alpha = ½$, that depends on frequency. If there is a change in the *RC*-EDP design topology due to geometric irregularities that will result in making new fractal impedances of different fractional orders, α. Such irregularities (heterogeneities) are introduced by changing the *RC*-EDP layers topology, which is a *static* one since it does not depend on time. The combination of such heterogeneities defines the basis of design and fabrication of *RC*-EDP in order to meet the desired functional capabilities and performance characteristics.

Suppose that a real *RC*-EDP device represents a "bad" multilayered capacitor, it will then be evident that its performance will be influenced by the number of its layers and their structure, layers configuration, layers thickness, cuts, overlapping, and their electrical contacts (terminals). The design features of such static heterogeneities, as classified as shown in Fig. 4.1, form the basis of the *RC*-EDP film design, which enjoy certain capabilities and performance characteristics.

A.Kh. Gil'mutdinov et al., *Fractal Elements and their Applications*, Analog Circuits and Signal Processing, DOI 10.1007/978-3-319-45249-4_4

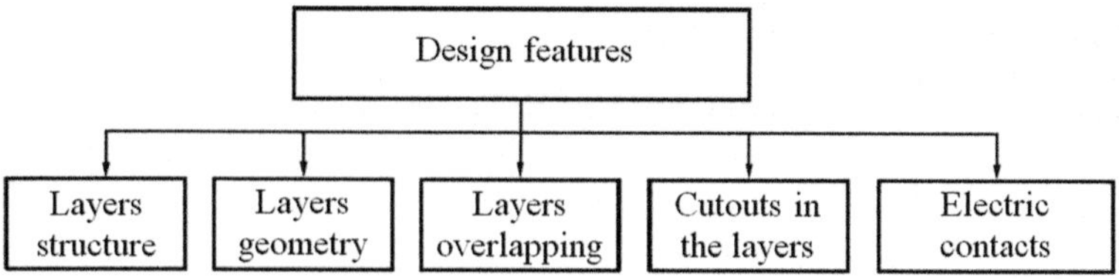

Fig. 4.1 Classification of the design features that determine the *RC*-EDP film design steps

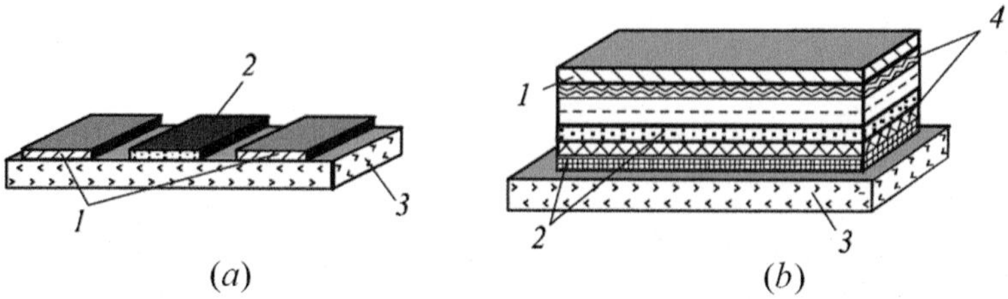

Fig. 4.2 *RC*-EDP design: (**a**) single-layered (planar); (**b**) multilayered; (*1*) conductor; (*2*) resistor; (*3*) substrate; (*4*) dielectric

In the flowing sections we consider in detail how the *RC*-EDP film design can be embodied by introducing the static heterogeneities design features. Some of these design steps have already been described in the corresponding literature. The proposed classification allows one to acquire new design methods where some of these methods have already been developed.

4.2 *RC*-EDP Film Design of Multilayer Structure

The layers structure is defined by the number and the sequence of the layers (stripes) used to fabricate the *RC*-EDP Film, which are made of materials of different electrophysical properties. All *RC*-EDP devices can be divided into two groups; the single-layered and the multilayered groups. The single-layered *RC*-EDPs are made of parallel resistive and conductive stripes located on a dielectric substrate (see Fig. 4.2a). The distributed capacitance of the element appears between the two neighboring stripes and depends on the gap geometry, the stripes measurements and configuration, and on the equivalent dielectric permittivity of the medium around the stripes.

The multilayered *RC*-EDP structure is shown in Fig. 4.2b. The minimum number of layers a multilayered *RC*-EDP film can be made of is three. The maximum number of these layers depends on the features of the used technology, the element reliability, and on the actual necessity of making such an *RC*-EDP. Multilayered elements can also be used to increase the *RC*-EDP time constant by providing serial connection of the conducting and/or the resistive layers located in different planes.

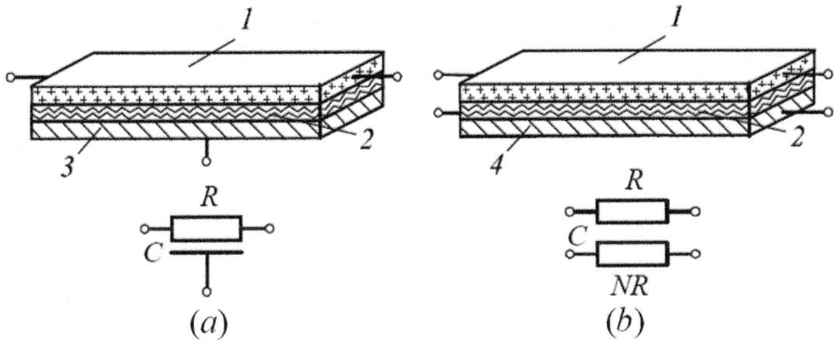

Fig. 4.3 Three-layered *RC*-EDP and their graphical symbols: (**a**) with the *R*-*C*-0 layers structure; (**b**) with the *R*-*C*-*NR* layers structure; (*1*) resistive layer *R*; (*2*) dielectric; (*3*) ideal conducting plate; (*4*) resistive layer *NR*

The three-layered *RC*-EDP design provides two possible layers structures: *R*-*C*-0 and *R*1-*C*-*R*2 (here *R* designates the film resistive layer, *C* represents the dielectric layer, and 0 denotes the ideal conducting layer). It should be noted now that the resistance of the conducting layer in real elements can be considered tentatively very low. Actually, every conducting layer has a finite resistance even if it is of very low value. The *RC*-EDP with the *R*-*C*-0 layers structure (hereinafter referred to as the *R*-*C*-0 EDP) and its graphical symbol are given in Fig. 4.3a. In addition, an *RC*-EDP with *R*-*C*-*NR* layers structure (hereinafter referred to as the *R*-*C*-*NR* EDP) and its graphical symbol are given in Fig. 4.3b. In this structure, the resistance of the lower layer is *N*-times larger than that of the top one, *NR*, over the whole surface. Such elements can be used as rejecting filters for large value of *N*.

There are two known four-layered *RC*-EDP structures: the first one has an *R*-*C*-*G*-0 layers and the second one has a 0-*C*-*G*-0 layers. Here *G* stands for the resistance of the surface layer that is covered with an ideal conducting layer to form one of the external terminals of the element. This is why an electric current exists through the *G*-layer transversely to its surface, and the corresponding transversal conductivity take the value $g_{sp} = 1/\rho \cdot d$; where ρ is the specific volume resistance of the layer material and *d* is the layer's thickness.

Structures of some four-layered *RC*-EDP and their graphic symbols are shown in Fig. 4.4. The first kind of the structure is used as a *feedback four-pole active RC-filters*. At certain values of the product $\gamma = RG$, these filters can possess zero sensitivity of the transfer function quality factor (*Q*-factor) to the variation of the active element gain $S_K{}^Q = (dQ/dK) \cdot (K/Q) = 0$. The application of the second structure depends on the variation of the dielectric layer, *C*, and on the thickness of the *G*-layers, which will be studied in Sect. 4.3.

We shall consider two basic types of five-layered *RC*-EDP that were studied and mentioned in the corresponding literature; the first type consists of 0-*C*1-*R*-*C*2-0 layers, while the second one consists of 0-*C*1-*R*1-*C*2-*R*2 layers. Figure 4.5a depicts a 0-*C*-*R*-*NC*-0 layers structure of an *RC*-EDP (a special case of the 0-*C*1-*R*-*C*2-0 structure where $C2 = NC1$; *N* is a scale factor).

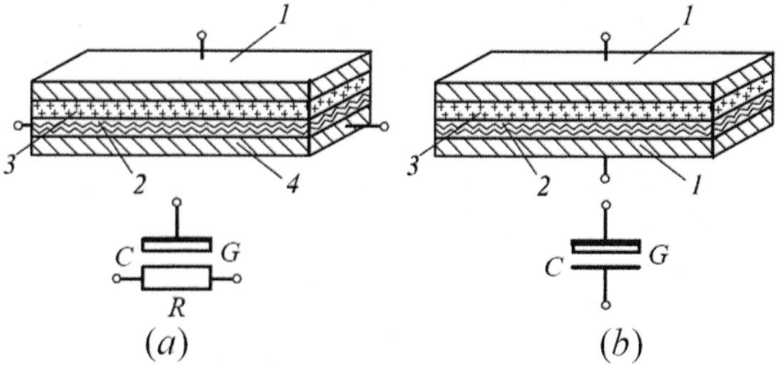

Fig. 4.4 Four-layered *RC*-EDP and their graphical symbols: (**a**) with the *R*-*C*-*G*-0 layers structure; (**b**) with the 0-*C*-*G*-0 layers structure; (*1*) conducting layer 0; (*2*) dielectric layer *C*; (*3*) resistive layer *G*; (*4*) resistive layer *R*

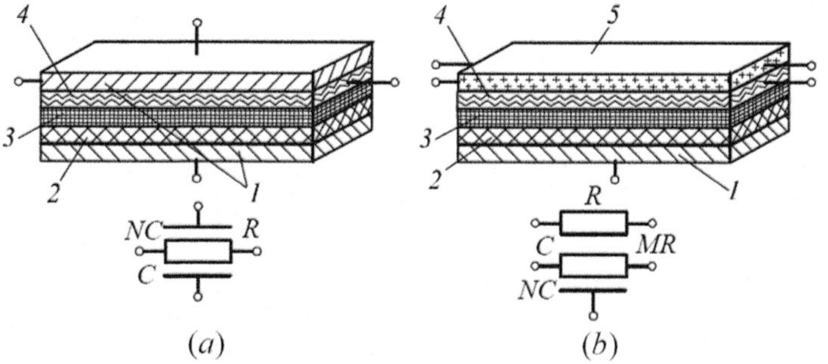

Fig. 4.5 Five-layered *RC*-EDP and their graphical symbols: (**a**) with the 0-*C*-*R*-*NC*-0 layers structure; (**b**) with the *R*-*C*-*MR*-*NC*-0 layers structure; (*1*) ideal conductor—layer 0; (*2*) dielectric layer *NC*; (*3*) resistive layer *R*; (*4*) dielectric layer *C*; (*5*) resistive layer *MR*

The *R*-*C*-*MR*-*NC*-0 layers structure *RC*-EDP shown in Fig. 4.5b is also known as the double-Kelvin transmission line. This line was used in active *RC*-filter circuits with transmission zeroes.

Notice that the *RC*-EDP devices with more than five layers have not been used for practical applications yet. Hence, the *RC*-EDP design can be classified according to the layers structure as shown in Fig. 4.6. This classification obviously makes it possible to synthesize numerous different designs of *RC*-EDP films.

4.3 Design Development of *RC*-EDP Films by Changing Layers Geometry

Notice that changing the characteristics or the performance of the *RC*-ELP-based circuits (lumped-parameter circuits) requires changing the number of elements, the circuit configuration, or the values of these elements, or, more often, all of the

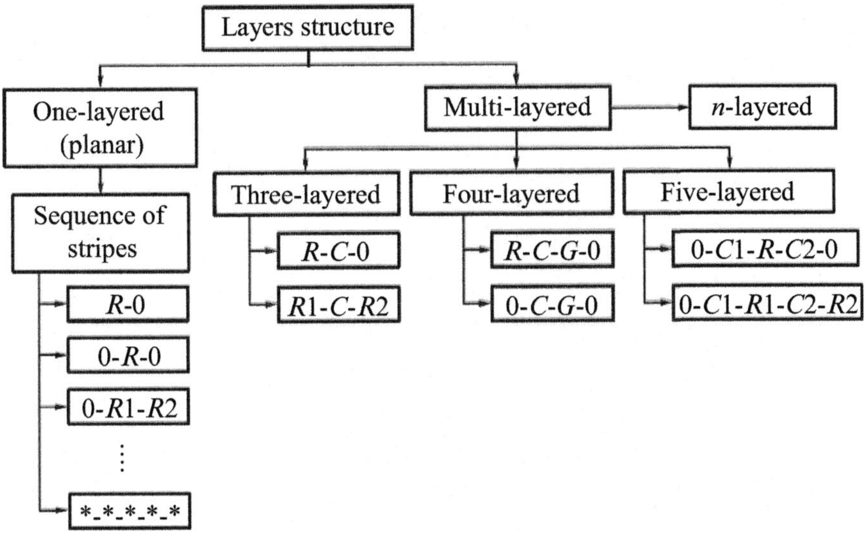

Fig. 4.6 Classification of the *RC*-EDP that differ in the layers structure

above. However, if there is an *RC*-EDP (distributed-parameter circuits) that is equivalent to some *RC*-ELP-based electrical circuit, then its performance and characteristics can be changed by varying the layers geometry, i.e., by introducing heterogeneities in their structures.

The *RC*-EDP with static geometric heterogeneities are the most studied today. The reason is that the topology of an *RC*-EDP layer and the rules of varying the layers' thickness are formed during the process of the element production, or when adjusting technological methods used.

From the analysis of previously published research, the design of *RC*-EDP films differs in layers configuration and thickness, in the process of shaping the elements capabilities, and in the classification of the *RC*-EDP layers geometry. This classification, as shown in Fig. 4.7, obviously makes it possible to synthesize *RC*-EDP film of different layers geometry.

Here b is the width of the resistive and the conducting layer strip; t is the thickness of the resistive layer; d is the thickness of the dielectric layer. The functions f_i, φ_i and Ψ_i are the rules (arbitrary in the general case) of varying the layers geometric parameters.

Several examples of the width variation formula of the resistive layer with the corresponding topologies of the one-dimensional heterogeneous *RC*-EDP resistive layer are given in Table 4.1. The constant parameters b_0, x_1 and k determine the topology of the resistive layer and the electrical performance characteristics of the *RC*-EDP.

Notice that, in a low-pass filter, replacing a one-dimensional homogeneous (ODHom) *RC*-EDP by a one-dimensional heterogeneous (ODHet) *RC*-EDP element where the width of the layers follows an exponential variation formula will

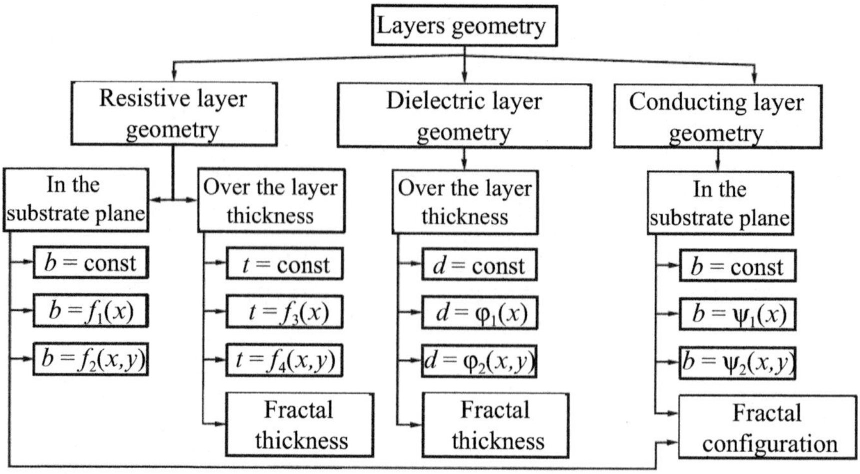

Fig. 4.7 Classification of the *RC*-EDP layers geometry

increase the slope of magnitude frequency response. If the same change is applied to a frequency-reject filter, a more symmetrical shape of the MFR will result near the rejection frequency. However, for *RC*-oscillators, it reduces their nonlinear signal output distortions.

The ODHet *RC*-EDP with a width variation that follows a Bessel formula is known as fractal elements, where its parameter variations change uniformly the phase frequency response of the *RC*-EDP input impedance.

The three-layered *RC*-EDP design depicted in Fig. 4.8a can also be classified as an ODHet *RC*-EDP. It is technically easier to implement and control the step law of the width variation $b(x)$. It increases the number of design parameters (amount of steps, geometry of every step) that can be adjusted to get the desired performance characteristics of the *RC*-EDP.

However, the one-dimensional approximation that is most widely used to analyze these designs provides proper precision only at small height differences between the neighboring steps. In this case, the step-type *RC*-EDP equivalent circuit can be represented as a set of connected ODHom *RC*-EDP where each of them reflects one step (see Fig. 4.8b). If the height of the neighboring steps differs significantly, then such kind of *RC*-EDP becomes an element with surface-distributed parameters, where its equivalent circuit will be two dimensional.

A known *RC*-EDP design example where the layers thickness varies according to a certain law is the 0-*C*-*G*-0-structured one-port four-layered *RC*-EDP depicted in Fig. 4.9a. It has two parallel conducting plates, where the resistive and the dielectric layers are located between them. The thickness of each of these two layers varies so that their total thickness remains constant along the element length. Figure 4.9b shows a similar *RC*-EDP structure, but differs in the thickness of its resistive layer G, while the width, $b(x)$, of the element varies according to the exponential law.

Table 4.1 One-dimensional heterogeneous *RC*-EDP

Name	**Resistive layer width variation law *b*(*x*)**	**The one-dimensional heterogeneous *RC*-EDP resistive layer topology**
Bessel	$b_0 \frac{x^{-k}}{x_1^{-k}}$	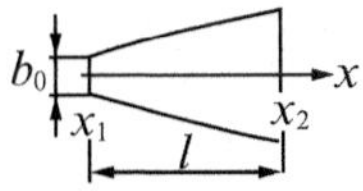
Exponential	$b_0 e^{-kx}$	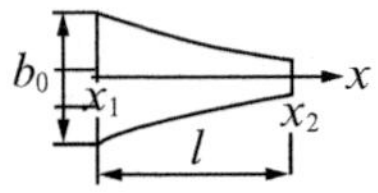
Linear	$b_0(1+kx)$	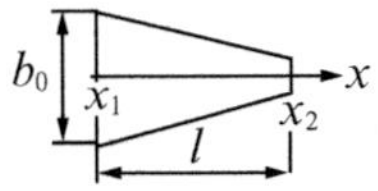
Parabolic	$b_0(1+kx)^2$	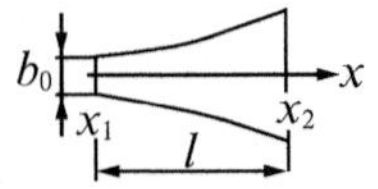
Hyperbolic secant	$b_0 \frac{\sec h^2 x}{\sec h^2 x_1}$	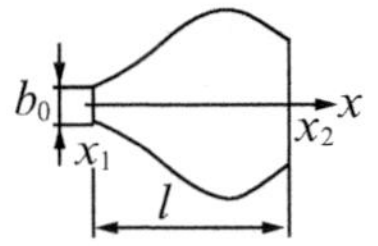
Hyperbolic cosecant	$b_0 \frac{\csc h^2 x}{\csc h^2 x_1}$	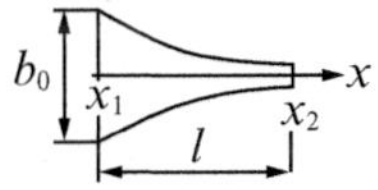
Trigonometric secant	$b_0 \frac{sc^2 x}{sc^2 x_1}$	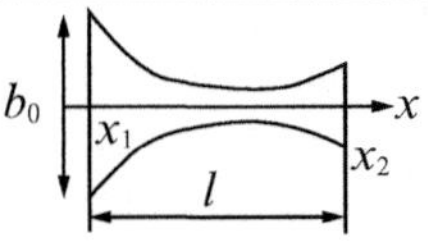
Trigonometric cosecant	$b_0 \frac{\csc^2 x}{\csc^2 x_1}$	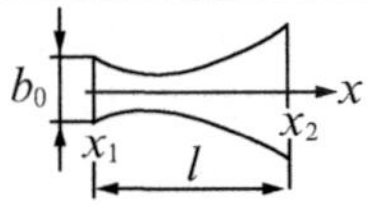

For example certain values of height, *h*(*x*), and width, *b*(*x*), of one-port *RC*-EDP devices have provided 30°, 45°, and 60° phase levels of an input impedance with ±1° tolerance over two decades, and a phase of 81° with tolerance of 9° over three decades.

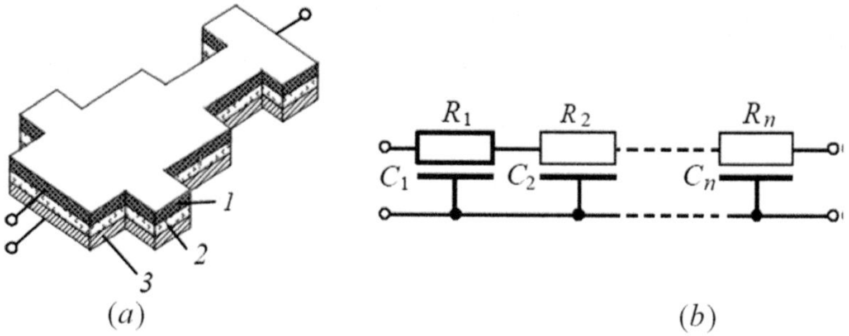

Fig. 4.8 Step-type *R*-*C*-0-structured one-dimensional heterogeneous *RC*-EDP: (**a**) general view of the design; (**b**) equivalent circuit: (*1*) ideal conductor; (*2*) dielectric; (*3*) resistive layer

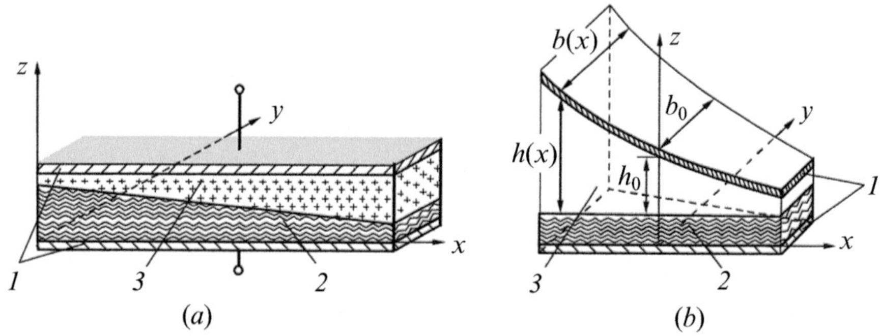

Fig. 4.9 Examples of the 0-*C*-*G*-0-structured *RC*-EDP with heterogeneous design parameters: (**a**) layers thickness heterogeneity; (**b**) layers thickness and width heterogeneity; (*1*) ideal conductor; (*2*) dielectric layer *C*; (*3*) resistive layer *G*

It is commonly known that the fractal geometry of the electrode-electrolyte interface surface allows synthesizing electrochemical elements with fractal impedance.

RC-EDP makes it possible to synthesize the fractal impedance via the fractal configuration of the resistive and/or the conducting layers by changing the fractal thickness (interface) of the resistive and the dielectric layers.

The fractal interface between the conducting and the dielectric layers can be created, for example, by making the *RC*-EDP lower plate of aluminum substrate with rough surface and the dielectric layer by an electrolytic anodizing process of this surface (the so-called aluminum-anodizing technology).

Figure 4.10 depicts an example of the *RC*-EDP with fractal configuration of the resistive layer made by means of photolithographic treatment of the metallic film covering the SiO_2 layer that was grown on the semiconductor substrate surface. The configuration of the conducting layer is an arrangement of the known regular fractal called "the Hilbert curve." It is interesting that the measured PFR of the experimental sample of the input impedance possessed a 45° uniform phase over four-decade frequency range.

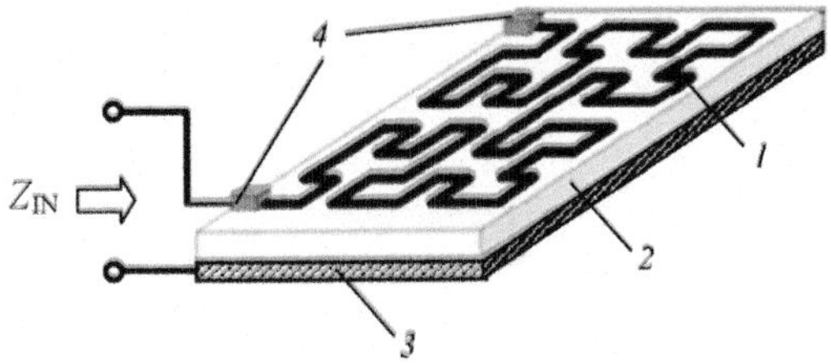

Fig. 4.10 *RC*-EDP with fractal configuration of the resistive layer: (*1*) resistive layer; (2) semiconductor substrate; (*3*) ideal conductor; (*4*) contacts

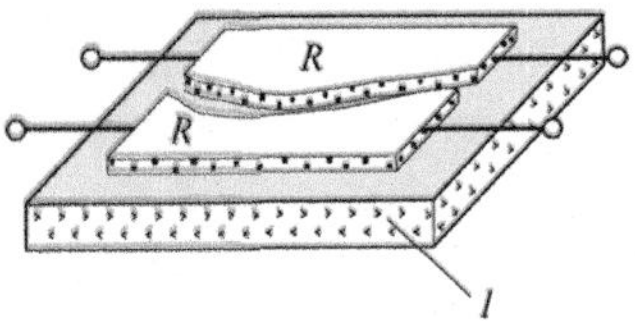

Fig. 4.11 *R*1-*C*-*R*2 EDP planar design: (*R*) resistive layer; (*1*) substrate

All options of *RC*-EDP with geometric heterogeneities that have been represented by the examples of three- and four-layered *RC*-EDP can be extended to *RC*-EDP with other numbers of levels. The classification given in Fig. 4.7 will change: the one resistive and one dielectric layers will be replaced with additional resistive and dielectric layers of the same geometric heterogeneities.

As for the planar *RC*-EDP design, the stripes shape and the gap between them can be varied according to the desired thickness laws. An example of the planar *RC*-EDP is depicted in Fig. 4.11, where the width of its stripes depends on the coordinates of the length and the constant gap, *d*, of the element. It is easy to imagine an *R*-*C*-0-structured *RC*-EDP with certain laws for the gap width variation $d = f_1(x)$. For example a resistive stripe width is denoted by $b_{res} = f_2(x)$ and the conducting stripe width is equal to $b_{cond} = f_3(x)$. In general, for multiple stripes of the *RC*-EDP, these rules can be set for all gaps and stripes. Such designs are obviously highly promising for the super-high-frequency applications, and they are the subject of investigation of several research areas.

4.4 Design Development of *RC*-EDP Film Using Layers Overlapping

In all the examples of the multilayered *RC*-EDP designs that have been considered before, the configurations of the resistive and conducting layers or of the two neighboring resistive layers were just the same; i.e., they did not go beyond the topology of each other. It is clear that if the variation laws of the geometric structures configuration of neighboring layers do not match, then the *RC*-EDP performance characteristics will differ from the desired ones. Therefore, there is another way to provide their desired performance characteristics by managing the degree of layers overlapping. Figure 4.12 depicts some examples of *RC*-EDP

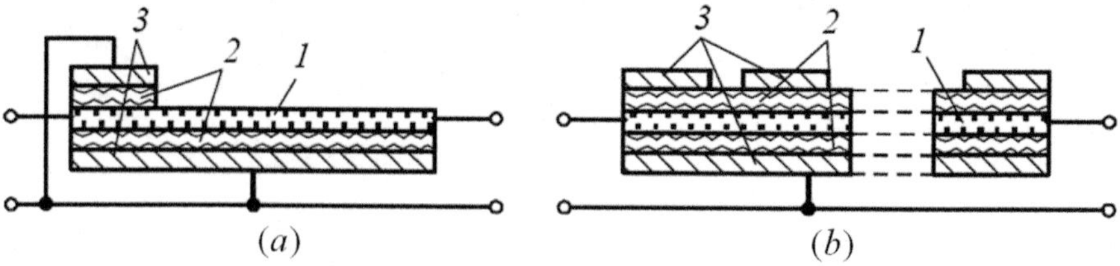

Fig. 4.12 Examples of *RC*-EDP with incomplete layers overlapping: (**a**) with high input capacitance; (**b**) with high slope of the MFR of the low-pass filter: (*1*) resistive layer; (*2*) dielectric layer; (*3*) conducting layer

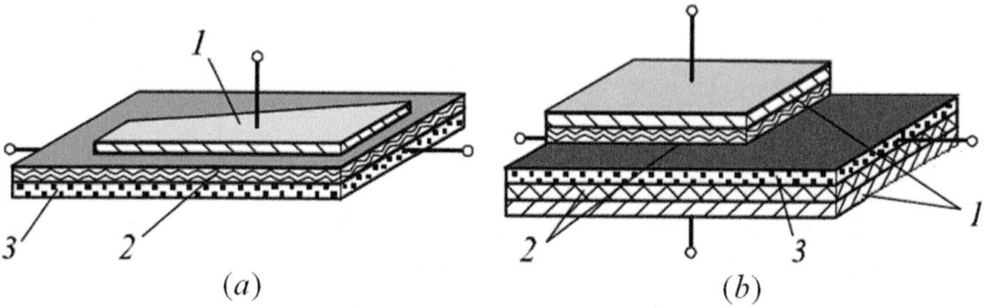

Fig. 4.13 Examples of *RC*-EDP with incomplete layers overlapping: (**a**) *R*-*C*-0 layers structure; (**b**) 0-*C*1-*R*-*C*2-0 layers structure: (*1*) conducting layer; (*2*) dielectric layer; (*3*) resistive layer

designs with incomplete layers overlapping that enhance the frequency characteristics of the devices.

Incomplete overlapping along two coordinates can also be fabricated (see Fig. 4.13). In this case, the electric potential distribution over the resistive layers is intrinsically two dimensional. Figure 4.13a shows the *R*-*C*-0-structured *RC*-EDP design where the conducting layer is completely within the borders of the resistive layer.

Figure 4.13b shows the 0-*C*1-*R*-*C*2-0-structured *RC*-EDP design where the upper conducting layer does not overlap with the resistive layer in two sides. It is obviously difficult to list all possible combinations of the layers incomplete overlapping, especially for the multilayered *RC*-EDP where the number of layers $n \geq 5$. Figure 4.14 gives the classification of all possible designs that differ in the manner of overlapping layers for the most known and studied three- to five-layered *RC*-EDP. The layers overlapping degree is also an efficient way to introduce the heterogeneities into the resistive layers.

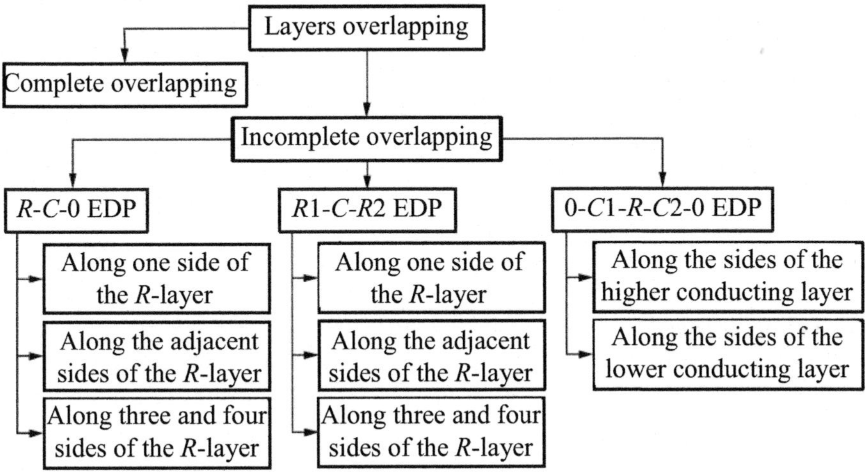

Fig. 4.14 *RC*-EDP designs with different forms of layers overlapping

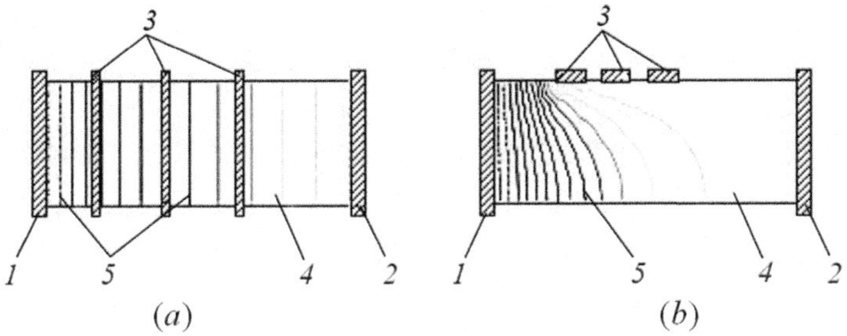

Fig. 4.15 *RC*-EDP designs with the resistive layers lead-outs: (**a**) the lead-outs do not change the potential distribution; (**b**) the lead-outs change the potential distribution: (*1*,*2*) *RC*-EDP terminals; (*3*) resistive layers lead-outs; (*4*) resistive layer; (*5*) equipotential lines

4.5 Design Development of *RC*-EDP Films via Contacts Combination

Another efficient method to design *RC*-EDP film structures of specific electrical properties is to use specified combination of electric contacts. This combination is formed to connect the element to the circuit and to provide desired distribution of an electric potential over resistive layers and to connect the layers or their parts.

In cases when it is essential to preserve the one-dimensional character of the *RC*-EDP, any additionally introduced contacts (they are often called the resistive layer lead-outs) should not distort the equipotential lines. This comes possible if the contact lead-out is made in the form of a narrow conducting stripe that crosses the whole of the resistive layer width (see Fig. 4.15a).

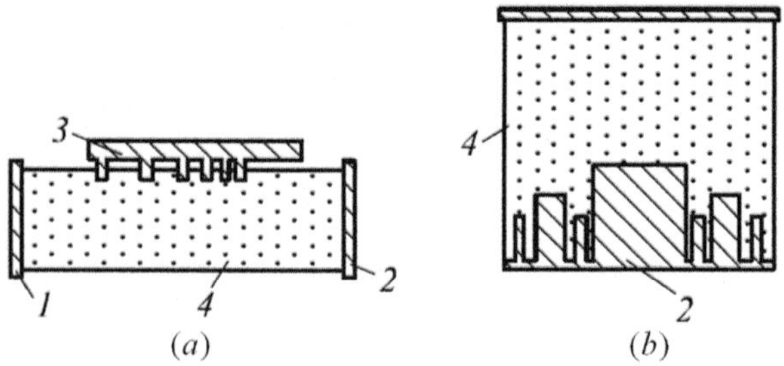

Fig. 4.16 Examples of the *RC*-EDP resistive layer topology: (**a**) with the comb-shape contacts; (**b**) with the fractal-shape contacts; (*1*,*2*) terminals; (*3*) resistive layers lead-out; (*4*) resistive layer

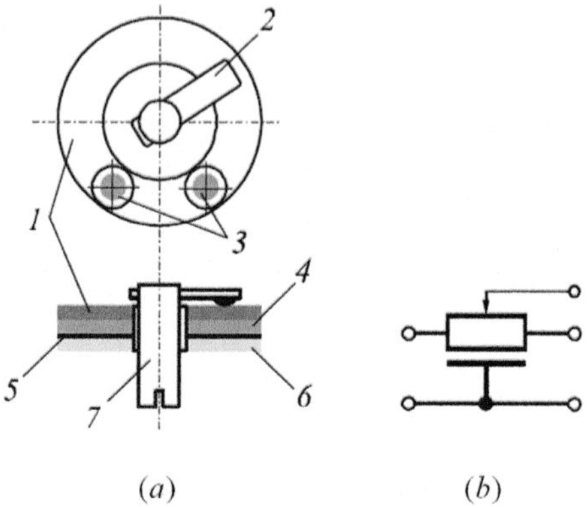

Fig. 4.17 Variable *RC*-EDP: (**a**) the design; (**b**) graphical symbol; (*1*) resistive layer; (*2*) moving contact; (*3*) terminals; (*4*) dielectric layer; (*5*) conducting layer; (*6*) substrate; (*7*) adjustment knob

The location of the lead-outs along the border (outline) of the resistive layer makes the potential distribution of the *RC*-EDP resistive layer a two-dimensional one (see Fig. 4.15b). The corresponding distribution of the electric field over the resistive layer depends on the location and geometry of the contacts.

Figure 4.16 shows some examples of resistive layer contacts with different geometry. Figure 4.16a depicts the comb-shape design of one *RC*-EDP contact with nonuniform law of comb spacing variation. Figure 4.16b, however, depicts the contact fractal geometry that conforms to the Cantor kernel regular fractal.

It is well known that variable resistors with movable contacts or capacitors with variable area of overlapping plates are widely used to perform various operations of adjustable radio-electronic devices. If one considers the *RC*-EDP as a hook-up component of the hybrid micro-assemblies, then applying a moving contact to the resistive layer of the *RC*-EDP will make it possible to vary the *RC*-EDP time constant and, hence, the frequency response of the circuits containing such device. An example of the *RC*-EDP design with a moving contact is given in Fig. 4.17. It is clear that numerous designs and types of variable (adjustable) *RC*-EDP can be designed using standard design and technology of variable resistors (different variation laws of resistance, coupled potentiometers, etc.) and variable capacitors.

The *RC*-EDP characteristics can also be changed stepwise by connecting the resistive layer lead-outs. Such connection can be done be soldering, welding, or electronic switches.

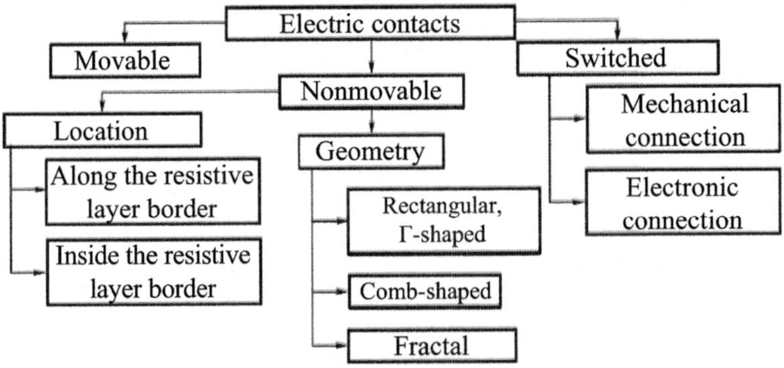

Fig. 4.18 *RC*-EDP electric contacts classification

Fig. 4.19 Examples of the *RC*-EDP designs with layers cutouts and their graphical symbols: (**a**) the Heizer *R*-*C*-0 structure; (**b**) the complementary *R*-*C*-0 EDP; (*1*) conducting layer; (*2*) dielectric; (*3*) resistive layer

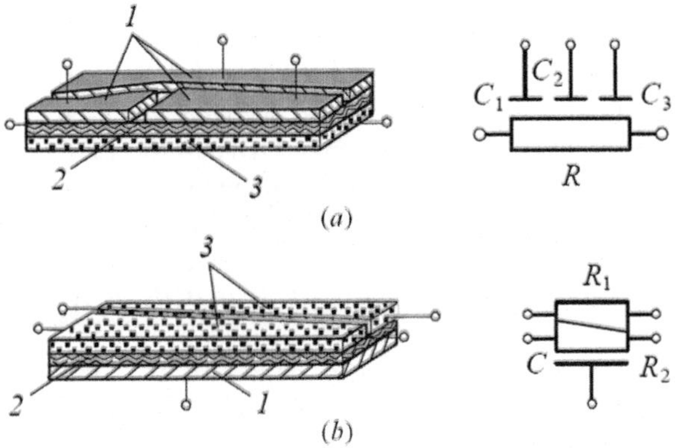

Summing up the known and the probable designs of the contacts that could be connected to the *RC*-EDP resistive layer, the following *RC*-EDP classification based on this index can be proposed as shown in Fig. 4.18.

It should be noted that switching contacts could be performed not just by means of the new design options but also by new classes of *RC*-EDP such as adaptive *RC*-EDP.

4.6 Design Development of *RC*-EDP Films via Layers Cutouts

The cutouts in the resistive and conducting layers of the *RC*-EDP film are another efficient design factor that could be used to synthesize them to achieve the desired characteristics. Figure 4.19 depicts the desired performance characteristics of *RC*-EDP which are provided by the corresponding laws of cutouts in the resistive and the conducting layers. Figure 4.19a represents the *RC*-EDP design of the

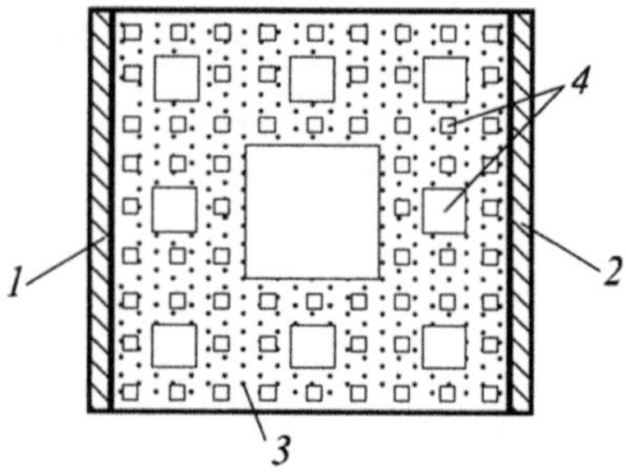

Fig. 4.20 *RC*-EDP with fractal structure of the resistive layer cutouts: (*1*,*2*) terminals; (*3*) resistive layer; (*4*) cutouts

conducting layer that consists of several areas separated with gaps. K.W. Heizer was the first to propose such structure. Under certain laws of gaps shapes, the performance characteristics of such *RC*-EDP models fractional-order rational transfer functions.

Figure 4.19b shows the *RC*-EDP resistive layer sliced into two parts to make the width variation of one resistive stripe, $b_2(x)$, complementary to the width variation of another resistive stripe, $b_1(x)$, such that $b_2(x) = b_0 - b_1(x)$. Here b_0 is the width of the resistive layer of the initial ODHom *RC*-EDP. Hence, that yields two heterogeneous *RC*-EDP layers over the area of one rectangular *R*-*C*-0-layers-structured ODHom *RC*-EDP. Setting the width variation law $b_1(x)$ and its parameters can provide the desired performance characteristics of the *RC*-EDP. The latter was called the "Complementary" *RC*-EDP (C*RC*-EDP).

The cutouts can be made in such a way to form the fractal layout in the resistive and/or the conducting layer. Figure 4.20 shows an example of such *RC*-EDP resistive layer cutouts that forms the "Sierpinski carpet" regular fractal.

It is known that the Sierpinski carpet Hausdorff dimension equals $D = \ln 8/\ln 3 = 1.89$. The corresponding index of the frequency-dependent fractional-order input impedance of the fractal element, based on an *RC*-EDP with the fractal cutouts structure, is $\alpha = 1/D = 0.5291$. The simulation results of the considered *RC*-EDP input impedance shows that when the fractal generation number increases, its PFR level tends to $\varphi_z = -47.355°$ and that corresponds to a fractional order of $\alpha = 47.355°/90° = 0.5262$ (see expression (3.33, *d*)). This is for the case of the Sierpinski carpet where $D = \ln 8/\ln 3 = 1.89$, which matches the known relation $\alpha = 1/D$ with 0.55% error.

The examples of the *RC*-EDP film designs that have been considered here on the basis of the proposed classification of the static heterogeneities do not cover all the cases of elements variations. The new design options can obviously be obtained by combining several static heterogeneities when fabricating a film. For example, we can fabricate the *RC*-EDP films using variable width and cutouts in the resistive and the conducting layers, or with incomplete layers overlapping and the cutouts fractal structure, etc. Finally, the most appropriate is determined by the ability of those devices to implement the desired performance and by production rate.

Practice

Develop the film *RC*-EDP designs that would combine several static heterogeneities described in Chap. 4.

Test Questions

1. Perform the classification of the static heterogeneities that determine the design basis of *RC*-EDP implementation.
2. What is the *RC*-EDP layers structure limited with?
3. Depict the classification diagram of the *RC*-EDP that differ in the layers structure.
4. Perform the classification of the *RC*-EDP that differ in the layers geometry. What designs are the most manufacturable to your mind?
5. Justify the possibility to vary the *RC*-EDP performance characteristics by means of incomplete overlapping of the layers.
6. Perform the classification of the *RC*-EDP electric contacts.

Chapter 5
Schematic Realization of *RC*-EDP Circuits

5.1 Preliminaries and Background

The design factors that determine the *RC*-EDP performance characteristics are based on their circuit diagram connection. Using the IC manufacturing technology, the synthesis of the input transfer functions via the one-dimensional homogeneous (ODHom) and/or a one-dimensional heterogeneous (ODHet) *RC*-EDP element can provide the layers topologies, which yield new forms of *RC*-EDP design methods. The combination of the design and schematic diagrams sufficiently enhances the use of the *RC*-EDP to be the basis of fractal elements and the active and passive radio electronic devices.

The basic factors that identify the process of using schematic diagrams to fabricate *RC*-EDP elements are classification in the following diagram given in Fig. 5.1

The following sections highlight the basics of utilizing the schematic factors to synthesize the frequency responses of the input and the transfer functions of the *RC*-EDP-containing circuits. Examples of developing the *RC*-EDP devices will be considered as well.

5.2 Characteristics of One-Dimensional (OD) Homogeneous and Heterogeneous *R-C-O* EDP Structure

The electrical characteristics of the *R-C*-0-structured EDP are clearly demonstrated via several examples of different forms of two-port ODHom circuits. Each circuit is assumed to have input impedance, Z_{ij}, and a frequency response of a voltage transfer function, T_{ijk}, where i, j, and k refer to the input, output, and the common

A.Kh. Gil'mutdinov et al., *Fractal Elements and their Applications*, Analog Circuits and Signal Processing, DOI 10.1007/978-3-319-45249-4_5

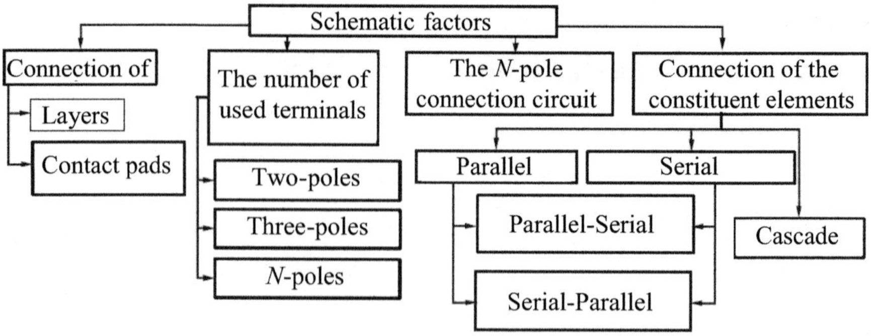

Fig. 5.1 Classification of schematic factors

terminals, respectively. Table 5.1 shows different options of those circuits that have different terminal wiring.

The *R*-*C*-0-layered ODHom *RC*-EDP has been used to develop seven ways of wiring circuit terminals. The application of these circuits covers a wide range of areas such as: high frequency operational amplifiers equalizer circuits (option 3), low-pass filters and phase shifters (option 4), frequency-independent attenuators with constant decay (option 7).

In the case of the *RC*-EDP connection in option 6, the magnitude of the transfer function a passive-element exceeds 1 (getting as high as 1.2) over some frequency range. This is the reason why this connection is used as a frequency-dependent feedback circuit for high-Q filters and *RC*-generators. As for the latter (i.e., *RC*-generator), such solution provides higher stability of the corresponding generator since the desired gain of the active element is less than one.

The schematics listed in options 1 and 2 both exhibit similar MFR and PFR frequency responses. Specifically, when the frequency increases, the magnitude of the input impedance, $|Z|$, decreases and φ_Z gradually rises from $-90°$ to $-45°$. As the frequency increases, the phase shift remains constant at $-45°$.

Similarly, the one-dimensional homogeneous *RC*-EDP of schematics 3 and 5 with a resistive coupling between their input terminals possesses almost identical magnitude and phase frequency responses. Unlike the previous case, here φ_Z gradually falls from 0° to $-45°$. As the frequency increases, the phase shift remains constant around $-45°$. Furthermore, one can see that the frequency responses between terminals two and four are the same as those taken between terminals one and three for the circuits of options 1, 2, and 3. In circuit theory, these types of circuits are called symmetrical.

Unlike the ODHom *RC*-EDP, option 6 represents a nonsymmetrical two-port configuration. It represents an ODHet *RC*-EDP structure. Hence, if an ODHet *RC*-EDP is used, then options 1–3 become nonsymmetrical. Therefore, the amount of different circuits of the two-port connections that would possess unique frequency responses increases almost twofold.

Table 5.1 Dependence of MFR and PFR of the *R*-*C*-0-structured EDP input impedances (*Z*) and transfer functions (*T*) on circuit connection and its terminals wiring

Option No.	Connection circuits	Circuit function Z or T	MFR and PRF
1	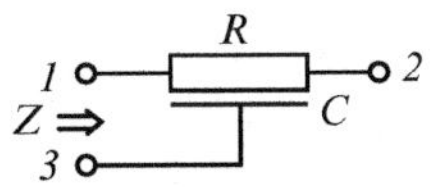	$\overline{Z}_{13} = \frac{\mathrm{cth}\theta}{\theta}$	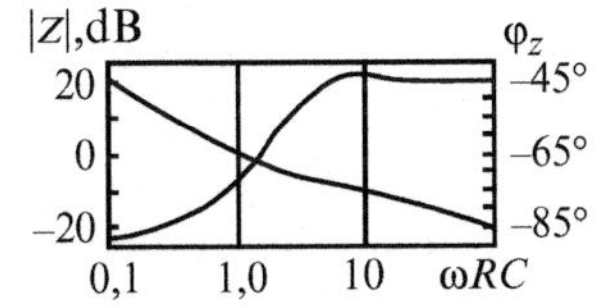
2	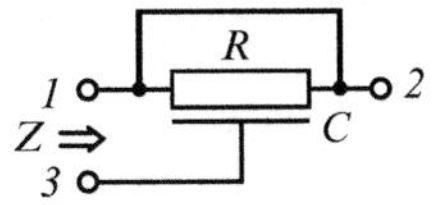	$\overline{Z}_{13} = \frac{\mathrm{cth}\frac{\theta}{2}}{2\theta}$	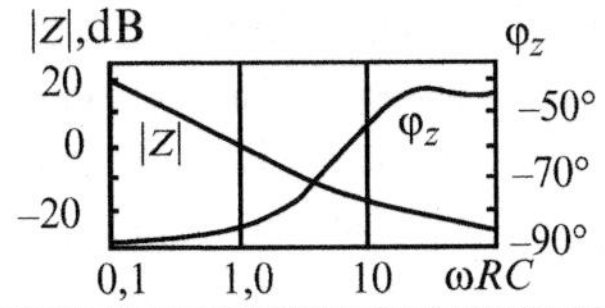
3	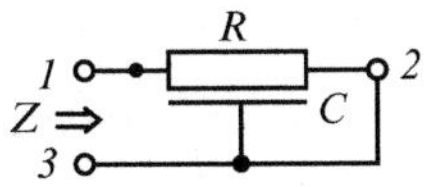	$\overline{Z}_{13} = \frac{\mathrm{th}\theta}{\theta}$	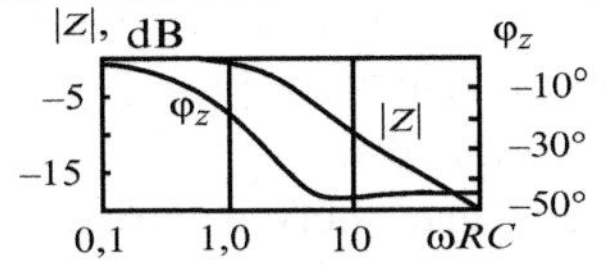
4	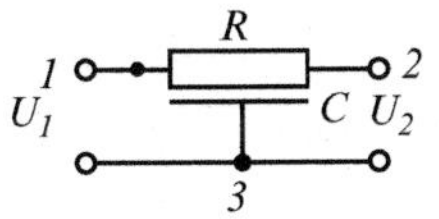	$T_{132} = \frac{1}{\mathrm{ch}\theta}$	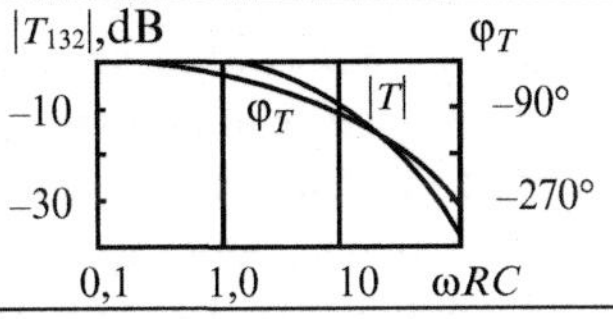
5	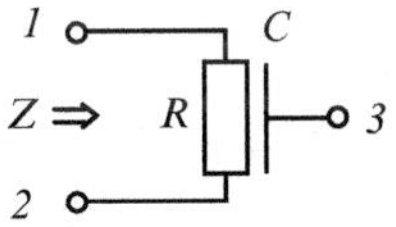	$\overline{Z}_{12} = \frac{2}{\theta}\mathrm{th}\frac{\theta}{2}$	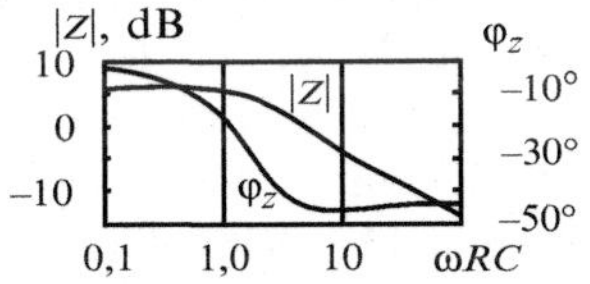
6	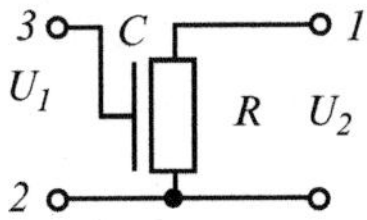	$T_{321} = 1 - \frac{1}{\mathrm{ch}\theta}$	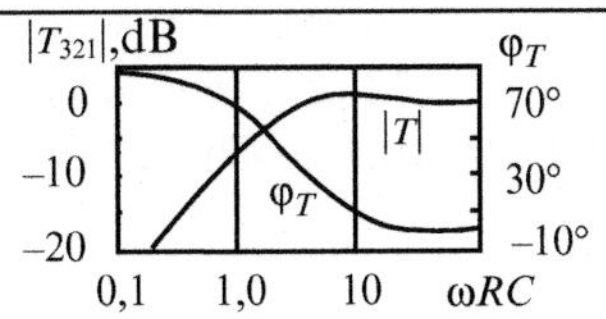
7	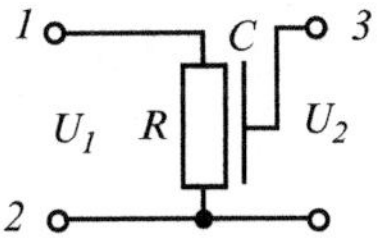	$T_{123} = 0.5$	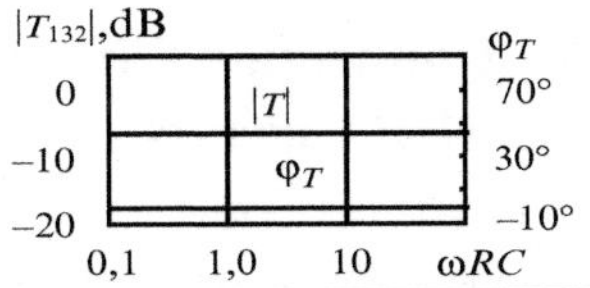

The same analysis of the transfer function frequency responses can be performed for the two-port connection of the *RC*-EDP. When the input and output are coupled via a resistor (option 4 for example), the transfer function MFR response is similar to that of a low-pass filter, while the absolute value of φ_T increases without limits unlike the ladder *RC*-circuits with limited amount of segments where φ_T approaches $n\pi/2$ (*n*: the number of segments).

The two-port wiring circuit of the ODHom *RC*-EDP is also symmetrical since simultaneous change of the input and the output ports does not influence the transfer function frequency responses. Thus, in general, using ODHet *RC*-EDP topology obviously yields nonsymmetrical circuit and creates one more version of the two-port connection with its own frequency response.

The two-port connection versions with capacitor-coupled input and output are nonsymmetrical. Their transfer function frequency responses strongly depend on the fact whether the input is of resistive or capacitive equivalence (and correspondingly the output is capacitive or resistive equivalence).

If the input signal is applied to the conductive plate of the *RC*-EDP, then the corresponding MFR response will be similar to that of a high-pass filter but with one significant difference; the passive circuit possesses an amplification gain factor greater than unity within some frequency range. Moreover, the *RC*-EDP PFR in this case also differs from the PFR of the corresponding ladder *RC*-circuit; i.e., φ_T changes its sign within some frequency spectrum and then tends to zero. Such extreme changes can be used to synthesize active *RC*-filters with a zero sensitivity of circuit Q-factor with respect to an active-element gain variation for a single-element frequency adjustment.

The application of the ODHet *RC*-EDP provides additional degree of freedom that makes it possible to control the parameters of the MFR and the PFR. After applying the signal to the resistive layer (option 7), the frequency response of the *RC*-EDP transfer function shows that they are similar to those of the resistive voltage divider, but unlike the latter, the coupling of the input and the output is of capacitive interface. Previous research revealed that the division factor of such divider could be varied when using the ODHet *RC*-EDP by specifying its width variation law (see Chap. 4) and its corresponding parameters.

5.3 Connection Circuits of OD *R-C-NR*-Structured EDP Circuits

Using the *R-C-NR*- or *0-C-R-NC-0*-layered *RC*-EDP configuration, instead the *R-C*-0-layered *RC*-EDP structure, increases the amount of terminals up to four. It is possible to synthesize 13 different types of two-port (four terminals) connection circuits based on the *R-C-NR* arrangement. These circuits are listed in Table 5.2 where the corresponding impedance expressions are also given.

Observe that the *R-C-NR*-layered EDP topology enables one to synthesize 15 different two-port circuits as shown in Tables 5.3 and 5.4. The corresponding expressions of their *z*-parameters are also listed for completeness.

Table 5.2 Two-port connection circuits of the *R-C-NR* EDP and their normalized impedances

No	Connection circuit	Input impedance (Z_{IN}/R)
1	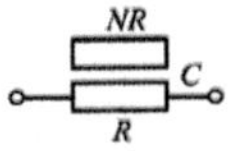	$\frac{1}{(N+1)\theta}\left(N\theta+2\,\mathrm{th}\frac{\theta}{2}\right)$
2	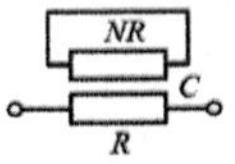	$\frac{(N+1)\mathrm{th}\,\theta/2}{\theta/2+N\mathrm{th}\,\theta/2}$
3	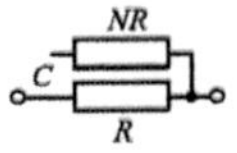	$\frac{1}{(N+1)\theta}(N\theta+\mathrm{th}\theta)$
4	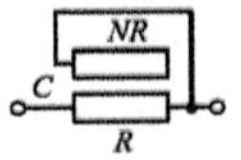	$\frac{(N+1)}{\theta}\left[\frac{N\theta\mathrm{ch}\theta+\mathrm{sh}\theta}{N\theta\mathrm{sh}\theta+(N^2+1)\mathrm{ch}\theta+2N}\right]$
5	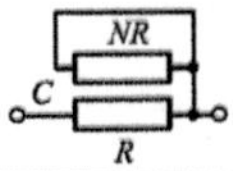	$\frac{(N+1)\mathrm{th}\theta}{\theta+N\cdot\mathrm{th}\theta}$
6	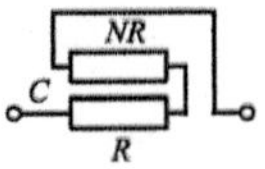	$\frac{(N+1)}{\theta}\mathrm{th}\theta$
7	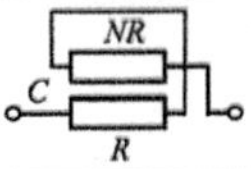	$\frac{(N+1)}{\theta}\left[\frac{(N-1)^2\mathrm{sh}\theta+2N\theta(\mathrm{ch}\theta+1)}{(N^2+1)\mathrm{ch}\theta+N(\theta\mathrm{sh}\theta+2)}\right]$
8	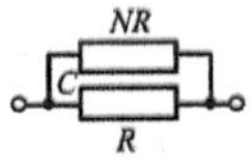	$\frac{N}{(N+1)}$
9	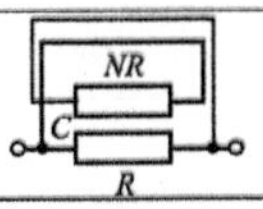	$N(N+1)\left[\frac{\mathrm{th}\theta/2}{(N-1)^2\mathrm{th}\theta/2+2N\theta}\right]$
10	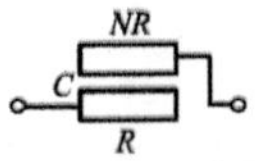	$\frac{1}{(N+1)\theta}\left[N\theta+\frac{N^2+1}{\mathrm{th}\theta}+\frac{2N}{\mathrm{sh}\theta}\right]$
11	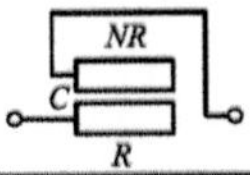	$\frac{(N+1)}{\theta\mathrm{th}\theta}$
12	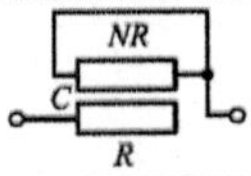	$\frac{(N+1)}{\theta}\left[\frac{\theta\mathrm{ch}\theta+N\mathrm{sh}\theta}{\theta\mathrm{sh}\theta+2N(\mathrm{ch}\theta-1)}\right]$
13	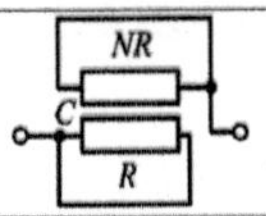	$\frac{(N+1)}{2\theta\mathrm{th}\theta/2}$

Table 5.3 Two-port connection circuits of the *R-C-NR* EDP with capacitive coupling

No	Connection circuit	Normalized matrix of *z*-parameters
1	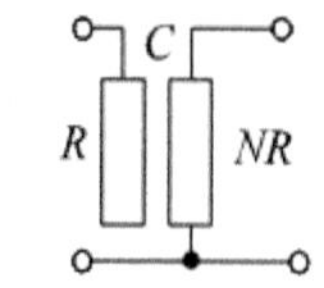	$\frac{N}{(N+1)\theta}\begin{bmatrix} \theta-2\mathrm{th}\frac{\theta}{2}+\frac{(N-1)^2}{N\mathrm{th}\theta} & \theta+(N-1)\mathrm{th}\frac{\theta}{2} \\ \theta+(N-1)\mathrm{th}\frac{\theta}{2} & \theta+2N\mathrm{th}\frac{\theta}{2} \end{bmatrix}$
2	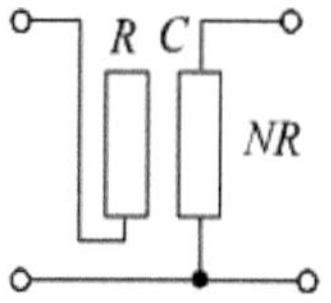	$\frac{1}{\theta}\begin{bmatrix} \frac{(N+1)}{\mathrm{th}\theta} & N\mathrm{th}\frac{\theta}{2} \\ N\mathrm{th}\frac{\theta}{2} & \frac{N}{(N+1)}\left(\theta+2N\mathrm{th}\frac{\theta}{2}\right) \end{bmatrix}$
3	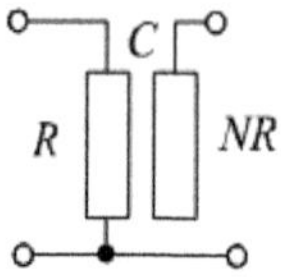	$\frac{N}{(N+1)\theta}\begin{bmatrix} \theta+\frac{2}{N}\mathrm{th}\frac{\theta}{2} & \theta-\frac{(N-1)}{N}\mathrm{th}\frac{\theta}{2} \\ \theta-\frac{(N-1)}{N}\mathrm{th}\frac{\theta}{2} & \theta-2\mathrm{th}\frac{\theta}{2}+\frac{(N+1)^2}{N\mathrm{th}\theta} \end{bmatrix}$
4	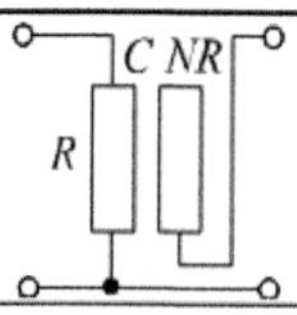	$\frac{1}{\theta}\begin{bmatrix} \frac{N}{N+1}\left(\theta+\frac{2}{N}\mathrm{th}\frac{\theta}{2}\right) & \mathrm{th}\frac{\theta}{2} \\ \mathrm{th}\frac{\theta}{2} & \frac{N+1}{\mathrm{th}\theta} \end{bmatrix}$
5	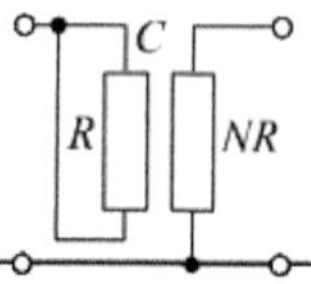	$\frac{1}{N+1}\begin{bmatrix} 2\theta\mathrm{th}\frac{\theta}{2} & -\theta\mathrm{th}\frac{\theta}{2} \\ -\theta\mathrm{th}\frac{\theta}{2} & \frac{\theta}{\mathrm{th}\theta}+\frac{1}{N} \end{bmatrix}$
6	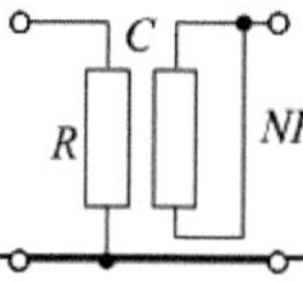	$\frac{1}{N+1}\begin{bmatrix} \frac{\theta}{\mathrm{th}\theta}+N & -\theta\mathrm{th}\frac{\theta}{2} \\ -\theta\mathrm{th}\frac{\theta}{2} & 2\theta\mathrm{th}\frac{\theta}{2} \end{bmatrix}$
7	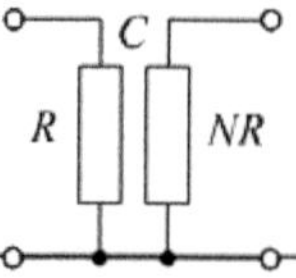	$\frac{1}{N+1}\begin{bmatrix} \frac{\theta}{\mathrm{th}\theta}+N & 1-\frac{\theta}{\mathrm{th}\theta} \\ 1-\frac{\theta}{\mathrm{th}\theta} & \frac{\theta}{\mathrm{th}\theta}+\frac{1}{N} \end{bmatrix}$
8	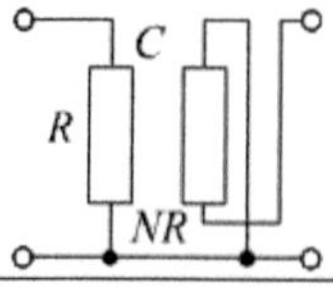	$\frac{1}{N+1}\begin{bmatrix} \frac{\theta}{\mathrm{th}\theta}+N & \frac{\theta}{\mathrm{sh}\theta}-1 \\ \frac{\theta}{\mathrm{sh}\theta}-1 & \frac{\theta}{\mathrm{th}\theta}+\frac{1}{N} \end{bmatrix}$

Table 5.4 Two-port connection circuits of the *R-C-NR* EDP with hard-line coupling

No	Connection circuit	Normalized matrix of *z*-parameters
1	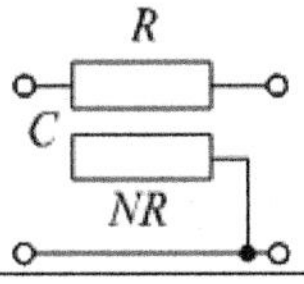	$\frac{1}{\theta}\begin{bmatrix} \frac{N}{(N+1)}\left(\theta-2\text{th}\frac{\theta}{2}\right)+\frac{N+1}{\text{th}\theta} & \frac{N}{\text{th}\theta}+\frac{1}{\text{sh}\theta} \\ \frac{N}{\text{th}\theta}+\frac{1}{\text{sh}\theta} & \frac{N+1}{\text{th}\theta} \end{bmatrix}$
2	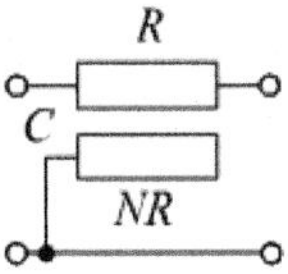	$\frac{1}{\theta}\begin{bmatrix} \frac{N+1}{\text{th}\theta} & \frac{N}{\text{th}\theta}+\frac{1}{\text{sh}\theta} \\ \frac{N}{\text{th}\theta}+\frac{1}{\text{sh}\theta} & \frac{N}{(N+1)}\left(\theta-2\text{th}\frac{\theta}{2}\right)+\frac{N+1}{\text{th}\theta} \end{bmatrix}$
3	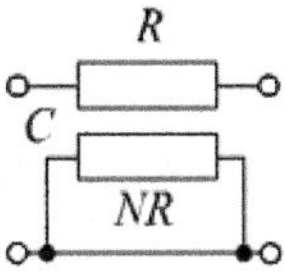	$\frac{1}{N+1}\begin{bmatrix} \frac{\theta}{\text{th}\theta}+N & -\frac{\theta}{\text{sh}\theta}-N \\ -\frac{\theta}{\text{sh}\theta}-N & \frac{\theta}{\text{th}\theta}+N \end{bmatrix}$
4	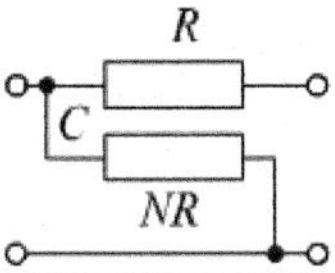	$\frac{1}{N+1}\begin{bmatrix} \frac{(N+1)^2}{N} & -(N+1) \\ -(N+1) & \frac{\theta}{\text{th}\theta}+N \end{bmatrix}$
5	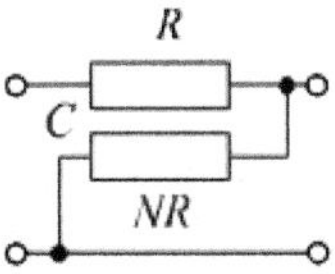	$\frac{1}{N+1}\begin{bmatrix} \frac{\theta}{\text{th}\theta}+N & -(N+1) \\ -(N+1) & \frac{(N+1)^2}{N} \end{bmatrix}$
6	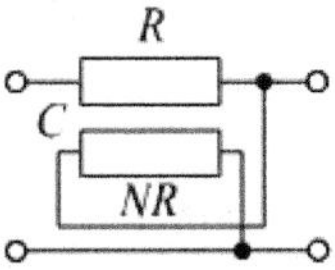	$\frac{1}{N+1}\begin{bmatrix} \frac{\theta}{\text{th}\theta}+N & 1-N-\frac{\theta}{\text{th}\theta/2} \\ 1-N-\frac{\theta}{\text{th}\theta/2} & \frac{2\theta}{\text{th}\theta/2}+\frac{(N-1)^2}{N} \end{bmatrix}$
7	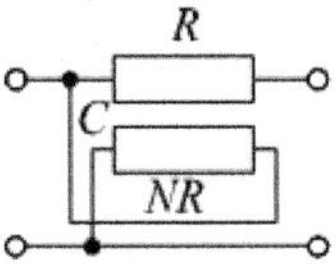	$\frac{1}{N+1}\begin{bmatrix} \frac{2\theta}{\text{th}\theta/2}+\frac{(N-1)^2}{N} & 1-N-\frac{\theta}{\text{th}\theta/2} \\ 1-N-\frac{\theta}{\text{th}\theta/2} & \frac{\theta}{\text{th}\theta}+N \end{bmatrix}$

Notice that for all circuits listed in Tables 5.3, 5.4, 5.5, and 5.6, the input voltage U_1 is applied to the left terminals; i.e., the input ports. The output voltage U_2 is measured from the right terminals; i.e., the output ports. Similarly, the input and output currents, I_1 and I_2, are defined accordingly.

Table 5.5 Connection circuits of the *R-C-NR* EDP in the form of transmission two ports

No	Connection circuit	Normalized matrix of *z*-parameters
1	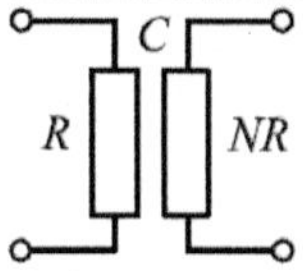	$\frac{N}{(N+1)\theta}\begin{bmatrix} \theta+\frac{2}{N}\text{th}\frac{\theta}{2} & \theta-2\text{th}\frac{\theta}{2} \\ \theta-2\text{th}\frac{\theta}{2} & \theta+2N\text{th}\frac{\theta}{2} \end{bmatrix}$
2	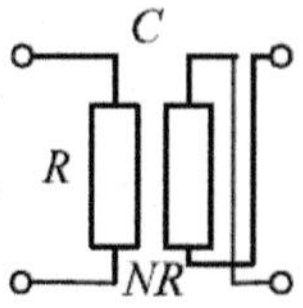	$\frac{N}{(N+1)\theta}\begin{bmatrix} \theta+\frac{2}{N}\text{th}\frac{\theta}{2} & -\left(\theta-2\text{th}\frac{\theta}{2}\right) \\ -\left(\theta-2\text{th}\frac{\theta}{2}\right) & \theta+2N\text{th}\frac{\theta}{2} \end{bmatrix}$
3	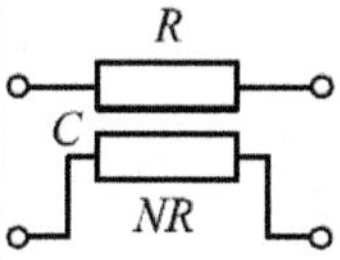	$\frac{N+1}{\theta}\begin{bmatrix} \frac{1}{\text{th}\theta} & \frac{1}{\text{sh}\theta} \\ \frac{1}{\text{sh}\theta} & \frac{1}{\text{th}\theta} \end{bmatrix}$
4	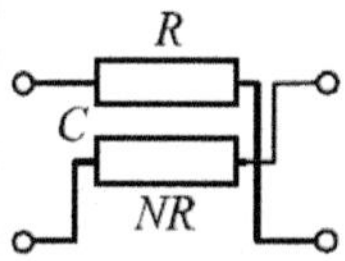	$\frac{N+1}{\theta}\begin{bmatrix} \frac{1}{\text{th}\theta} & -\frac{1}{\text{sh}\theta} \\ -\frac{1}{\text{sh}\theta} & \frac{1}{\text{th}\theta} \end{bmatrix}$
5	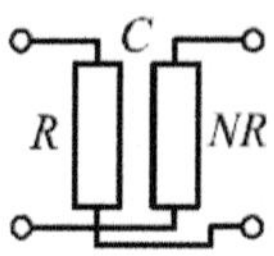	$\frac{N}{(N+1)\theta}\begin{bmatrix} \frac{N^2+1}{\text{th}\theta}+\frac{2N}{\text{sh}\theta}+N\theta & -\left(\frac{N^2+1}{\text{sh}\theta}+\frac{2N}{\text{th}\theta}-N\theta\right) \\ -\left(\frac{N^2+1}{\text{sh}\theta}+\frac{2N}{\text{th}\theta}-N\theta\right) & \frac{N^2+1}{\text{th}\theta}+\frac{2N}{\text{sh}\theta}+N\theta \end{bmatrix}$
6	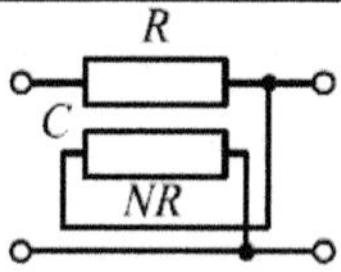	$\frac{N}{(N+1)\theta}\begin{bmatrix} \frac{N^2+1}{\text{th}\theta}+\frac{2N}{\text{sh}\theta}+N\theta & \frac{N^2+1}{\text{sh}\theta}+\frac{2N}{\text{th}\theta}-N\theta \\ \frac{N^2+1}{\text{sh}\theta}+\frac{2N}{\text{th}\theta}-N\theta & \frac{N^2+1}{\text{th}\theta}+\frac{2N}{\text{sh}\theta}+N\theta \end{bmatrix}$

Table 5.5 shows six circuits of the *R-C-NR* EDP in the form of transmission two ports. Observe that the multiplying factor, *N*, of the *R-C-NR*-based two ports (connection circuits 1, 2, and 3 given in Table 5.4) provides a rejection-shaped frequency responses as depicted in Table 5.6. The frequency response listed in Table 5.6 indicates that it is basically possible to synthesize a complete signal processing unit using a single electronic device.

Table 5.6 Performance parameters and characteristics of the *R-C-NR* EDP-based rejection filters

<table>
<tr><th>Connection circuit</th><th>Parameters</th><th>Magnitude and Phase Frequency Response</th></tr>
<tr><td>1 and 2 from table 5.4</td><td>N=0,0866
ω(1+N)RC=19,8</td><td>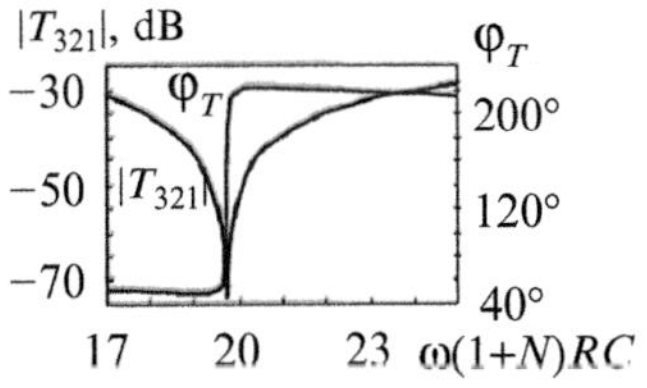
</td></tr>
<tr><td>3 from table 5.4</td><td>N=0,218
ω(1+N)RC=30,8</td><td>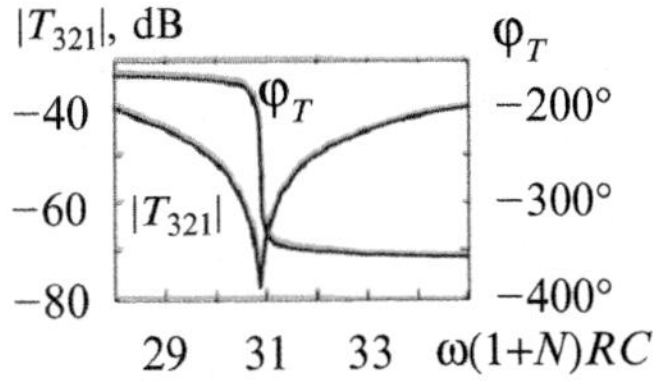
</td></tr>
</table>

5.4 Connection of an *0-C-R-NC-0*-Structured EDP Circuits

The *0-C-R-NC-0* topology allows one to synthesize 14 different two-port connection circuits and 15 different four-port EDP circuits as given n Tables 5.7, 5.8, 5.9, and 5.10.

Using the ODHet *RC*-EDP turns the symmetrical circuits into a nonsymmetrical one (except for the cases when the *RC*-EDP geometric or physical parameters heterogeneity is located symmetrically with respect to the input and the output terminals). The corresponding number of circuits' terminals may double.

5.5 Effect of Number of *RC*-EDP Terminals on Circuits Connections

It is clear that an increase in the number of *RC*-EDP terminals increases the number of connection circuits that has unique performance characteristics.

Figure 5.2 depicts the relationship between the number of possible versions of circuits' connections, N, and the number of terminals, n, of the initial *RC*-EDP (with undefined connection circuit). The notation kP represents the number of ports used in the circuits connections, where k is the number of ports, $k = 1, 2\ldots$

One can see, for example, that the *RC*-EDP with five terminals possesses about 500 two-terminal and four-terminal versions and about 60 versions each of five

Table 5.7 Two-port connection of the *0-C-R-NC-0*-structured EDP with hard-line coupling

Circuit number	Connection circuits	Circuit number	Connection circuits
1	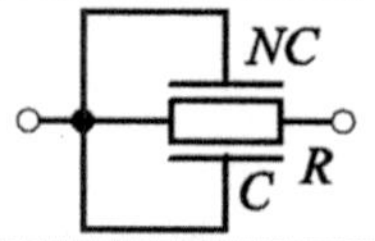	4	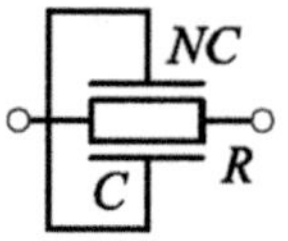
2	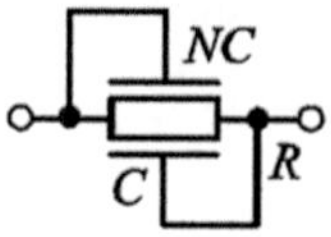	5	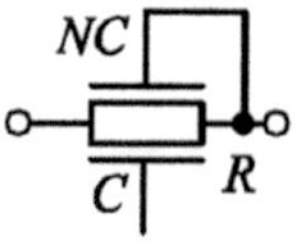
3	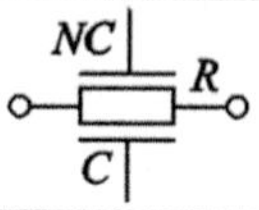		

Table 5.8 Two-port connections of the *0-C-R-NC-0*-structured EDP with capacitance coupling

Circuit number	Connection circuits	Circuit number	Connection circuits
1	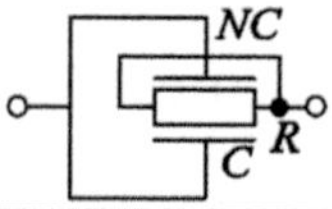	6	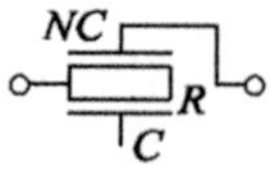
2	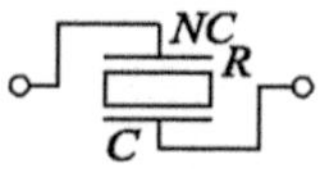	7	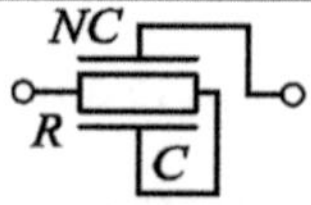
3	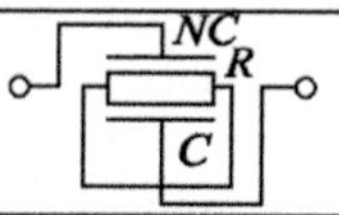	8	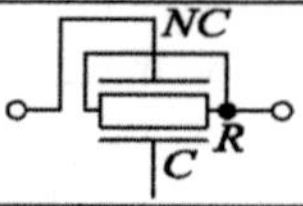
4	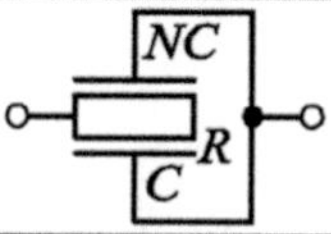	9	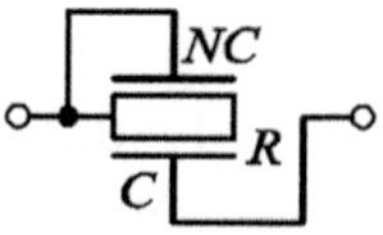
5	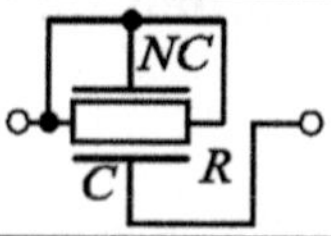		

terminals. Moreover, the *RC*-EDP with six terminals provides as much connection circuits as 1000 of two-terminal configuration.

Such great variety of the derived circuits that can be made out of multiport elements (multilayered *RC*-EDP actually fall into this category) have the potential

Table 5.9 Four-terminal connection of the *0-C-R-NC-0*-structured EDP with hard-line coupling

Circuit number	Connection circuits	Circuit number	Connection circuits
1	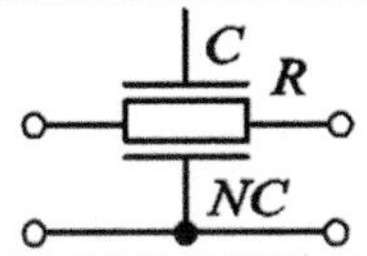	4	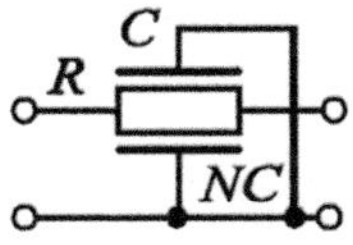
2	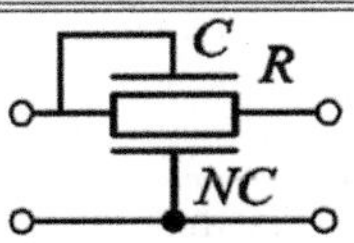	5	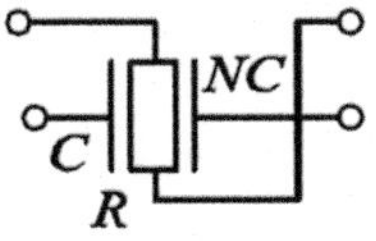
3	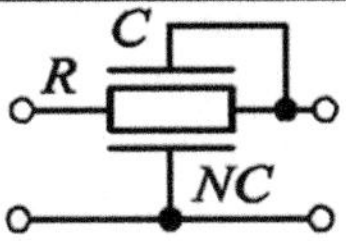		

Table 5.10 Four-terminal connection of the *0-C-R-NC-0*-structured EDP with capacitive coupling

Circuit number	Connection circuits	Circuit number	Connection circuits
1	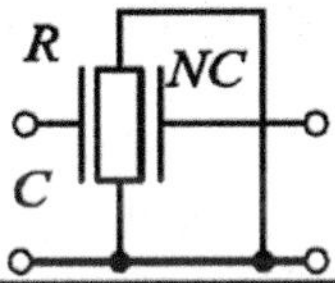	6	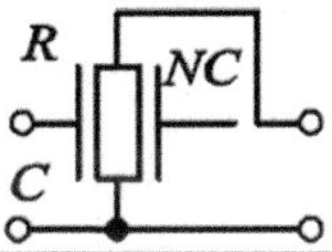
2	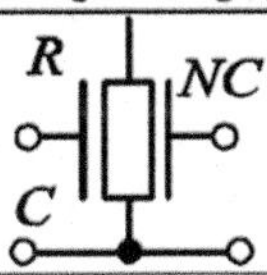	7	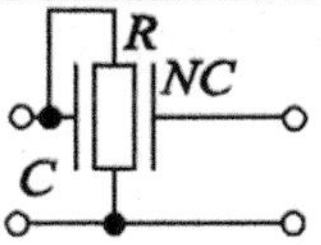
3	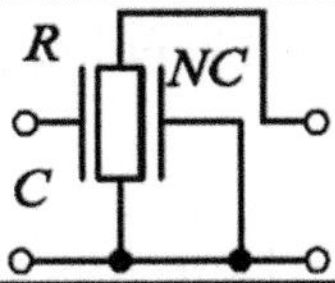	8	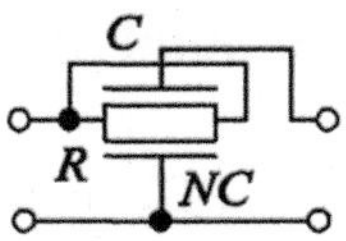
4	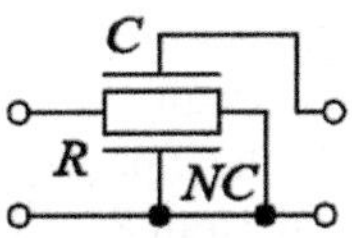	9	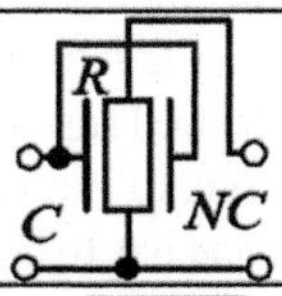
5	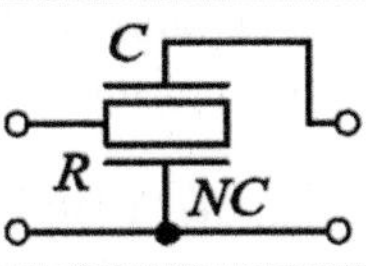	10	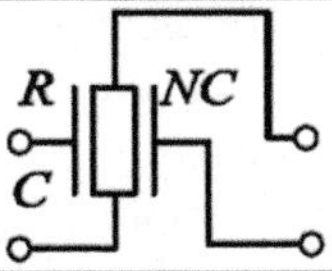

Fig. 5.2 Dependence of the number of possible versions of circuit connections, *N*, on the number of terminals, *n*, of the initial *RC*-EDP

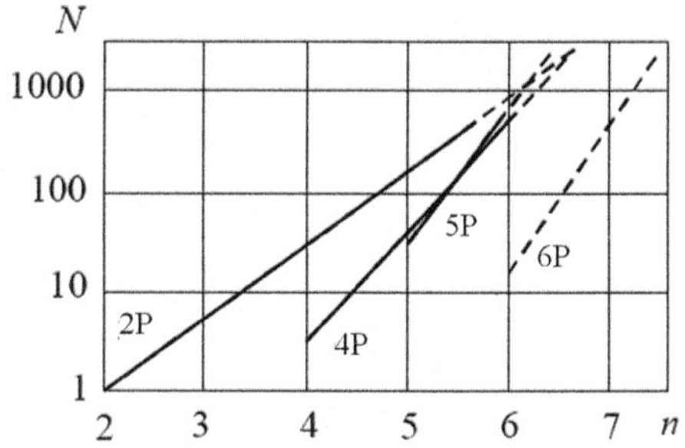

to be the basis to synthesize the *RC*-EDP devices that meet specified frequency response. This can be accomplished by choosing the optimal connection version. This requires a database that would comprise information on various connection circuits, their frequency-dependent parameters, and on their voltage variation due to the static or dynamic heterogeneous characteristics of the chosen *RC*-EDP design structure.

Figure 5.3b, c depicts an illustrating example of two forms of two-terminal connection of six-port *RC*-EDP circuit using ODHom *NR-C-R-MC-KR*-structure (here *N*, *M*, and *K* are the positive real numbers) with the corresponding input impedance PFR curves.

As a comparison, Fig. 5.3a shows the known connection of the *R*-*C*-0-layered ODHom *RC*-EDP circuit. The PFR of its input impedance is constant and equals to $-45°$. It can also be concluded from Fig. 5.3b, c that it is possible to synthesize fractal elements with input impedance phase response that vary from $-40°$ to $-35°$ with $\pm 1°$ tolerance over two frequency decades. Notice that to expand the operating frequency range and to reduce PFR fluctuations within this range, one can solve an optimization problem to find the optimum parameters scaling factors, *N*, *M*, and *K*, and their spacing of the initial *RC*-EDP.

5.6 Compound Structure of *RC*-EDP Circuits

Integrating the ODHom and the ODHet *RC*-EDPs into a single structure provides further enhancement of the *RC*-EDP circuits to achieve desired frequency response. The connection layout of these distributed-parameter elements can be synthesized to model a desired input transfer function of the circuit.

For example, to synthesize a circuit that meets the input impedance MFR shown in Fig. 5.4a, one has to compose three cascaded ODHom *RC*-EDP elements of equal length but of different widths. These circuits can be fabricated as step-shaped *RC*-EDP devices as depicted in Fig. 5.4b. The terminals of this device are allocated at its narrowest edge.

Another example of generating a compound *RC*-EDP circuit is shown in Fig. 5.5. The synthesized circuitry comprises of four ODHom *RC*-EDP combined

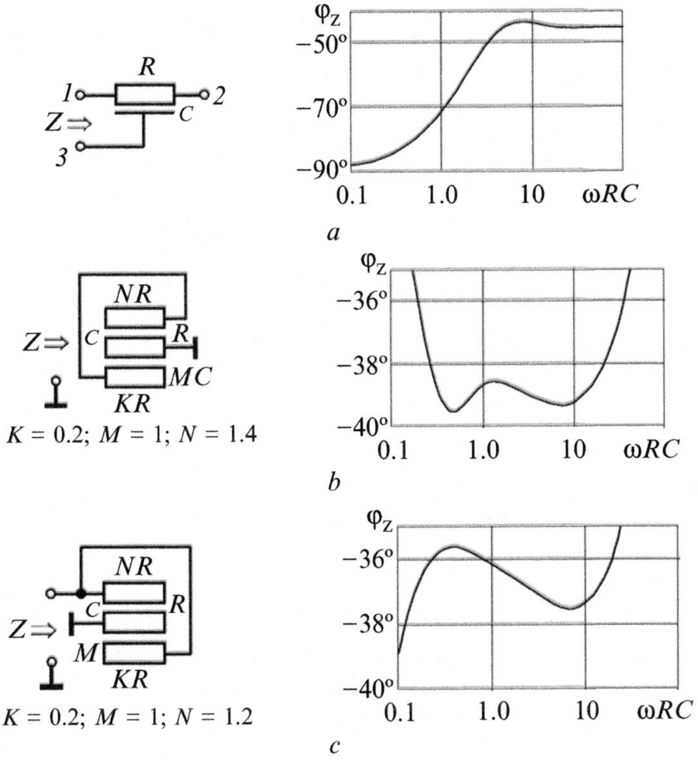

Fig. 5.3 Examples of the *RC*-EDP two-terminal connection circuits and their input impedance PFR: (**a**) with *R-C-0* layer structure; (**b**, **c**) with *NR-C-R-MC-KR* layer structure

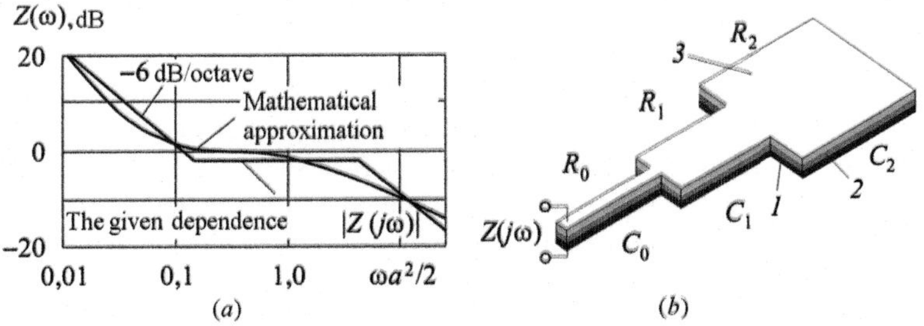

Fig. 5.4 An example of *RC*-EDP synthesis to satisfy desired input impedance MFR: (**a**) the desired input impedance MFR; (**b**) the *RC*-EDP synthesized design; (*1*) conducting layer; (*2*) dielectric layer; (*3*) resistive layer

as two pairs in such a way that each pair forms two serially connected circuits. The fact that both circuits could have the same resistive layer makes it possible to create an *integral* functional element with distributed parameters. The design structure of one section of such element is depicted in Fig. 5.5b.

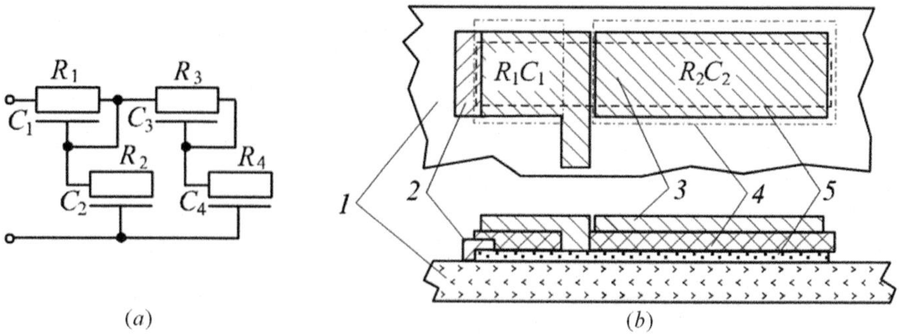

Fig. 5.5 Compound structure of *RC*-EDP design: (**a**) ODHom *RC*-EDP-based circuit; (**b**) *RC*-EDP design structure; (*1*) substrate; (*2*) contact pad; (*3*) conducting plate; (*4*) dielectric layer; (*5*) resistive layer

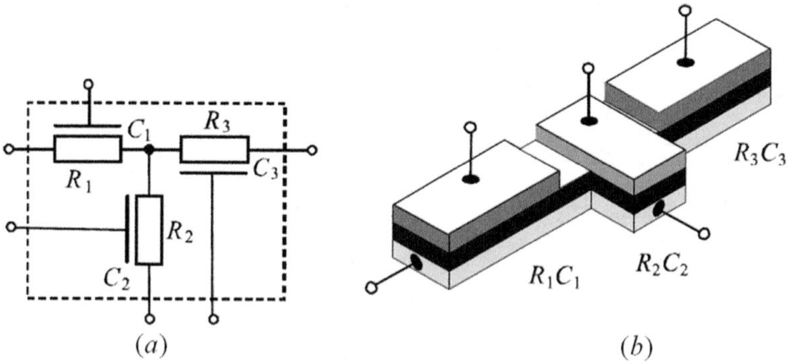

Fig. 5.6 Example of compound *RC*-EDP design structure: (**a**) equivalent circuit; (**b**) *RC*-EDP design structure

Figure 5.6 provides another example of synthesizing new fractal devices by combining several *RC*-EDPs. Figure 5.6a shows the circuit diagram of an active *RC*-filter using three ODHom *RC*-EDP elements of equal time constants $\tau = R_1C_1 = R_2C_2 = R_3C_3$, which can be considered as a frequency-dependent feedback circuit. The design structure of this circuit can be implemented in the form of a single functional multiport element with six terminals as shown in Fig. 5.6b.

Now consider a more general case of synthesizing a desired frequency response using compound *RC*-EDP structure comprised of four units of ODHom layered-*RC*-EDP to form the new structure *NR-C-R-MC-KR*. The synthesis process is performed to meet desired characteristics of the circuit frequency response in compliance with the terminal connections of the ODHom *RC*-EDP components.

Figure 5.7a, for example, depicts the equivalent circuit of a two-port with $16 \pm 0.5°$ constant phase of uniformity of the input impedance PFR over four frequency decades (Fig. 5.7b). Here $\overline{\omega}$ denotes the normalized frequency.

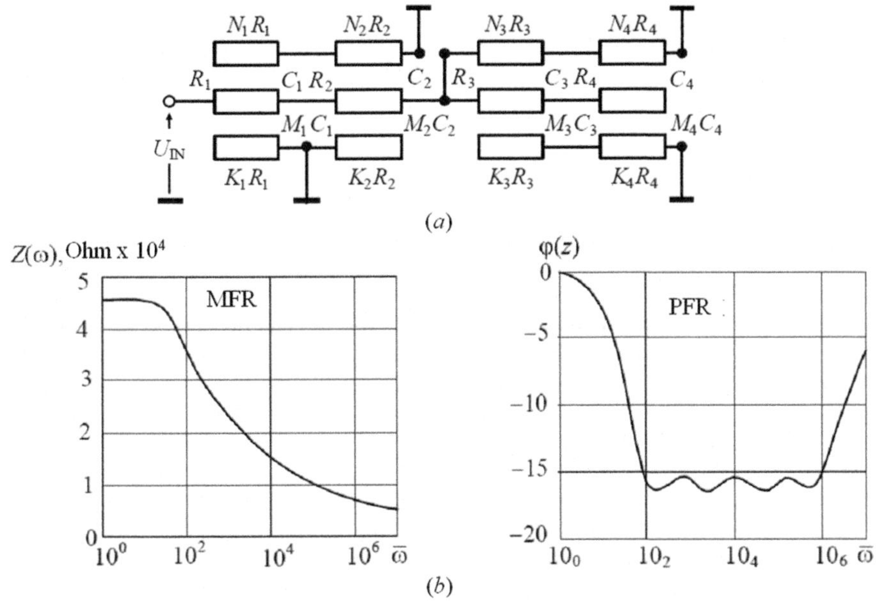

Fig. 5.7 Compound *RC*-EDP: (**a**) equivalent circuit; (**b**) magnitude and phase frequency responses (MFR and PFR) of the input impedance

There are three layers in this ODHom *RC*-EDP design. The parameters of these circuits can be uniquely identified as shown in Fig. 5.7a. Notice that N_i define the ratios of the upper and the central resistive layer specific resistances, K_i denote the ratios of the bottom and the central resistive layer specific resistances, M_i represent the ratios of the upper and the bottom layer dielectric specific capacities, and i is the number of the ODHom *RC*-EDP elements.

The frequency response (MFR and PFR) of the two-port circuit of this example shown in Fig. 5.7b was obtained using the following parameter ratios of the ODHom *RC*-EDP:

(a) The upper-central coefficient ratios: $\boldsymbol{N} = \{0.476; 0.33; 0.33; 0.33\}$,
(b) The upper-lower coefficient ratios: $\boldsymbol{M} = \{30; 30; 30; 30\}$,
(c) The bottom-central coefficient ratios: $\boldsymbol{K} = \{12.86; 1; 1; 1\}$.

Observe that the resistances, R_i, and the capacitance, C_i, of the layers are the same for all sections of the ODHom *RC*-EDP circuit. As we can see, the fractal element has been synthesized by connecting several ODHom *RC*-EDP circuits where the parameters of the first section differ from the others.

The design structure of the equivalent circuit can be done in several ways.

First method: combine three ODHom *RC*-EDP of equal layer parameters into a single structure to have desired contact pads and insulating areas according to the equivalent circuit and then connect this structure to the other parameters of the separate first ODHom *RC*-EDP one.

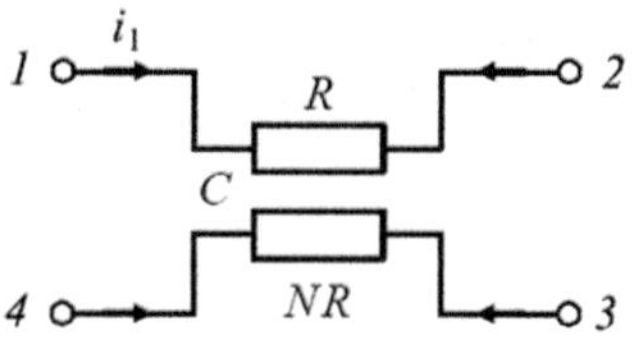

Fig. 5.8 *R-C-NR*-layered EDP

Second method: the upper and the bottom resistive layers of the whole of the *RC*-EDP are made of materials sensitive to some outside fields that could change their electrical and physical properties, thus changing the layers' resistance.

A common practice of today's IC includes the development of programmable logic and analog devices (PLD and PAD). Such approach can also be used to create compound *RC*-EDP that meets the desired performance characteristics. This requires connecting sets of separate *RC*-EDP terminals (bipolar or MOS technology), which could be programmed by one of the following methods:

- Mask programming: it is a photolithographic process to make the connection lines between the terminals using a set of photographic masks;
- Fusing: connecting the links that short-circuit the *RC*-EDP terminals by passing short electric current pulses through these links (the fuse technology);
- Antifusing technology: forming the links between terminals by applying the breakdown voltage to the insulator;
- Create links by programming the MOS switches, which accumulate charges with the newly formed conducting channels.

Practice

1. Let the *Y*-matrix of the *RC*-EDP, depicted in Fig. 5.8, be given as follows:

$$[Y]=\frac{1}{(1+N)R}\begin{bmatrix} \frac{\theta}{\mathrm{th}\theta}+N & -\frac{\theta}{\mathrm{sh}\theta}-N & \frac{\theta}{\mathrm{sh}\theta}-1 & 1-\frac{\theta}{\mathrm{th}\theta} \\ -\frac{\theta}{\mathrm{sh}\theta}-N & \frac{\theta}{\mathrm{th}\theta}+N & 1-\frac{\theta}{\mathrm{th}\theta} & \frac{\theta}{\mathrm{sh}\theta}-1 \\ \frac{\theta}{\mathrm{sh}\theta}-1 & 1-\frac{\theta}{\mathrm{th}\theta} & \frac{\theta}{\mathrm{th}\theta}+\frac{1}{N} & -\frac{\theta}{\mathrm{sh}\theta}-\frac{1}{N} \\ 1-\frac{\theta}{\mathrm{th}\theta} & \frac{\theta}{\mathrm{sh}\theta}-1 & -\frac{\theta}{\mathrm{sh}\theta}-\frac{1}{N} & \frac{\theta}{\mathrm{th}\theta}+\frac{1}{N} \end{bmatrix}$$

   Use this matrix to derive the expression of the *RC*-EDP impedance for different options of two-port connections (see Table 5.2). With the help of mathematical software, build the input impedance logarithmic MFR and PFR for the chosen connection variant.
2. Use the *Y*-matrix from exercise 1 to derive the expressions of the *Y*-matrixes of the *R-C-NR*-layered EDP two-port connection circuits (see Tables 5.3 and 5.4).

With the help of mathematical software, build the two-port transfer function logarithmic MFR and PFR for the chosen connection model (the left port is input, the right one is output).

3. Depict the variants of two-port connection of the *NR*-*C*-*R*-*MC*-*KR*-layered EDP. Suggest a method to estimate the amount of such connections.

Test Questions

1. Name the schematic factors that make it possible to vary the *RC*-EDP performance characteristics without changing the structure of the element.
2. Name the possible applications of different variants of two-pole and two-port connection of the *R*-*C*-0-layered *RC*-EDP.
3. Depict the electric circuit diagrams of the rejecting filters based on the *R*-*C*-*NR*- and 0-*C*-*R*-*NR*-0-layered EDP.
4. Name the advantages and the disadvantages of the compound *RC*-EDP.
5. Name the methods of connection of ODHom *RC*-EDP in a compound *RC*-EDP.

Chapter 6
Design and Fabrication of *RC*-EDP Circuits

6.1 Preliminaries and Background

The *RC*-EDP production techniques are sufficiently influenced not only by their performance characteristics such as operating frequency range, production precision, capability of parameters adjustment, stability, and overall dimensions but also by the nature and types of the mathematical models that are supposed to simulate the real design structures when analyzing the *RC*-EDP-based circuits.

The *RC*-EDP devices can be made both as parts of integrated circuits and as discrete components. The *RC*-EDPs included in integrated circuits are made using the same production technologies of the corresponding integrated circuits. Figure 6.1 represents the classification of the different design-and-fabrication methods of *RC*-EDP.

When the *RC*-EDPs are made as discrete components, their production process is initially performed as batch fabrication method similar to film or semiconductor production technologies. The master die is cut into separate "chips," where the latter are equipped with terminals or SMD-contact pads and then sealed. Such manufacturing process enhances the technology effectiveness of the end product and the reproducibility and stability of their performance characteristics. However, it is quite possible that the *RC*-EDP from different production runs could possess different performance characteristics (temperature resistance coefficient (TRC), temperature capacitance coefficient (TCC), timing stability etc.).

6.2 Thin-Film *RC*-EDP Fabrication Methods

The first *RC*-elements with distributed parameters were manufactured as thin-film *RC*-structures since the minimum thickness of the films is (hundreds of Angstroms) best complied with one of the fundamental requirements of *RC*-elements; i.e., the

A.Kh. Gil'mutdinov et al., *Fractal Elements and their Applications*, Analog Circuits and Signal Processing, DOI 10.1007/978-3-319-45249-4_6

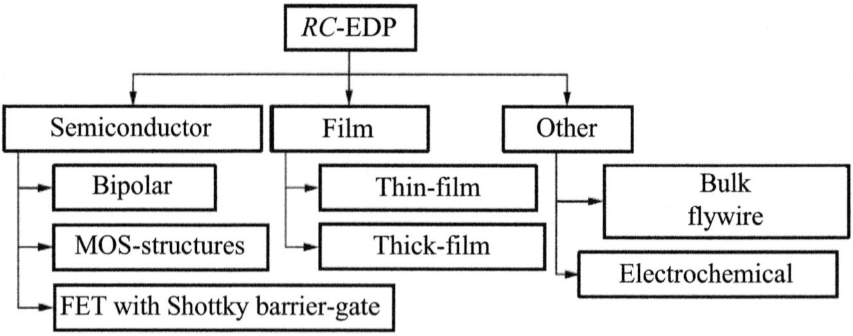

Fig. 6.1 Design-and-technology options of *RC*-EDP

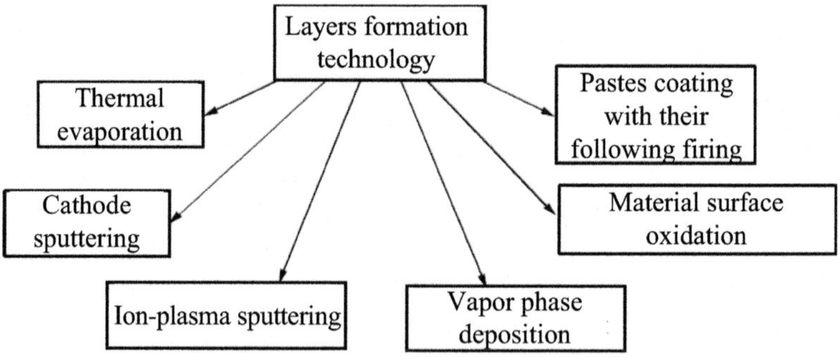

Fig. 6.2 Classification of *RC*-EDP layers formation technology

film cross section should be much less than its length as shown in Fig. 3.30, which depicts the three-layered formation technology of *RC*-EDP structures.

The classifications of the layers and pattern formation technologies that could be used to produce the *RC*-EDP films are given in Figs. 6.2 and 6.3. Each method will be explained in the subsequent discussion.

Figure 6.4, on the other hand, shows the schematic diagram of a vacuum deposition chamber that performs *thermal evaporation* of materials in vacuum to create films. The vacuum deposition chamber is comprised of a cap *1*, bottom plate *2* with elastomeric sealing *3*. Within the chamber, there are heated substrate holder *4* with substrate *5*, removable mask *6*, door *7*, and film material thermal evaporator *8*. Prior to material deposition, 10^{-3}–10^{-4} Pa operating vacuum is established within the chamber. The particles of the evaporated material go through the holes of the mask, which is pressed to the substrate and deposit on the latter, forming a film layer pattern to reproduce the pattern of mask holes.

This method is especially appropriate to make metallic conducting and resistive films; it provides sufficient reproducibility of electrophysical performance characteristics. The method is distinct for high film deposit rate, high purity propagation medium for the film material from the evaporator to the substrate. Applying

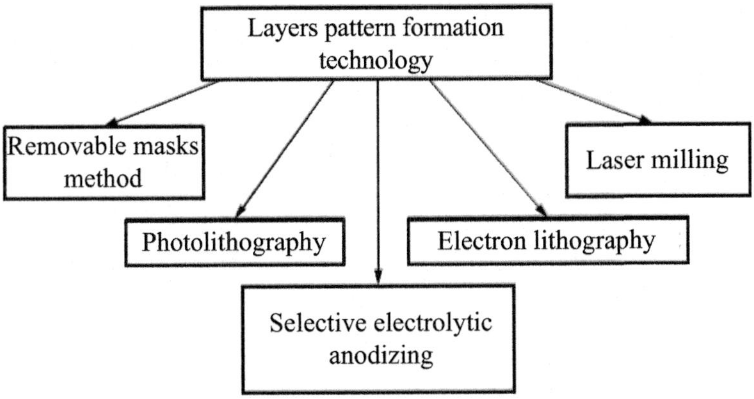

Fig. 6.3 Classification of *RC*-EDP layers pattern formation technology

Fig. 6.4 Vacuum deposition plant schematic diagram

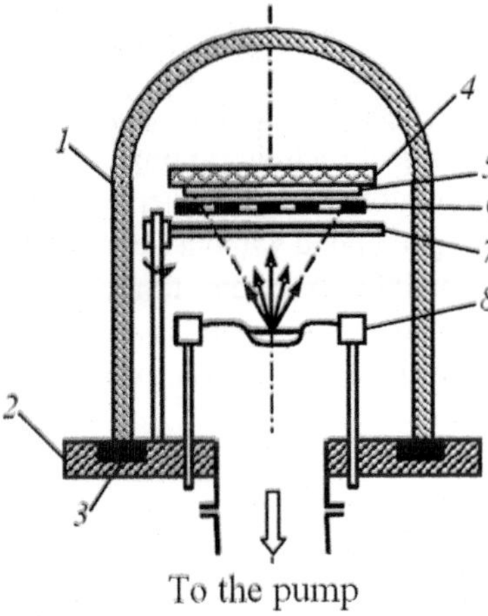

electron beam evaporators makes it possible to evaporate and deposit films of high-melting on the substrates, corrosion-resisting, and temperature-stable metals such as W, Ta, Ti, and Mo. The following summarizes the basics methods of fabricating *RC*-EDP films.

6.2.1 Cathode Sputtering Process

In case of *cathode sputtering* process, the target of the sputtered material is the cathode and the substrate holder is the anode. Applying high voltage between them causes glow discharge in the inert gas (physical sputtering) or in the reactive gas

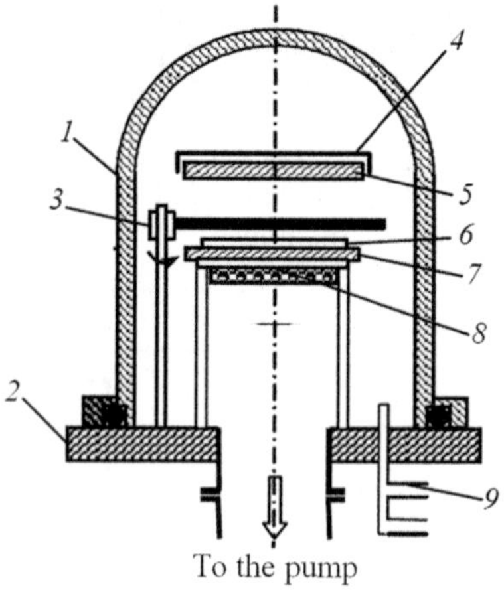

Fig. 6.5 Cathode sputtering plant schematic diagram: (*1*) cap; (*2*) bottom plate; (*3*) door; (*4*) grounded shield; (*5*) cathode; (*6*) substrate; (*7*) anode; (*8*) substrate heater; (*9*) gases supply

(reactive sputtering). The schematic diagram of a cathode sputtering film deposit plant is shown in Fig. 6.5.

Compared to thermal sputtering, the cathode sputtering is a slower process and it has more difficulties to produce pure films of active metals. However, reducing the medium pressure, in general, accelerates this process and creates conducting films with good bond without intermediate layers. The cathode sputtering method makes it possible to create films of high-melting metals such as W, Ta, Ti, and Mo. Moreover, when oxygen or nitrogen is introduced to the inert medium, it becomes possible to get oxides and nitrides of these materials.

6.2.2 Ion-Plasma Sputtering Process

This process is a version of cathode sputtering. Here, gas discharge plasma ions bombard a special target resulting in deposit of the material atoms. If the gas discharge is caused by high-frequency voltage, then the target can be dielectric and it becomes possible to create oxide layers.

6.2.3 Oxidation of the Conducting Surface or Resistive Material

This method allows acquiring dielectric films used to create the *RC*-EDP distributed capacitance or to protect the surface of the conducting or resistive layer from the environment. Oxidation can be performed under high temperature either in the atmosphere or with oxygen or by means of electrolytic anodizing process.

Electrolytic anodizing is a well-known method, since 1817, used to get oxide films. It is used to form oxide films of metals and semiconductors surfaces. It is performed at the anode in an electrolytic medium. The method is widely used, for example, to grow tantalum oxide (Ta_2O_5) and aluminum oxide (Al_2O_3).

The thickness and properties of the oxide film created by electrolytic anodizing technique depend on several factors such as the content, the concentration, the temperature of the electrolyte, the electrodes voltage, the film current, and the duration of the process. Making the *RC*-EDP based on tantalum pentoxide and titanium dioxide allows getting dielectric layers with specific capacitance of $C_o = 1000\ldots2000$ pF/mm^2, while the metals of Ta and Ti can be used to create the conducting and resistive layers.

6.2.4 Vapor Phase (VP) Films Deposition

This process is a relatively new one that appeared as a result of tremendous advancement in applied chemistry. Vapors of the initial compounds (halogenides, hydrides, carboniles, and organo-element compounds (OEC)) are placed into the chamber where they reach heated-up substrates to certain temperatures. These vapors react with the substrates to form solid film-forming substance, and the gaseous by-products are removed from the substrates zone.

The method is based on various synthesis reactions to replace the hydrolysis and pyrolysis processes. Applying electrical voltage, light, electron beam, and other kinds of exposure additionally energizes the reacting molecules that intensify the chemical reaction at lower temperature. The most commonly used method to intensify gas-vapor-phase film deposition today is gas discharge plasma.

The following gas-vapor-phase film deposition processes can be recommended to make *RC*-EDP.

6.2.5 Monosilane Pyrolytic Decomposition Process

This process occurs at temperatures from 300 to 400 °C and produces silicone dioxide as a by-product according to the following chemical reaction:

$$SiH_4 + 2O_2 \rightarrow SiO_2 \downarrow + 2H_2O.$$

Low temperatures make it possible to use this process to make films on the thin-film *RC*-EDP plate, which is generally made of aluminum. In this case, the SiO_2 layer operates as a good dielectric for the *RC*-EDP devices.

6.2.6 High-Melting Metal Films Deposition

The number of carbon monoxide ligands in a metal carbonyl complex is described by a Greek numeral, followed by the word *carbony*. The carbon monoxide ligand may be bound terminally to a single metal atom or bridging to two or more metal atoms. Carbonyls degrade at temperatures from 250 to 320 °C according to the following reactions:

$$W(CO)_6 \rightarrow W\downarrow + 6CO\uparrow,$$
$$Mo(CO)_6 \rightarrow Mo\downarrow + 6CO\uparrow.$$

These processes can be used to create in *RC*-EDP stable resistive layers of high-melting materials with high-stable specific resistance. Furthermore, it does not require using expensive cathode or ion-plasma sputtering equipment.

6.2.7 Polycrystalline Silicon Films Deposition

This process occurs as a result of pyrolytic decomposition of monosilane at temperatures from 540 to 700 °C as follows:

$$SiH_4 \rightarrow Si\downarrow + 2H_2\uparrow$$

Introduction of impurities (doping) can significantly vary polycrystalline silicon films conductivity. It means that the same equipment can be used to produce both conducting plates of the *RC*-EDP and their resistive layers with different specific resistance.

In cases when there is a need to reduce the temperature of the process or it is hard to get films with the required content and properties, gas-vapor-phase deposition is performed in the discharge plasma medium. Plasma makes it possible to perform chemical reactions at 200...400 °C lower temperatures.

The given examples do not cover all the capabilities of the method. Further development of applied chemistry and, in particular, of organometallic compounds chemistry, will make it possible to create films for thin-film elements in easier ways than the conventional vacuum methods provide.

6.2.8 Photolithography

Photolithography is the basic method to create the layers pattern today. The most modern machines of industrial optical lithography provide 1 μm level resolution. 35...50 μm resolution is quite enough for the hybrid integrated circuits (HIC) with

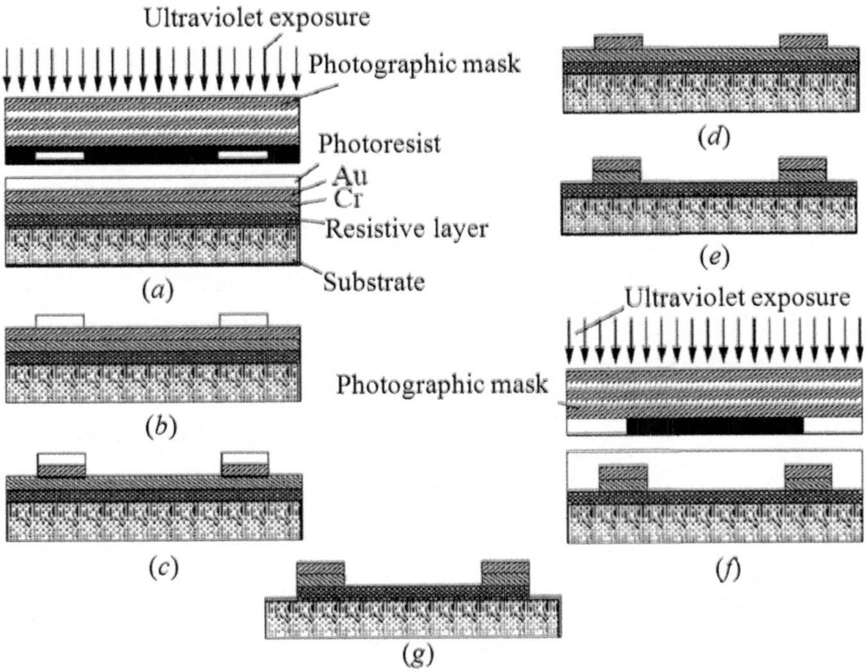

Fig. 6.6 Schematic diagrams of the double lithography method: (**a**) photoresist exposure; (**b**) photoresist development; (**c**) selective etching of gold; (**d**) photoresist removal; (**e**) selective etching of chrome; (**f**) photoresist application and exposure; (**g**) selective etching and removal of the photoresist

hook-up wiring elements. Therefore, the basic problem in creating the layers pattern of the *RC*-EDP-type multilayered structures is to choose such etching agents that would be inert to the lower layers film materials.

This problem can also be resolved by combining the lift-off and lift-on photolithography of chemical and ion-plasma etching.

Figures 6.6 and 6.7 show schematic diagrams of couple of examples of thin-film elements production process. Figure 6.6 represents the stages of the double photolithography process technology of creating the resistive layer with Cr–Au contact system. Figure 6.7 shows a method of lift-off photolithography process applied to form dielectric layer pattern onto the previously created resistive layer. Completion of the *RC*-EDP production requires applying a solid conducting layer and its photolithographic modeling.

As we can see, the described methods require significant amount of operations even to form a pattern of one layer. An easier way to make the *RC*-EDP layers pattern is to use removable metallic masks for every operation of layer creation. The design of such bimetallic mask is shown in Fig. 6.8.

However, such masks do not allow resolutions less than 50 μm due to the finite mask's thickness, improper adherence of the mask and the substrate, and some

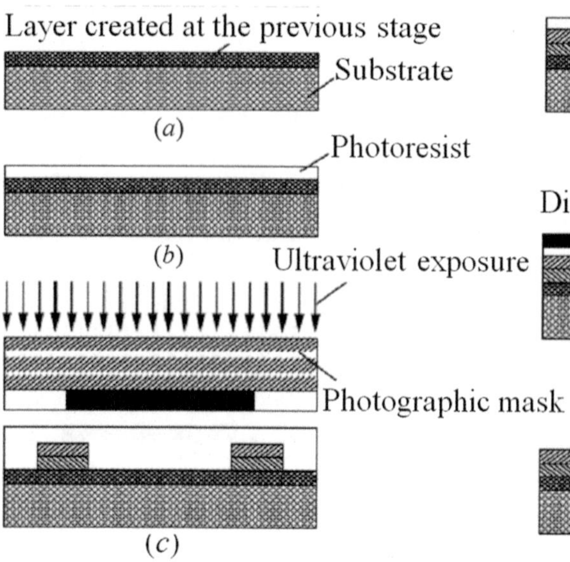

Fig. 6.7 Schematic diagrams of the lift-off lithography method: (**a**) initial workpiece; (**b**) photoresist application; (**c**) photoresist exposure; (**d**) contacts creation of photoresistive mask; (**e**) dielectric layer deposition; (**f**) removal of the photoresist with the layer deposited on it

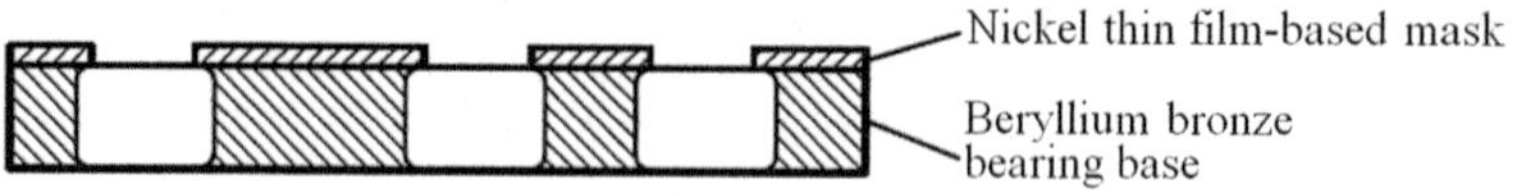

Fig. 6.8 Bimetallic removable mask design

other reasons. Besides, layers patterns alignment requires introducing significant alignment tolerance, which yields an increase in the structure's overall dimensions, thus introducing lumped parasitic elements. Meanwhile, the method of removable masks is relatively simple. It is not related to the "wet" chemical processes and can be recommended to manufacture many small products.

The so-called tantalum technology of *RC*-EDP fabrication has shown an excellent performance. This technology can be described as follows: the solid tantalum film covers the dielectric substrate by sputtering the tantalum target using photolithography to form a resistive layer pattern. Then an aluminum solid film is thermally sprayed over it using photolithography to form the conducting line that electrically connects all the parts of the tantalum film.

With photographic mask on, the substrate is put into an electrolytic bath where all open tantalum areas are subject to electrolytic anodizing. This process is conducted to adjust the surface resistance of the resistive layer and to form the dielectric layer of Ta_2O_5 used in the *RC*-EDP.

After anodizing, the photomask is removed and the solid aluminum film is deposited. The third photolithography results in the *RC*-EDP plate pattern. The

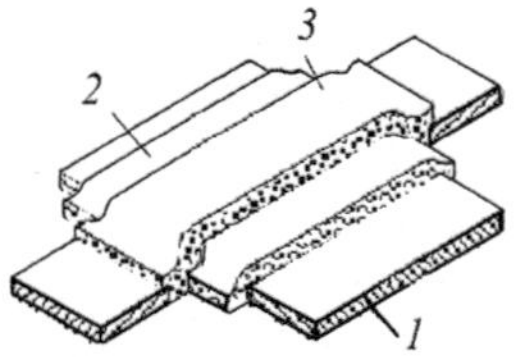

Fig. 6.9 Thick-film *RC*-EDP: (*1*) plate; (*2*) dielectric; (*3*) resistor

tantalum-made *RC*-EDP device shows high stability of its parameters, significant specific capacitance, and high breakdown voltage.

Thick-film technology is the dominant technique to manufacture hybrid integrated circuits today. In particular, 70 % of the whole of the manufactured hybrid integrated circuits outside Russia today are such kind of ICs. Design development of the thick-film hybrid integrated circuits is stimulated by demand in the radio-electronic components, where it possesses high levels of power and operation voltages. As for the thin-film circuits, they should rather be used to produce high-speed mini-devices of the control-and-measurement and radio-electronic equipment intended to operate within centimeter and millimeter wavelength ranges.

Thick-film hybrid integrated circuits are made as follows: dielectric, resistive, and conducting pastes are deposited on the dielectric substrate by screen printing with subsequent firing of every layer at high temperature (500–1000 °C). The thick film can be from units to several tens of micrometers thick.

The thick-film *RC*-EDP design structure is depicted in Fig. 6.9. It should be noted that, as a rule, the resistive layer is deposited on the dielectric layer because the resistive layer firing temperature is lower than that of the dielectric layer.

Precious metals (gold, silver, platinum, and palladium) are the most commonly used materials to make conducting pastes. These metals can be used both in pure form and in the form of alloys.

Resistive pastes are made of ruthenium oxide in most cases. They provide the surface resistance value that ranges from 3 Ω/mm^2 to 10 $M\Omega/mm^2$. The dielectric pastes comprise ferroelectric powders (Barium Titanate $BaTiO_3$ for instance) that possess high relative dielectric constant ε. The corresponding specific capacitances vary from 40 to 100 pF/mm^2.

It seems promising to make hybrid integrated circuits based on polymeric thick-film pastes. The latter have many advantages such as wide range of specific surface resistances, low-temperature hardening, low cost, and availability to produce the conducting, resistive, and dielectric layers.

Firing is an important stage of thick-film HIC production technology. Substrates of the HIC are usually subject to firing after each screen printing in order to provide strong bond of pattern and substrate. The substrate is heated up at a certain heating rate, and then exposed to a constant temperature for certain time period, and then is cooled down at a certain cooling rate. The high yield of the produced ICs requires not just maintaining a precise high temperature control but also a stable timing for each stage.

Infrared furnaces are used to enhance the manufacturing productivity. These furnaces differ from the conventional ones by the fact that they heat up the substrates by radiation using several heaters at a very high temperature (about 2000 °C) in spite of moving them through zones of specified temperature levels. The heating process of infrared furnaces is three to four times faster than that of the conventional ones but more difficult to figure out the combination of those heaters that would provide the desired parameters of the resistors.

Microwave ovens were created few years back. They can be considered as perfect candidates to use heat radiation for film production. They provide a perfect environment for a homogeneous distribution of heat over the working chamber. Microwave heat treatment may be carried out in a vacuum, and in oxidizing, inert, or reducing environment.

The conventional materials of the thick-film (also used to produce the *RC*-EDP) today can be replaced with organometallic compounds.

Organometallic compounds are highly soluble in organic solvents that are usually used to produce the thick-film HIC pastes. After firing, organometallic compounds form a very smooth thin film on the substrate surface. This film externally resembles the films that are made by vacuum deposition. The thickness of the screen-printed organometallic film is 150–200 nm (whereas the conventional thick-film technology provides a paste layer thickness of 10–40 μm).

Organometallic films can be deposited on the substrate in different ways: dipping, sputtering, and screen printing. Drying and firing processes of the organometallic compounds and the thick-film pastes are similar. However, the higher content of organic materials in the organometallic compounds involves requirement in good ventilation of the furnaces.

The maximum firing temperature is not as important here as in case of thick-film pastes. Its value just should be high enough to provide decomposition of the organic matter and high adhesion of the film to the substrate. As a rule, depending on the substrate type, the organometallic films firing temperature lies within the 500–850 °C temperature range.

The organometallic compounds enjoy sufficient advantages over the thick-film pastes: they are cheaper; the range of the substrates based on these materials is significantly wider (e.g., aluminum oxide, aluminum nitride, or polyimides). Furthermore, chemical or plasma etching of the fired organometallic film provides the pattern resolution comparable with photolithography. This makes it possible to increase the chip density in comparison with the conventional thick-film technology.

The range of materials used to create layers by the film technology is enormous, and it is constantly fabricated using new materials. So, let us consider some examples and the general description of parameters of the film resistors and capacitors made by the thin-film technology and the thick-film technology.

Comparison of the film resistors and capacitors performance characteristics is given in Tables 6.1 and 6.2. The information is from various Russian and foreign references.

The microwave made *RC*-EDP devices are formed in planar structures (see Fig. 4.2a). Their stripe-to-stripe capacitance can reach tenth of picofarads and the stripes resistance can reach to tens of Ohms. Hence, the specific time constants the

Table 6.1 Film resistors performance characteristics

Parameters	Thick film	Thin film
Resistances range, *R*	3 Ω...20 MΩ	10 OΩ...100 kΩ
Rating accuracy, %	0.5...5	0.1...0.5
Temperature resistance coefficient, 10^{-6}/°C	±100	±25
Operating temperatures range, °C	−55...+325	−55...+125
Maximum working voltage	100 V	50 V
Dissipation power, mW	125...500	100...200

Table 6.2 Film capacitors performance characteristics

Parameters	Thick film	Thin film
Specific capacitance, nF/cm^2	10...250	10...200*
Rating accuracy, %	0.5...5	0.1...0.5
Dielectric dissipation factor (at 1 kHz, 25 °C), %	0.5...2.0	0.04...1
Temperature capacitance coefficient, 10^{-6}/°C	±1000	±100
Operating temperatures range, °C	−55...+125	−55...+125
Maximum operating voltage	1000 V	100 V

Remark: Specific capacitance marked with asterisk (*) is measured for tantalum pentoxide (Ta_2O_5)- and titanium dioxide (TiO_2)-based capacitors

film *RC*-EDP can vary from 2×10^{-11} s/mm for a planar *RC*-EDP structure to 1 s/cm for a thick-film multilayered *RC*-EDP one.

The parameters of the film resistors and capacitors given above allow estimating the time constants and the operating frequency ranges of the thin- or thick-film *RC*-EDPs.

6.3 Performance Analysis and Parameter Tuning of *RC*-EDP Circuits

The film *RC*-EDP manufacturing methods are characterized by the electrophysical parameters and geometric errors of the fabricated elements. Unlike the resistors and capacitors of lumped parameters, where the rated values of these elements are accurately controlled, the *RC*-EDP distributed parameters enjoy a number of performance properties; for example, the ODHom *R*-*C*-0-EDP structures are characterized by the total resistance (impedance) and total capacitance that specify the *RC*-EDP time constant. As for the ODHet *R*-*C*-0-EDP structures, in addition to the above two mentioned properties, the width of the *RC*-EDP varies according to a *certain law* and their layers electrophysical properties vary over the element length. The *RC*-EDP devices with more complex structures (*R*-*C*-*NR*, 0-*C*-*R*-*NC*-0, ... etc.) possess more performance properties.

The key point of the two-dimensional *RC*-EDP adjustment depends on the variation of the structure design parameters. Changing the resistance and/or the capacitance may cause a variation of the electrical field distribution over the resistive layers, which also depends on the degree of layers overlapping. It provides changes in MFR and PFR performance of the fabricated devices. We shall not consider the adjustment strategies, but we shall consider just two techniques that can be used to vary the film *RC*-EDP performance parameters and characteristics.

Two basic adjustment methods are practically applied today to the electrical circuit elements and their functions: static and dynamic adjustments.

Static adjustment means changing the rated values of the circuit's elements. It is performed for discrete *RC*-EDP element that is not connected to the circuit. Such kind of adjustment can only be practically applied to the ODHom *RC*-EDP since the corresponding performance characteristics are specified by just two parameters (R and C), which can be tuned without influencing the electrical potential distribution of the resistive layer.

Applying the static adjustment to the ODHet *RC*-EDP is limited since it is not possible to change layers' width or varying the electrophysical properties of the layers over the *RC*-EDP length.

Dynamic adjustment means tuning the *RC*-EDP parameters during circuit operation. The *RC*-EDP parameters are varied in such a way that the desired output performance and accuracy are achieved. Dynamic adjustment can successfully be applied to both the ODHom and ODHet designs as well as to the two-dimensional designs of the *RC*-EDP elements.

In most cases of the ODHom *RC*-EDP applications (*RC*-filters, *RC*-oscillators, etc.) it is more important to provide an accurate time constant, τ_{RC}, for static adjustment. The value of the *RC*-EDP time constant depends on the specific resistivity of the layer, ρ_o, specific capacitance, C_0, and on its structural geometry. The value of the ODHom *RC*-EDP time constant is calculated from the following expression:

$$\tau_{RC} = RC = \rho_o C_o l^2$$

where l is the length of *RC*-EDP element.

It follows that the time constant τ_{RC} of the ODHom *RC*-EDP can be adjusted by both varying the electrophysical characteristics of the *RC*-EDP layers materials (ρ_o, ε) and the length of the *RC*-EDP.

It should be pointed out that discrete (static) adjustment of resistors and capacitors may not be suitable for all *RC*-EDP deterministic adjustment. The adjustment process should result in only changing the time constant. However, if the time constant variation causes a change in the distribution of the electrical field, it may yield an unpredictable effect of using such *RC*-EDP element in a certain device. The *RC*-EDP time constant adjustment methods are systematized according to the classification diagram shown in Fig. 6.10.

Electrical adjustment of the resistive layer resistance is basically used for thin-film resistors. The process means to pass, for a short-time, high-current pulses

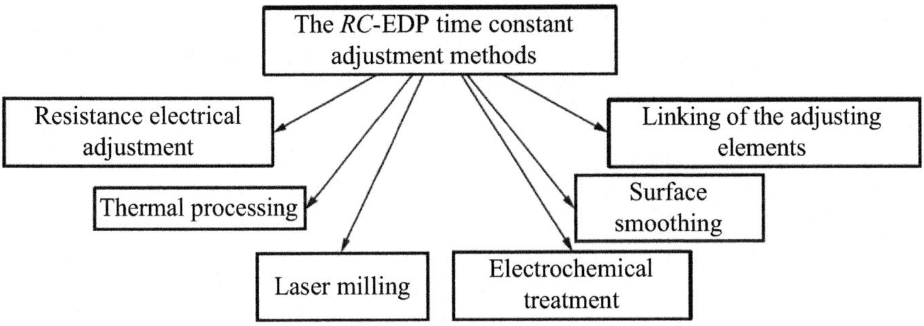

Fig. 6.10 Classification of time constant adjustment methods of one-dimensional *RC*-EDP structures

through the *RC*-EDP resistive layer. It results in changing the film structure and, consequently, its resistance.

Thermal processing is basically used to adjust the thin-film resistive layers resistance by means of exposing the *RC*-EDP to high temperatures in an oxygen or inert gas medium. In the first case, the high temperature (which however does not exceed the film evaporation temperature) causes oxidation of the resistive layer surface and an increase in the resistance of the film. In the second case, as a rule, the high temperature induces recrystallization of the film material when its structure becomes more regular causing thus reduction of the film resistance.

Laser milling can be used both for the thin-film and thick-film resistive layers. The resistance changes will be due to the spread-out of the laser beam on the film. Certain power of the laser radiation causes partial evaporation of the film material and growth of its resistance.

Electrochemical treatment is usually applied to such materials as tantalum that are used in the tantalum technology. Here, electrolytic anodizing creates the dielectric layer. The thickness of the conducting tantalum layer that serves as a resistor is correspondingly reduced, thus increasing the resistor resistance.

Surface smoothing can be applied basically to the thick-film resistors that possess relatively large thickness and that are actually thermally untreatable. Smoothing is done with tiny abrasive wheels. Felts, fine-grained sand papers, and diamond-covered disks are used here as abrasive materials.

Linking (connecting and disconnecting) the adjustment elements is the only adjustment method, which does not vary the materials electrophysical properties. The elements adjustment can be made in the form of linked plates located on the common resistive layer or in the form of the linked resistors located on the common plate. This adjustment method causes an increase of the *RC*-EDP overall dimensions.

It is practically reasonable to use partitioned *RC*-EDP in cases when the same element is intended to operate in devices with close parameter values; i.e., when using a set of electronic filters with different cut-off frequencies.

It is clear from the previous discussion that the resistive layer is basically subject to treatments. Thus, it is apparent that the availability of the resistive layer for a certain type of treatment ultimately determines the corresponding adjustment technology. In addition, the proposed adjustment methods vary the resistance in different ways: some of them just increase the resistance (smoothing, electrochemical treatment, laser milling) and others, depending on the treatment conditions, can reduce it (resistance electrical adjustment, thermal processing).

Dynamic adjustment of the device parameters within the desired ranges provides variation not just for the *RC*-EDP time constant, but also for the resistive layer surface electrical potential distribution. This can be done by changing the value and the configuration of the layers' overlapping area (see Fig. 4.14), the location of the taps of the resistive layer or the size and the location of the terminals relatively to the resistive layer (see Fig. 4.18), and by introducing cut-outs into the resistive and the conducting layers.

Variation of the overlapping area can technically be done by removing part of the conducting or the resistive layer by laser cutting or chemical etching of the layer material.

Variation of the location of the taps of the resistive layer or the size and the location of the terminals of this layer is done with laser cutting that reduces the conducting contact effective width. Laser cutting at one side of the contact change both the location and the size of the tap. Laser cutting at two sides of the contact just reduces the contact size.

Laser cutting can vary the distribution of the electrical potential not just over the *RC*-EDP top resistive layer but also over the layers located on the substrate surface (e.g., in case of the *RC*-EDP with such layer structures as *R-C-NR*, 0-*C-R-NC*-0 etc.). To provide this, it is required to create special adjustable areas of the resistive layer that would overlap the *RC*-EDP overlapping areas.

As long as the technical methods of adjustment are irreversible, the adjustment strategy should be verified using the test samples or the computer-simulated proper mathematical models.

Test Questions

1. Specify the types of the integrated circuits that differ in design-and-technology features and describe them in brief.
2. Specify the methods to create film layers of different materials.
3. Describe the thermal vacuum evaporation technology used to create films.
4. Describe the cathode evaporation technology used to create films.
5. Describe the vapor-phase deposition technology used to create films.
6. Describe the *RC*-EDP manufacturing photolithographic methods routing sequences.
7. Describe the techniques of the *RC*-EDP layers pattern forming by means of the contact masks and the moving masks methods.
8. Describe the thick-film *RC*-EDP manufacturing technologies.
9. Specify the design and technological errors of the *RC*-EDP manufacturing that should be taken into account when using the mathematical models of these elements.
10. Describe the *RC*-EDP parameters adjustment methods.

Chapter 7
Nonlinear Structure of *RC*-EDP Circuits

7.1 Presuppositions

In the previous chapters, we analyzed the possibilities of creating different *RC*-EDP design solutions to obtain the desired performance characteristics. The electro-physical performance characteristics of *RC*-EDP films are assumed constant during operation. The properties of the *RC*-EDP layers should be stationary and independent of external interference to guarantee stable and robust operation, which can only be achieved for fixed parameters of the *RC*-EDP-based devices.

The modern electronics, however, widely uses resistive and dielectric materials that vary during operation, which significantly vary their performance characteristics due to the effect of external fields (electrical, magnetic, thermal, mechanical, etc.). Hence, this feature can be capitalized to vary the *RC*-EDP performance characteristics during operation.

The effect of external fields on the *RC*-EDP can be performed in two ways: over the whole surface of the element (length) and at local areas of the element's surface (along the length). In the first case, if the intensity of the external field is the same on all areas of the *RC*-EDP, then its time constant will change and the frequency response will not. However, if the field intensity varies along the length of the *RC*-EDP (for example, temperature, voltage, illumination gradients), then such *RC*-EDP will be heterogeneous, and that will cause variation of both the time constant and the frequency response of the *RC*-EDP.

In the second case, local effect of external or internal fields on the properties of the *RC*-EDP layer materials can be performed, in particular, by managing systems' electrodes and by introducing the shading masks located on the element's surface. It yields heterogeneities in the *RC*-EDP that cause variations of the element's performance parameters and characteristics, even under the constant intensity of external effects.

Hereinafter the heterogeneities of the electro-physical properties of the layer materials and the *RC*-EDP geometry that are formed by external field's intensity but

A.Kh. Gil'mutdinov et al., *Fractal Elements and their Applications*, Analog Circuits and Signal Processing, DOI 10.1007/978-3-319-45249-4_7

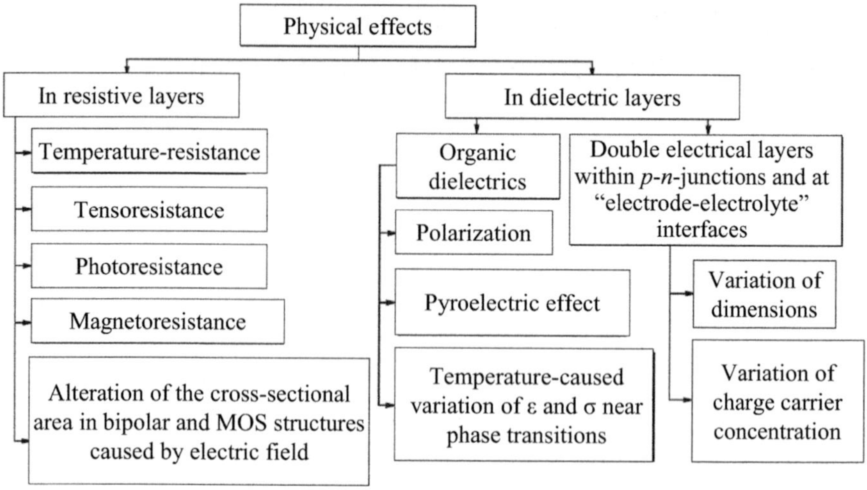

Fig. 7.1 Physical effects that generate the *RC*-EDP dynamical heterogeneities

not during their production, which can vary with time, will be called the *dynamical heterogeneities.*

Figure 7.1 represents a classification of the physical effects that can be used to generate the dynamical heterogeneities.

If the resistive and/or the dielectric layer of the *RC*-EDP is linearly dependent on the control action value, then such element will be called the *parametric RC*-EDP (an *RC*-EDP with time-variant performance characteristics and parameters).

If the nonlinearities in the performance of the *RC*-EDP layers depend on the value of the control action, then these elements will be called *nonlinear RC*-EDPs. Connecting *RC*-EDPs in a certain way and having them operate in a certain mode can provide signal amplification, selection, rejection, mixing, and other functions.

7.2 External Field Effect on the Fabrication of *RC*-EDP Resistive Parameters

7.2.1 Thermoresistive Effect

Thermoresistive effect refers to the variation of the resistance of a material caused by changes of its temperature. This effect is used as the major principle of operation of thermistors (thermoresistors with negative thermoresistor coefficient (TRC)) and posistors (thermoresistors with positive TRC). Both types of thermoresistors are made of semiconductors whose TRC variation ranges between (−65 °C; +70 °C).

The dependence of thermoresistors on temperature variations significantly affects their performance characteristics. Experiments proved that most negative

TRC thermoresistors produced in Russia could be approximated by the following expression, which is valid for the entire range of the operating temperatures or its part and practically provides acceptable accuracy:

$$R_{\mathrm{T}} = A \cdot e^{\frac{B}{A}}, \tag{7.1}$$

where R_{T} is the value of the thermoresistor resistance at temperature, T, K; A denotes a constant that depends on the physical properties and the thermoresistives' geometric layers; and B is the temperature-sensitivity coefficient (units in °C or K), which depends on the physical properties of the material used to fabricate *RC*-EDP elements.

If one assumes A to be constant, then for two different temperatures T_2 and T_1, (7.1) yields

$$R_{\mathrm{T}_2} = R_{\mathrm{T}_1} \cdot e^{B\frac{T_1 - T_2}{T_1 \cdot T_2}}. \tag{7.2}$$

Equation (7.2) allows one to calculate the value of the resistance of the thermoresistor at any temperature T_2 (at the operating range of temperatures) if the value of the constant B is known as well as the sample resistance at some temperature T_1.

The value of B is found experimentally by measuring the thermoresistor resistance at two temperatures T_1 and T_2. Solving for B from (7.2) yields

$$B = \frac{2.3026\ \Delta \log R}{\Delta(1/T)}, \tag{7.3}$$

where $\Delta \log R = \log R_{T_2} - \log R_{T_1}$, and $\Delta(1/T) = \Delta(1/T_2) - \Delta(1/T_1)$.

The TRC parameter, α_{T}, is determined in the common way as follows:

$$\alpha_{\mathrm{T}} = \frac{1}{R} \cdot \frac{\mathrm{d}R}{\mathrm{d}T},$$

and then it follows from (7.1) that

$$\alpha_{\mathrm{T}} = -\frac{B}{T^2}.$$

The resistance of many types of posistors over a wide range of temperature (several tens of degrees centigrade) strongly obeys an exponential law as follows:

$$R_{\mathrm{T}} = A \cdot e^{\alpha T},$$

where A is the constant, and α is the TRC at temperature T °C in absolute units.

Thin-film thermoresistive layers (film thickness of several nanometers) are created by depositing the materials on the substrate by cathode sputtering. As a result, the process is run in the medium with partially controlled pressure of oxygen to provide the oxide films except the metallic ones. However, the content of the deposited film cannot be controlled with proper accuracy since the oxide targets with different contents produce films that are chemically similar. The oxygen stoichiometry of the deposited film is easily adjusted by heating this film in the oxygen medium after deposition.

Noble metals (gold, palladium, or platinum) are usually used to make the contacts. Electrical properties of the thin-film thermoresistors often sufficiently differ from the properties of the bulk thermoresistors made of the same material. Specific layers' resistance depends on the film deposition conditions as well as on the oxygen content of the film. The single-film TRC is usually higher than that of the bulk thermoresistors; but their stability is much less.

One of the most commonly used methods to make thick-film thermoresistors is to screen-print the paste made of metal (Mn, Ni, Co, Cu, Fe) oxide mixture on the aluminum oxide substrate with the subsequent firing in the oxygen medium. The oxide paste can also comprise powder frit glass that forms the protective glaze cover on the film surface and, at the same time, increases adhesion of the film to the substrate. The thick-film thermoresistive layers acquired in such a manner have micrometer thickness. Variation of the material content and the thermoresistor size makes it possible to obtain the resistance values from 1 to $10^6\,\Omega$ at room temperature and with a TRC from -2 to $6.5\,\%/^\circ$C. Metallized electrodes are created at high-temperature processing using thick films made of such noble metals as platinum, gold, or palladium.

Posistors are also made of semiconductor ferroelectric ceramics. They provide a wide range of resistance variation (about 10^6 times).

7.2.2 Magnetoresistive Effect

Magnetoresistive effect means that the external magnetic field affects the ferromagnetic material where it changes the scattering electrons in different probability directions. In its turn, it causes changes to the electric resistance.

Film magnetoresistors have become widely used in recent years. The magnetic-sensitive element of these devices is made of ferromagnetic films (Ni-Co or Ni-Fe alloys). The maximum value of the thin-film magnetoresistor magnetic resistance refers to the zero external magnetic field; that is, the magnetic field reduces the device resistance.

The thin-film magnetoresistors are produced by a batch-type process similar to the thin-film hybrid integrated circuit production technique. The magnetic-sensitive material film is deposited on the substrate by electron-beam sputtering of a specially composited target. The pattern of the future magneto-sensitive element is created with photolithography or electron lithography.

The basic performance parameters of the magnetoresistors are the following:

R_0: initial resistance of the magnetoresistor at zero magnetic field.

Table 7.1 Parameters of thermoresistor films

Film material content	Resistance rating at zero induction, Ohm	Allowable deviation of the resistance rating, %, not exceeding	Relative magnetic sensitivity, %, at least (in the field with inductance, mT)	Temperature resistance coefficient at zero induction, %/°C
Nickel-cobalt	1200	+40	2.2 (20)	0.36
Nickel-iron	3500	+15	1.7 (20)	0.34

R_B: magnetoresistor resistance under the effect of magnetic field.

R_0/R_B: magnetoresistive ratio.

γ: relative magnetic sensitivity, which represents the relative variation of the magnetoresistor resistance caused by a certain induction magnetic field and defined as follows:

$$\gamma = \frac{(R_B - R_0)}{R_0} 100\%$$

where *f* is the cutoff frequency, the frequency of the modulated sinusoidal magnetic flux when the sensitivity of the magnetoresistor magnetic gets to 0.707 of the non-modulated flux sensitivity.

Typical values of some performance parameters of the thermoresistor films are given in Table 7.1.

The relative magnetic sensitivity, γ, of the magnetoresistor depends on the magnetic inductance, B, and the environment temperature, T_{env}.

Magnetoresistors are suitable for high-speed operations. Experiments showed that specially designed magnetoresistors could operate in a magnetic flux modulated with frequencies that exceed 10 GHz.

7.2.3 *Photoresistive Effect*

Photoresistive effect appears in semiconductor or dielectric materials. It means a change of the material electrical conductivity under an infrared, visible, or ultraviolet radiation.

The mechanism of photoconductivity can be explained as follows: the blacked-out semiconductor contains small amount of mobile charge carriers (electrons or holes) caused by thermal energy; that is, the semiconductor possesses an initial conductivity, σ_0, called dark conductivity:

$$\sigma_0 = q\left(n_0\mu_n + p_0\mu_p\right),$$

where q is the elementary electronic charge; n_0 and p_0 are the electron and hole concentrations of the mobile charge carriers within the semiconductor in the state of equilibrium, respectively; and μ_n and μ_p represent the mobility of the charge carriers.

Light induces growth of the mobile charge carriers. There are several possible mechanisms of their generation:

- Transfer of electrons from the valence energy band to the conductivity energy band, i.e., increase of the intrinsic conductivity
- Transfer of electrons from the valence energy band to the free impurity levels, i.e., increase of hole conductivity
- Transfer of electrons from the impurity levels to the vacant energy band, i.e., increase of electron conductivity

Hence, when exposing the semiconductor to the light radiation, the mobile charge carrier concentration increases by Δn and Δp with the consequent sufficient growth of its conductivity:

$$\sigma_0 = q\left[(n_0 + \Delta n)\mu_n + (p_0 + \Delta p)\mu_p\right].$$

The change of the semiconductor electrical conductivity caused by a light effect is called *photoconductivity*, σ_{Ph}, where

$$\sigma_{Ph} = \sigma - \sigma_0 = q\left(\Delta n_0\mu_n + \Delta p_0\mu_p\right).$$

Variation of the illumination intensity changes the semiconductor photoconductivity.

After the light is on, it takes the intensity of the charge carriers some time to get to a steady-state value that corresponds to the applied illumination intensity. It increases exponentially in time as described by

$$\Delta n(t) = \alpha\beta\tau N\left(1 - e^{-t/\tau}\right),$$

where N is the amount of photons that fall onto a unit area per second, α is the absorption coefficient that characterizes the energy absorbed by the semiconductor, β is the number of charge carriers appeared due to absorption of one photon, and τ is the lifetime of the nonequilibrium charge carriers.

If the duration of illumination is long enough: $t > (3\text{–}5\tau)$, then the nonequilibrium charge carrier concentration reaches its steady state. When electrons and holes are generated in pairs as a result of electron transition from the valence energy band to the conduction energy band (semiconductor intrinsic energy absorption), the number of the nonequilibrium holes equals the number of the nonequilibrium electrons as follows:

$$\Delta n_{\mathrm{st}} = \Delta p_{\mathrm{st}} = \alpha\beta\tau N.$$

In case of impurity absorption, one type of charge carriers are mainly generated, thus forming either electron or hole photoconductivity. This conductivity also changes in transition modes according to the exponential law. When removing illumination, then the nonequilibrium charge carrier concentration at $\Delta n \ll (n_0 + p_0)$ obeys the following expression:

$$\Delta n(t) = \Delta n_{\mathrm{st}} e^{-t/\tau}.$$

Gradual change of σ_{Ph} when switching the illumination on and off is called photoconductivity relaxation.

Photoresistors are often made of sulfides, selenides, and tellurides of different elements as well as the $A^{III}B^{V}$-type compounds. The infrared range of the operating photoresistors are made of PbS, PbSe, PbTe, and InSb, while the visible and near-ultraviolet ones are made of CdS.

Today, lead sulfide, PbS, is one of the most studied and developed materials used to create photosensors with maximum photosensitivity wavelength of $\lambda_m = (1.9–2.8)$ μm at 293 K. In recent years, the method of PbS film chemical deposition on various dielectric substrates has become a frequent practice. It demonstrates higher productivity and economy than the vacuum deposition method. The produced polycrystalline films are the two-phase system where the low-resistance *n*-type crystallites are surrounded with the opposite conductivity layers. Potential energy barriers appearing at the crystal boundaries facilitate efficient spatial separation of the photo-carriers. It results in a significantly increased photo-carrier lifetime and, consequently, the photosensitivity of the photoresistors. The same mechanism takes place for lead selenide, PbSe, that has a maximum photosensitivity wavelength of $\lambda_m = (3.6–3.8)$ μm at 293 K.

All research and development surveys in this field recently focused on the technically controlled methods to synthesize PbS and PbSe films. The variations of the growth conditions made it possible to control and change the performance characteristics of these films according to the changes in the photosensitivity critical wavelength λ_{c0}, response delay τ, and dark resistance R_D. It was found that the distribution and the density of nucleation centers of the crystallites (along with their size) that make the film were completely determined by the substrate material, its surface treatment, PbS and PbSe, and the oxygen-containing oversaturation phases.

The technology of an industrial production of lead-sulfide layers makes it possible to create photoresistive layers that have a time delay, $\tau = 1 \times 10^{-4}$ (293 K), and cause small pink or 1/*f*-type noises for the fast-response photoresistors. Such technology provides a fabrication of one-element devices whose dimensions vary from 6×6 mm^2 to 10×10 mm^2, while producing 40–100 μm multielement chains and matrices with up to 3000 different topologies. The heterogeneity of layers over an area of 25×25 mm^2 does not exceed 1–4 %. Similar processes are well established for PbSe and they provide manufacturing steps of the photoresistors with small and large amounts of elements (up to 256) and $\tau = 2$–5 µs.

7.2.4 Tensoresistive Effect

Tensoresistive effect means a change of the conducting and the semiconducting resistance caused by mechanical stress. Tensoresistors today are the basic strain sensors used to study the physical properties of materials and engineering structures. They are also used to produce many force, weight, pressure, motion, torque, and vibration sensors.

Film tensoresistors are made by supplementing the tensosensitive material with its subsequent deposition on the substrate. Such elements are made of nickel-based alloys (constantan, manganin, nichrome).

Tensoresistive effect in semiconductors is caused by, first of all, the change of charge carrier energy spectrum with deformation that varies the carriers' concentration and their effective mobility. This fact induces high sensitivity of the semiconductor tensoresistors (by one or two orders greater as compared to the metallic tensoresistors). However, the metrological performance characteristics (accuracy, linearity, temperature, and temporal stability) of the semiconductor tensoresistors are worse than those of the metallic ones. The semiconductor tensoresistors are made by mechanical treatment of semiconductor monocrystals (usually silicon) or by epitaxial growth of doped silicon on a monocrystal elastic element. The silicon-on-sapphire (SOS) technology has become the common practice and today it is used to produce the pressure sensors.

However, unlike heat, magnetic, or electromagnetic fields that can be localized at certain areas of the thermoresistive, photoresistive, or magnetoresistive layers, the mechanical deformation fields are impossible to localize in the small-sized tensosensitive layers. Therefore, the authors suggest that the tensoresistive effect will be used to create the *RC*-EDP only by making the measurement sensors in the form of distributed measurement media.

The MOS-structured *RC*-EDP also provides dynamical variation of their performance parameters. The resistor of the *RC*-EDP design structure depicted in Fig. 3.32 is implemented by the induced channel that appears under the effect of the electrical field applied to the metallic gate of the MOS structure (the *RC*-EDP plate). Varying this voltage causes a change in the channel width and, consequently, a change in the resistance of the *RC*-EDP resistive layer. If the voltage between the terminals of the resistive layer is constant, the current through the channel will cause variation of the resistor resistance along the channel length. Hence, it becomes possible to make the ODHet *RC*-EDP element with controlled heterogeneity of the resistive layer of the specific resistance.

7.3 External Field Effect on the Fabrication of *RC*-EDP Dielectric Layer Materials

While the conductors possess the metallic bonding based on sharing of all electrons, the insulators have the ionic, covalent, or molecular bonds that come from bound states of electrons. As a result, the electric field applied to the insulator actually does not induce conductivity, i.e., a transfer of the free electrons. It just causes shift of the bound electrons, i.e., *electrical polarization*. The latter involves almost all particles that make the insulator and that change their mutual position, while very few charged particles, that are relatively free, transfer the electrical charges through the insulator, thus causing an electrical conductivity.

Different effects on the insulator such as exposing to electrical, magnetic, or mechanical fields and variation of the insulator illumination or temperature cause various internal physical changes. The majority of these changes are caused by polarization which only suites insulators.

Dielectric material polarization induced with different fields causes appearance and variation of charges within the material structure. If the insulator geometry remains unchanged, the variation of the charge will result in variation of this layer's specific capacitance and, consequently, variation of the *RC*-EDP time constant. Hence, the relationship between the different external fields and the so-called polarization-driven electrical response of insulators will be analyzed first. Table 7.2 gives the information on insulators' properties in the "action–response" manner. Responses to electrical and mechanical fields and temperature variation are considered. Magnetic fields regulate the electrons' magnetic moments in the substance and do not induce sensible electrical response in most of insulators (non-ferromagnetic).

It should be noted that many interesting phenomena of physical and technical application of insulators (especially in electronics) appear due to combined action of several factors (i.e., more than one) on the insulator, for example, the combined effect of light and electrical field, or light and mechanical stress.

Electrical action on insulators causes a number of reversible and irreversible physical phenomena such as electrical and mechanical, and thermal actions. The first electrical response that should be noted is *polarization*, which imparts a specific electrical moment also called polarization, P_n. In a first approximation, the polarization is proportional to the electrical field, or $P_n = \varepsilon_0 \chi_{nm} E_m$, where E_m is

Table 7.2 Insulators' electrical responses to external field effect

Action	Electrical response
Electrical field E_m	Polarization $P_n = \varepsilon_0 \chi_{nm} E_m$
	Electrical conductivity $j_n = \sigma_{nm} E_m$
	Electrical breakdown
Mechanical stress X_{kl}	Direct piezoelectric effect $P_i = d_{iki} X_{kl}$
Temperature variation ΔT	Pyroelectric effect $E_i = q_i \Delta T$

the electric field, $\varepsilon_0 = 8.854 \times 10^{12}$ F/m is the dielectric constant, and χ_{nm} is the dielectric susceptibility constant.

Polarization is a reversible phenomenon since the insulator usually gets back to the non-polarized state after the electrical field is ceased. P_n and E_m are not proportional to each other under strong electrical fields due to the dielectric nonlinearity.

Electrical conductivity is the second important phenomenon, which appears in all insulators due to the effect of electric field. Small electric field induces electric current with a density proportion to the field itself, i.e., $j_n = \sigma_{nm} E_m$, where σ_{nm} is a specific cubic conductance (or just conductance). The electrical conductivity is a transfer phenomenon.

7.3.1 Pyroelectrical Effect

Pyroelectrical effect means the appearance of electrical charges on the polar dielectric surface due to a change of its temperature. Figure 7.2 depicts the pyroeffect mechanism by a simple model of a one-dimensional polar crystal comprised of a series of dipoles (polar molecules).

Every such dipole (denoted with an arrow) possesses a spontaneous electrical moment. If thermal motion is absent (an idealized case when $T = 0$ K), then all dipoles are strongly oriented and generate the maximum spontaneous polarization (see Fig. 7.2a). With an increase in temperature $T_1 > 0$ K, then for $T_2 > T_1$ a random thermal motion causes, first, partial disordering of the dipoles and, second, a thermal expansion of the crystal (see Fig. 7.2b,c). Both of these mechanisms reduce the spontaneous polarization, P_s, with temperature increase (see Fig. 7.2d). If the temperature variation is small enough, then the direct proportionality between ΔP and ΔT can be defined as $\Delta P = k\Delta T$, where k is a pyroelectric constant coefficient. Pyroelectric coefficients of various pyroelectric materials may significantly differ.

Ferroelectric materials are the most important pyroelectrics. They are represented by such compounds as triglycine sulfate and its isomorphs, lithium

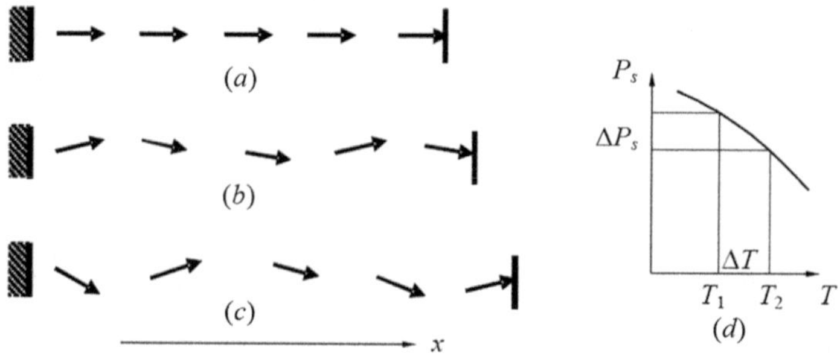

Fig. 7.2 Pyroelectric effect mechanism model: (**a**) $T = 0$ K; (**b**) $T_1 > 0$ K; (**c**) $T_2 > T_1$; (**d**) P_s polarization

niobate and tantalate (they are polarized with the current bias in the course of the crystal growth), ferroelectric phase potassium nitrate thin films (near 450 K), ceramic lead titanate, and lead zirconate titanate with various mixtures.

Pyroelectric detectors of thermal and electromagnetic radiation are highly sensitive and require no special cooling unlike the semiconductor detectors. Pyrodetectors feature low noise factor and a quite broad-spectrum band that makes it possible to detect radiations of 10^{10}–10^{20} Hz frequencies (from the centimeter waves to the X-rays and even the γ-rays).

Unlike the electret (substance with "weak" electric dipoles) residual polarization, spontaneous polarization signifies thermodynamically stable conditions of polar dielectrics. Spontaneous polarization, P_s, shows itself in the most interesting manner in ferroelectric materials where it can change its direction that is influenced by the electrical field. Hence, ferroelectric materials can be classified as a subclass of pyroelectrics where spontaneously polarized state is flexible and sufficiently changes under the action of electrical field, temperature, pressure (Fig. 7.3), and other factors.

When depolarizing, the electrical field actually remains the absolute value of P_s and changes its direction (see Fig. 7.3a). When the temperature or the pressure exceeds the critical values, P_s disappears (see Fig. 7.3b,c); for this reason ferroelectric materials can be called nonlinear pyroelectrics.

The poly-domain structure of ferroelectric crystals is extremely favorable since a mono-domain crystal generates an electric field in the environment (due to the fact that it is an electret).

High polarizing capability of the poly-domain ferroelectrics causes significant dielectric permittivity and motion in domains induced by electrical field that trigger the nonlinear properties of ferroelectrics.

Ferroelectric materials are commonly used to make electrical capacitors. Their high dielectric permittivity allows making capacitors with significant specific capacitance (such as ferroelectric ceramics with smeared phase transition, ferroelectric semiconductors). Temperature-caused variation of ε and σ near the phase transition makes it possible to create the temperature-controlled and temperature-compensating elements.

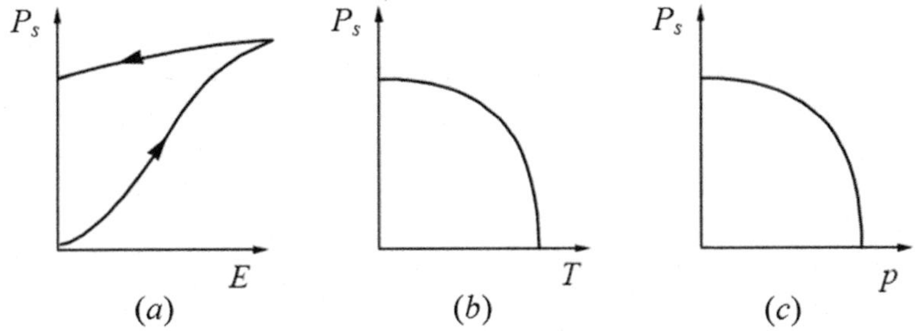

Fig. 7.3 Variation of spontaneous polarization of ferroelectric materials caused by (**a**) electric field; (**b**) temperature; (**c**) pressure

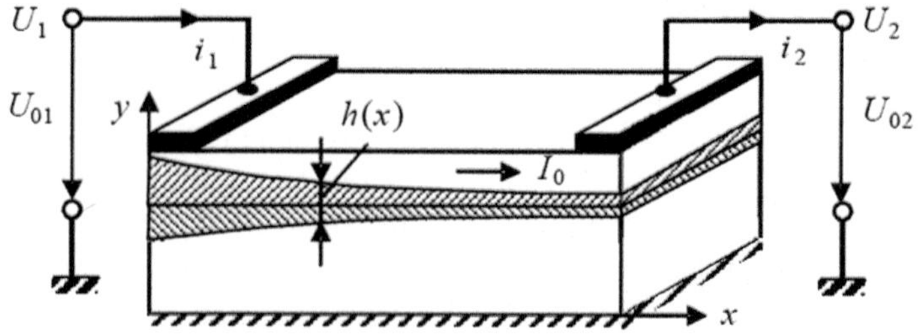

Fig. 7.4 Heterogeneous semiconductor *RC*-EDP

Dielectric nonlinearity, i.e., variation of the capacitance of the capacitor, due to the controlled electric field, is used to make electrically adjustable capacitors.

The variation of the *RC*-EDP layers' parameters due to changes in electric fields can be performed for the *RC*-EDP comprised of the semiconductor diffused layer and the barrier capacitance of the reverse-biased *p-n* junction (see Fig. 7.4).

Direct current, I_0, flowing through the resistive layer along the *p-n* junction generates differential voltage along this layer $\Delta U(x)$, which is added to the *p-n* junction reverse bias as follows: $U_{rev} = U_{02} - U_{01}$, i.e., $U_{rev} + \Delta U(x)$. The differential voltage, $\Delta U(x)$, induces a change in the thickness of the *p-n* junction depletion layer and, consequently, of the *RC* structure line parameters along the junction length. This is why the given structures are heterogeneous. The electrical parameters of this element can be controlled by adjusting the reverse voltage, U_{rev}, and the differential voltage, $\Delta U(x)$, which depends on the DC current, I_0.

It seems promising to make the controlled *RC*-EDP by combining the ferroelectric materials with other materials (for example, photoconductors and semiconductors). Assume that there is a "conducting plate–ferroelectric–photoconductor" structure. If the conductivity of the photoconductor is locally changed (for example, by illuminating some of its surface areas), then the resistance of the photoconductor-illuminated areas will significantly decrease and the applied voltage will almost be applied to the ferroelectric layer. The latter thereby is locally depolarized, thus creating a spatial heterogeneous capacitance along the distribution of the photoconductor layer illumination.

"Ferroelectric–semiconductor" structure is another example of this kind of hybrid structures. Direct contact of ferroelectric and semiconductor materials makes it possible to control the semiconductor electrical properties by varying the ferroelectric polarization. In particular, such "sandwiches" are used to make adaptive resistors that are widely specified by the short pulses of the driving voltage. It requires no further power consumption to maintain the specified conductivity.

The examples given above do not cover all of the physical effects of the materials that could be used to make parametric *RC*-EDP of desired characteristics. Practical implementations of the considered capabilities, however, require more studies in material science, and the fabrication technology and the physics of film and semiconductor layers.

7.4 Nonlinear *RC*-EDP Circuits

It is apparent that the distribution of the *p-n* junction-based semiconductor *RC*-EDP can be considered nonlinear for large signals (the signal voltage and the reverse voltage are of the same order). In this case, the voltage and current signals cause changes of the *RC*-EDP line parameters along the junction, thus influencing the electrical parameters of the entire element.

The multilayered *RC*-EDP films may also have nonlinear properties if the electro-physical characteristics of their layers nonlinearly depend on the field that is influencing them. Ferroelectrics, in particular, are typical materials that enjoy the aforementioned properties.

Reoriented polarization is the major distinctive feature of ferroelectrics and it is reflected in their definition. Figure 7.5a depicts a dielectric-hysteresis loop, which represents a clear nonlinear relationship. Just like it occurs in ferromagnetic

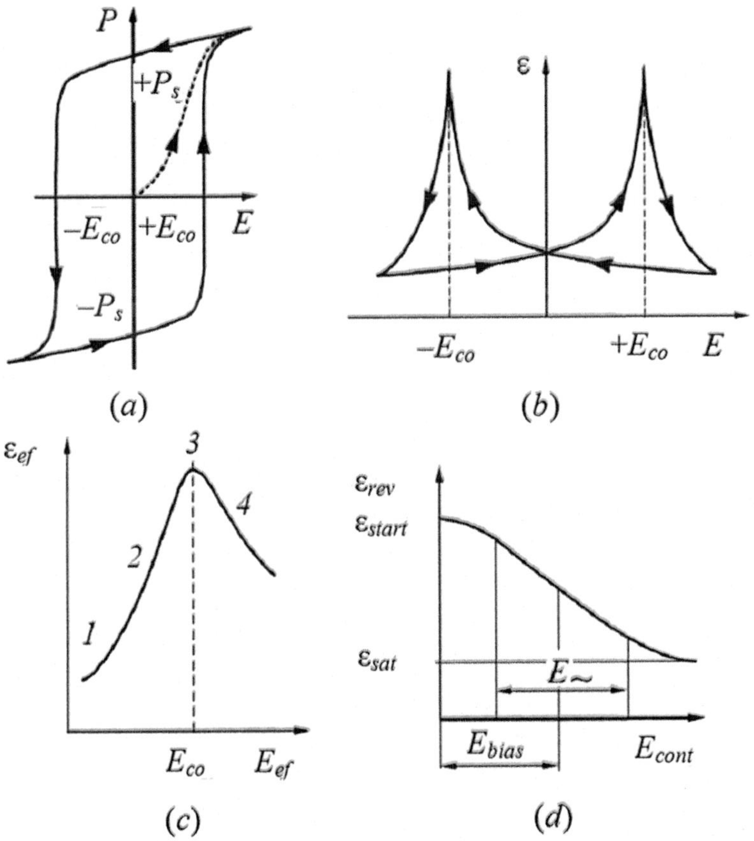

Fig. 7.5 Basic properties of ferroelectrics: (**a**) dielectric hysteresis; (**b**) dynamical nonlinearity; (**c**) effective nonlinearity; (**d**) inverse nonlinearity

materials, hysteresis here is specified with field reorientation as follows: first, the domains get oriented with growth of the electric field, E (dashed line), then the process saturates, and $P(E)$ growth rate decreases.

The reduction of the electric field causes a decrease of the just induced polarization; therefore, at $E = 0$, only spontaneous polarization, P_s, remains. An inversion of the field sign (see Fig. 7.4a) initially influences the induced polarization (the linear part of the hysteresis loop); but as soon as the field gets the coercive (breakdown) value $-E_{co}$, the spontaneous polarization rapidly changes from $+P_s$ to $-P_s$. Further variation of the field repeats the cycle so that $P(E)$ dependence has the form of a loop.

The relative dielectric permittivity obeys the law of the polarization due to the field ratio as follows: $\varepsilon = 1 + P/\varepsilon_0 E$, where $P = P_{ind} + P_{or}$ (P_{ind} is the induced polarization; P_{or} is the orientational (domain) polarization). Since ferroelectrics have $\varepsilon \approx P/\varepsilon_0 E \gg 1$, then, as depicted in Fig. 7.5b, the dependence of $\varepsilon(E)$ exhibits two spikes at $-E_{co}$ and $+E_{co}$. Hence, *the capacitance of a ferroelectric capacitor gets peak values two times per one period of an applied sine voltage.*

The effective dielectric permittivity, ε_{ef}, depends on the effective intensity of the electric field as shown in Fig. 7.5c. It has a peak that is not as sharp as that of the peaks of ε depicted in Fig. 7.5b. Observe that region *1* of $\varepsilon_{ef}(E)$ is linear for weak (small) electrical fields since the field intensity is not sufficient to cause domain depolarization; and the dielectric permittivity, in this case, is specified with P_{ind}. Region *2*, however, reveals a jump of ε_{ef} because P_{ind} is added to P_{or}. When almost all domains are involved into depolarization, $\varepsilon_{ef}(E)$ gets its peak in region *3*. Then, in region *4*, the dielectric permittivity decreases as the contribution of P_{ind} becomes insignificant, while P_{op} remains constant as the field intensity E keeps growing.

It is inappropriate to describe the ferroelectrics effective nonlinearity with series due to significant variation of $\varepsilon_{ef}(E)$. In real practice, this nonlinearity is characterized by the effective coefficient, $K_{ef} = \varepsilon_{max}/\varepsilon_{start}$. For example, the nonlinearity effective coefficient of the nonlinear capacitors (ferroelectric capacitors) could get to $K_{ef} = 20$. The effective nonlinearity of the ferroelectric capacitors is utilized to build frequency multipliers (here the higher order harmonics of the current are selected), and voltage suppressors (here the capacitance sharply rises with the voltage while the AC resistance decreases).

In addition to the effective nonlinearity, the ferroelectric capacitors possess reversible nonlinearity that is also used to build electronic devices. This reversible nonlinearity means variation of ε_{rev} that is exposed to the alternating field and the controlling electrical biasing E_{cont} (see Fig. 7.5d). The controlling field "strictly" orients the domains so that they do not influence a depolarization. It causes a reduction of ε_{rev} and its saturation at ε_{sat} level. The reversible nonlinearity coefficient of the ferroelectric capacitors gets up to 10. This coefficient is a ratio of the start dielectric permittivity and its saturation value: $K_{rev} = \varepsilon_{start}/\varepsilon_{sat}$. The reverse nonlinearity is used to electrically adjust the capacitance of the ferroelectric capacitor.

One may capitalize the dependence of $\varepsilon_{rev}(E_{cont})$, which can be used, for example, to perform amplitude modulation of the high-frequency signal. To provide linearity of the modulation characteristics, it makes sense to apply constant biasing E_{bias} in addition to the controlling alternating field. Unlike the effective nonlinearity, the reversible one can be used for a wider frequency range on the condition that the high-frequency signal voltage is small. In this case, the controlling field causes anisotropy of ε and saturates the polarization. If the nonpolar (paraelectric) phase is used, the reverse nonlinearity can be utilized also in the microwave frequency range.

Practice

1. Suggest some temperature-controlled *RC*-EDP designs.
2. Suggest some magnetic field-controlled *RC*-EDP designs.
3. Suggest some illumination-controlled *RC*-EDP designs.

Test Questions

1. Define the differences between the parametric elements with lumped parameters (R, L, C) and the parametric *RC*-EDP distributed ones.
2. Give the definition of the parametric *RC*-EDP.
3. Give the definition of the nonlinear *RC*-EDP.
4. Can the nonlinear *RC*-EDP be considered the parametric circuit elements?
5. Explain the physical mechanism of the thermoresistive effect.
6. Explain the physical mechanism of the magnetoresistive effect.
7. Explain the physical mechanism of the photoresistive effect.
8. Illustrate the relation between the dielectric polarization under the action of the controlling fields and variation of the *RC*-EDP performance parameters and characteristics.
9. Explain the physical mechanisms of controlling the parameters of the bipolar- or MOS-technology-based *RC*-EDP.
10. Give some examples of combined application of different physical effects capable of controlling the *RC*-EDP performance parameters.

Chapter 8
Applications of Fractal Elements

It was noted in Chap. 3 that due to the absence of single design of fractal *RC*-EDP elements, the corresponding fractal impedance transfer functions were recently realized using certain circuits based on the multi-section *RC* elements with lumped parameters (*RC*-ELP). These circuits were adequate only for certain number of sections and within certain frequency ranges.

The *RC*-EDP elements, as compared to the *RC*-ELP circuits, reduce the number of elements required to implement fractal impedances. This is a great advantage in itself. They also provide much more degrees of freedom (in the form of multiple design and schematic parameters with their combinations), to synthesize the functional radio-electronics (including the fractal ones) devices with the desired performance characteristics.

One way to apply the *RC*-ELP-based circuit theory is to replicate the physical processes such as heat transfer, or substances and charges in various continuum media. The authors suppose that the developed theory of the two-dimensional heterogeneous *RC*-EDP could make it possible to synthesize more accurate models of these processes. In its turn, this will allow to figure out more exact descriptions of physical, chemical, biological, and other types of systems.

This chapter considers the known examples of the models and the devices where the fractal elements are used (either based on the *RC*-ELP or in the form of the fractal impedance *RC*-EDP). The scientific and technical fields where the *RC*-EDP could significantly enhance the operation accuracy, parameter identification and control of the real objects and processes will be considered as well.

A.Kh. Gil'mutdinov et al., *Fractal Elements and their Applications*, Analog Circuits and Signal Processing, DOI 10.1007/978-3-319-45249-4_8

8.1 Application of Fractal Elements and Devices in Signal Processing

The term "fractal" has recently appeared in the literature and deals with the devices that comprise fractal elements or RC-EDP-based circuits that employ fractal impedances (fractional-order filters, fractional-order oscillators). Let us consider some of these fractional-order devices described in the literature.

The "fractal" or fractional-order low-pass filter is a typical one that requires fractal lumped or distributed elements to make such device. It is a voltage divider, one arm of which is a resistor and the load element is a fractional-order capacitor ("fractal" capacitor). As it was noted in Chap. 3, the current, $i(t)$, through the fractional-order capacitor (FOC), C_F, is proportional to the fractional derivative of voltage, $u_c(t)$, across its terminals; that is,

$$i(t) = C_F \frac{d^\alpha u_c(t)}{dt^\alpha}. \tag{8.1}$$

Then the impedance of the fractional-order capacitor FOC can be expressed as follows:

$$Z_{CF} = (\omega^\alpha C)^{-1} e^{\frac{-j\alpha\pi}{2}}. \tag{8.2}$$

The complex transfer function of the voltage divider will be

$$T(j\omega) = \frac{jZ_{CF}}{R + jZ_{CF}}. \tag{8.3}$$

Substituting from (8.2) into (8.3) with the subsequent necessary transformations makes it possible to derive the following MFR and PFR of the fractional low-pass filter:

$$|T(\omega)| = \frac{1/RC}{\sqrt{(\omega^\alpha \cos(\alpha\pi/2) + 1/RC)^2 + (\omega^\alpha \sin(\alpha\pi/2))^2}} \tag{8.4}$$

and

$$\varphi(\omega) = -\text{arctg}\left(\frac{\omega^\alpha \sin(\alpha\pi/2)}{\omega^\alpha \cos(\alpha\pi/2) + 1/RC}\right). \tag{8.5}$$

The graphs of the magnitude and phase frequency responses (MFR and PFR) from (8.4) and (8.5), respectively, are shown in Fig. 8.1 for $\alpha = 1.5$.

The graphs show that when $1 < |\alpha| < 2$ the transfer function MFR grows like a second-order low-pass filter with a pole Q-factor $Q_{pole} > 1$.

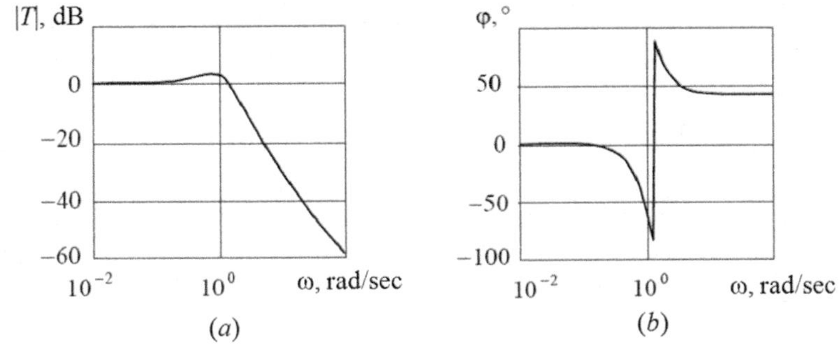

Fig. 8.1 Frequency responses of the fractional-order low-pass filter for $\alpha = 1.5$: (**a**) MFR and (**b**) PFR

The works of W. Ahmad, R. El-Khazali, and A. C. Elwakil (2001) consider the fractional-order self-oscillator made on the basis of a classical Wien-bridge oscillator circuit by replacing the conventional capacitors with the fractal ones. This work shows that the frequency of such an oscillator can be adjusted by varying the fractional order, α, without changing the value of the fractal capacitors.

As long as there is no complete physical implementation of the fractal capacitor as depicted in (8.2), it can be approximated by an *RC*-EDP with small resistance of the resistive layer.

Fractional-order oscillator (FOO) using *RC*-EDP-based fractal impedances was proposed by A. Oustaloup instead of the FOC as early as 1981. He built the fractional-order oscillator using an operational amplifier with frequency-dependent negative feedback approximated by a fractal impedance of two-port equivalent network based on the Foster second form 5-segment circuit (the Foster circuit diagram is depicted in Fig. 3.11). It was found that the oscillating signal distortion factor was minimum when $\alpha = 0.458$. The device with $\alpha = 0.5$ was then considered and exhibited higher linearity of the modulation frequency within the low-frequency band while maintaining a high value of a modulation index.

Today all the known *RC*-ELP-based circuits can be replaced with the *RC*-EDP-based circuits with a fractional order, α, that takes values between $0 < |\alpha| < 1$. Consequently, it is obvious that such replacement will cause a reduction of the device size and an increase of their reliability, and provide any value of order, α, index within the specified range for a broad frequency band. This can be achieved, not by increasing the discrete elements, but by introducing a heterogeneous-distributed *RC* medium. These heterogeneities are made by either specific fabrication methods or exposing these devices to controlling fields of different physical nature (see Chap. 7).

8.2 Application of the *RC*-EDP to Model Fractal Processes and Objects

8.2.1 Presuppositions

Objects and processes that are described by fractional-order differential equations are very common. This fact was proven by various published works (for example, check out the wide reference list of the following works: A. A. Potapov. Fraktaly v radiofizike i radiolokatsii (Fractals in Radiophysics and Radiolocation) (In - Russian)-Moscow, Universitetskaya kniga, 2005.—pp. 848; V. V. Uchaykin. Metod drobnykh proizvondykh (The fractional derivatives method) (In - Russian)—Ulyanovsk: Artishok, 2008—pp. 512).

For example, the current-voltage relationship of electrochemical systems that exhibit unsteady-state diffusion obeys fractional-order differential equation. This is caused by the rough surface of the electrodes or the heterogeneous parameters of the electrochemical reaction.

Fractional-order differential equations can be used to describe abnormal diffusion in heterogeneous media. Fractional-order derivatives make it possible to explain the memory effect in insulators and describe the electrical conductivity processes in complex composite materials such as plastics and cements.

In hydrodynamics for example, when water percolates through a porous barrier, the flow rate is proportional to a fractional-order derivative of the dynamical pressure at the water-border interface. In deformable solid mechanics, fractional-order derivatives are used to simulate viscoelastic behavior of various materials.

Experts have been describing the behavior of different natural systems by means of electrical models for many decades. Such models can either be a one-dimensional and reflects the prevailing flow of processes or a multidimensional one that describes set of objects and processes. The structure of such electrical models and their elements reflects the structure of the simulated objects, their internal processes, and parameters. For this reason, such kind of models are used to study the complex nature of physical, chemical, and biological systems, and the properties and structure of composite materials, etc.

Notice that a quite number of prevailing models of these systems are based on the *RC*-ELP. When simulating the transfer functions of such processes (diffusion, heat propagation, percolation, etc.), the equivalent circuits often comprise Warburg impedances that are expressed in the following way:

$$Z_W(\omega) = \frac{\sigma}{\sqrt{\omega}}(1 - j),$$

where σ is the Warburg constant.

This implies that the equivalent circuits based on the existing conventional electronic components cannot always describe the processes accurately, which take place in real physical media for a broad frequency band.

A number of distinctive examples of physical processes and objects of fractal nature will be considered in subsequent sections. These models have been developed as a result of experimental data analysis.

8.2.2 Alumina Electrolytic Capacitor Model

Alumina electrolytic capacitors (AEC) are widely used to build power supplies. The design structure of an AEC is depicted in Fig. 8.2a.

A detailed image of the AEC structure of an electrochemical system is given in Fig. 8.3a. This formation has three distinctive basic areas that determine most of the capacitor's parameters (capacitance, series resistance, operating voltage, etc.). These areas are the anode aluminum foil *1* with Al_2O_3 film *2* covering its surface, an electrolyte *6* with a separating paper *5* saturated by the electrolyte, and the cathode aluminum foil *4* that performs hard-line coupling of the electrolyte and the external terminal of the capacitor.

In real electrolytic capacitors, the total length of the aluminum stripes is always much greater than their width and particularly than their thickness. In addition, the electro-physical properties of the aluminum foil and the insulator should be as one-dimensional and isotropic as possible. Therefore, the R_1-Y-R_2 line structure and its graphical symbol are correspondingly depicted in Fig. 8.2b, and Fig. 8.2, which can be used as an equivalent circuit of the *RC*-EDP devices.

Aluminum oxide (Al_2O_3) thin film is the capacitor's insulator. The operating electrolyte is its second plate. To increase the capacitance, the anode foil surface is etched to make micropores in the material.

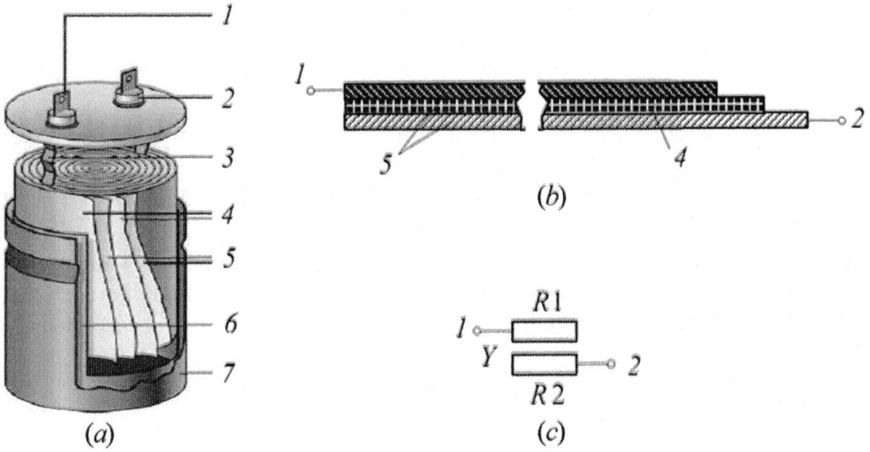

Fig. 8.2 Alumina electrolytic capacitor: (**a**) design structure; (**b**) capacitance layer structure; (**c**) R_1-Y-R_2-line graphical symbol: (*1*) positive terminal; (*2*) negative terminal; (*3*) connecting lead; (*4*) insulator; (*5*) aluminum stripes; (*6*) aluminum casing; (*7*) plastic insulator

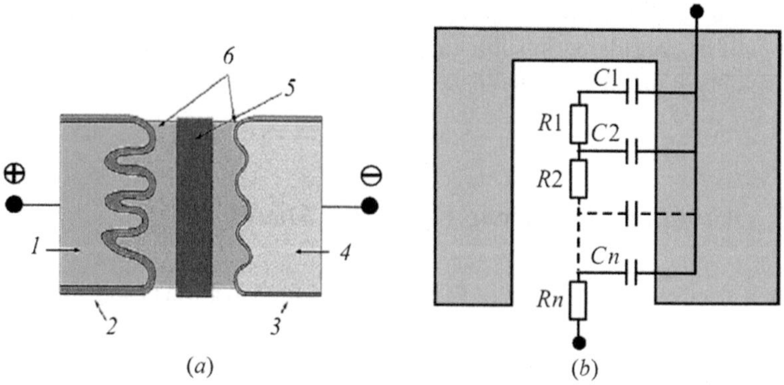

Fig. 8.3 AEC: (**a**) electrochemical system structure (cross section); (**b**) equivalent circuit of a pore in a porous electrode: (*1*) anode foil; (*2*) oxide Al_2O_3 foil; (*3*) natural oxide film of the cathode foil; (*4*) cathode foil; (*5*) separate paper saturated with electrolyte; (*6*) electrolyte

Ion transfer processes in electrolyte located in pores today are simulated with a model in the form of a ladder *RC* circuit with lumped parameters (De Levi model) as shown in Fig. 8.3b.

Therefore, the transversal conductance between the anode and cathode foils of the AEC electrochemical system can also be simulated by an *RC*-EDP-containing circuit.

8.2.3 Lithium-Ion Polymeric Battery Model

A simplified structure of a lithium-ion polymeric battery is given in Fig. 8.4. It contains a positive electrode (Li_xCoO_2) and a negative electrode ($Li_{x-1}C_6$) placed into the ionic conductive electrolyte, which is carbonic acid solution of lithium salt ($LiPF_6$).

P. L. Moss studied the impedance spectrum of lithium-ion polymeric batteries and in 2008 derived the following equivalent circuit as depicted in Fig. 8.5

The resistor R_s here simulates the electrolyte resistance. The first *Foster form-I* *RC* ladder circuit ($R_1C_1, \ldots, R_nC_m$) simulates the migration of the lithium ions through the multilayer surface films of the electrodes. The resistor R_{CT} and capacitor C_{DL} simulate the parameters of the dipole layer on the electrode borders, and the solid-state diffusion of lithium ions is approximated by the Warburg element $Z_W(\omega)$.

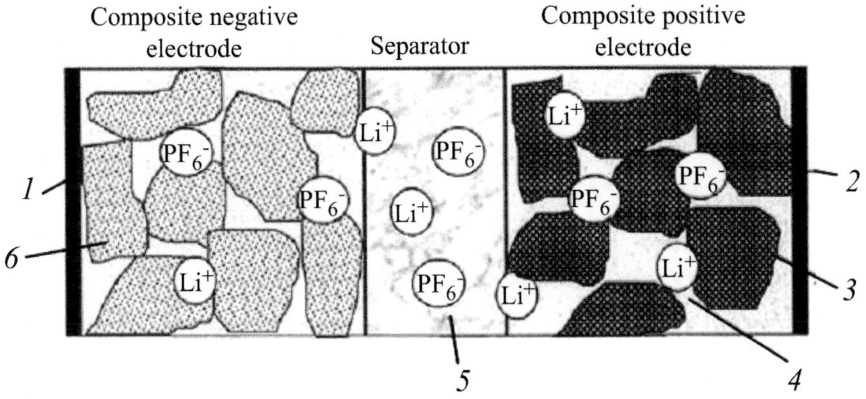

Fig. 8.4 Lithium-ion polymeric battery structure: (*1*) current collector (Cu); (*2*) current collector (Al); (*3*) active material (Li_xCoC_2); (*4*) filling agent, bonding agent, electrolyte; (*5*) electrolyte; (*6*) active material ($Li_{x\text{-}1}C_6$)

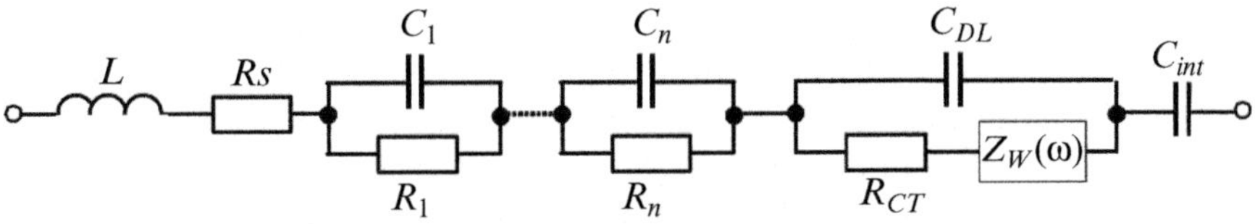

Fig. 8.5 Equivalent circuit of the lithium-ion polymeric battery input impedance

8.2.4 Modeling of Composite Heterogeneous Materials

Heterogeneous materials that comprise conducting and insulating phases represent a large class of artificial and natural materials. Such systems as polycrystalline and amorphous materials including ceramics, polymers, composition materials, wet cement, and electronic and ionic conductors have been studied to reveal the fact that all of them possess similar frequency dependence of conductivity and dielectric permittivity.

In 2006, C. R. Bowen and D. P. Almond studied the frequency response of the two-dimensional *RC* circuit input impedance that was intended to simulate a heterogeneous microstructure with conducting and insulating areas. As an example, Fig. 8.6a displays a ceramic composite image (Al_2O_3–TiO_2) obtained using a scanning electron microscope. This image can be approximated with a two-dimensional *RC* circuit.

The gray areas in the image represent Al_2O_3 phase and are simulated with capacitors C_{Al2O3}. The light areas signify TiO_2 phase and are simulated with resistors R_{TiO2}.

The equivalent circuit of an arbitrary cross section of the sample material is a two-dimensional *RC*-EDP-based grid as shown in Fig. 8.6b. In this circuit, the

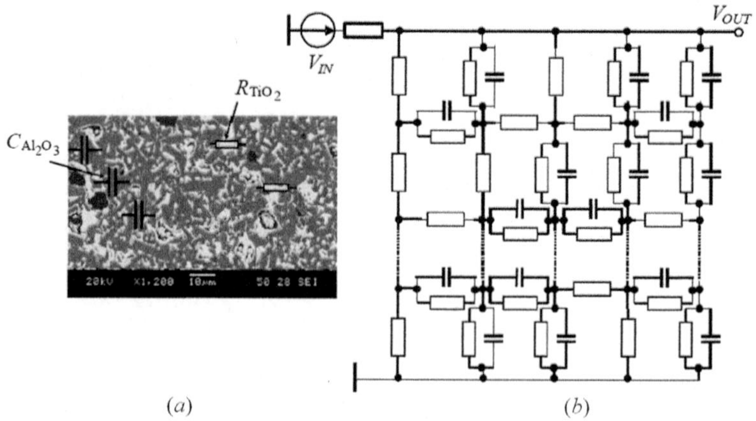

Fig. 8.6 (**a**) Microstructure of a composite Al_2O_3–TiO_2 material, and (**b**) its two-dimensional electrical equivalent circuit

heterogeneous materials are represented either by purely resistive branches or by capacitive branches bypassed with resistors. The measured frequency response of the sample impedance was used to perform random choice of the type of every branch and the values of their elements in the circuit synthesis.

One published work of the authors of the present book also shows the relation between the heterogeneous medium conductivity frequency response and the medium structure. The heterogeneous medium was synthesized based on the *R-C-G*-0-layered *RC* medium (see Chap. 4). The objective of the experiment is to demonstrate that the constant phase shift φ_{Zc} (as well as α) of the two-pole-connected *R-C-G*-0-layered EDP phase frequency response (PFR) of the input impedance is strongly influenced by the parameters of its resistive and conducting layer topology.

In the experiment, the area of the initial two-dimensional homogeneous *RC*-EDP was divided into 400 finite elements with 20 × 20 grid. The following three values of α were specified within the two decades' frequency range: $\alpha = 0.4$ ($\varphi_{Zc} = -36° \pm 1°$), $\alpha = 0.5$ ($\varphi_{Zc} = -45° \pm 1°$), and $\alpha = 0.59$ ($\varphi_{Zc} = -53° \pm 1°$). Genetic algorithm (GA) was used to perform the synthesis.

Examples of the synthesized structures of the heterogeneous medium with the corresponding input impedance PFR are depicted in Figs. 8.7 and 8.8. In the images of the *RC*-EDP topologies, the light squares symbolize the removed resistive layer (0-layer), the dark ones represent the removed conducting layer (*R*-layer), while the gray ones define the *R-C-G*-0-layered structure.

Since the genetic algorithm, GA, integrated by the synthesis program is based on random search methods, the topologies synthesized with this method will vary under the same initial conditions. Hence, in order to estimate some integral performance characteristics, ten models were synthesized for every specified value of α. The results of statistical processing of the corresponding parameters at 0.95

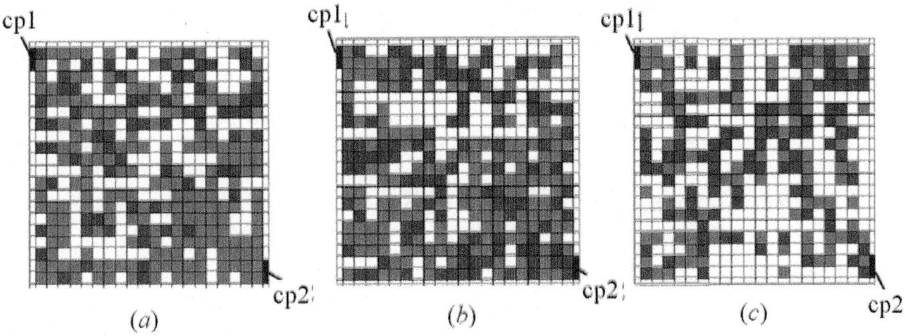

Fig. 8.7 Topology of two-dimensional homogeneous *RC*-EDP synthesized to model input impedances with phases: (**a**) $\varphi_{Zc} = -36° \pm 1°$; (**b**) $\varphi_{Zc} = -45° \pm 1°$; (**c**) $\varphi_{Zc} = -53° \pm 1°$

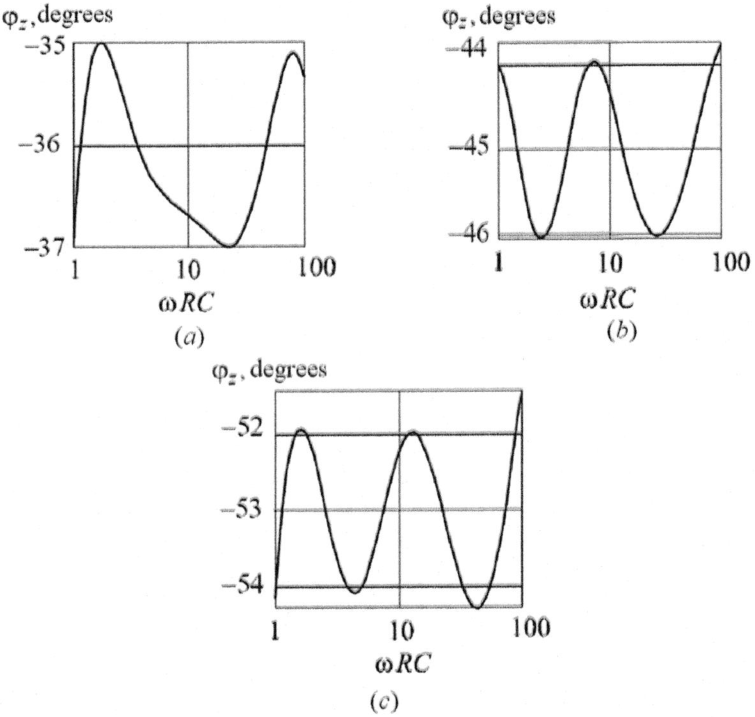

Fig. 8.8 The PFR curves of the two-dimensional homogeneous *RC*-EDP that corresponds to the topologies in Fig. 8.7: (**a**) Fig. 8.7a; (**b**) Fig. 8.7b; (**c**) Fig. 8.7c

confidence probability are given in Table 8.1 (the numbers are rounded up to the closest integer number of the finite elements).

Here $\bar{N}_R$, $\bar{N}_{RC}$, and $\bar{N}_0$ are, respectively, the average number of the *R*-layered, the *R-C-G*-0-layered, and the material-free-layered finite elements that are included into the percolating cluster (a structure consisting of the *R*- and *R-C-G*-0-type finite

Table 8.1 Dependence of the two-dimensional *RC*-EDP structural parameters on α

α	$\bar{N}_{\mathrm{R}} \pm \Delta$	$\bar{N}_{\mathrm{RC}} \pm \Delta$	$\bar{N}_{0} \pm \Delta$
0.4	47 ± 4	228 ± 10	125 ± 8
0.5	94 ± 7	132 ± 7	174 ± 8
0.59	76 ± 6	86 ± 6	202 ± 8

elements that form a path to electrical current flow between contact pads cp_1 and cp_2 of the two-dimensional homogeneous *RC*-EDP) (see Fig. 8.7); and Δ is the confidence intervals for the mean values related to every type of the finite elements.

It is evident from the table that there is a statistically significant relationship between the value of the fractional-order power index, α, of the frequency-dependent input impedance and the parameters of the *RC*-EDP-based model structure. For example, an increase of α causes a decrease of both the amount of the *R*-*C*-*G*-0-layered finite elements and the density of the occupied area of the *RC*-EDP material. In addition, Fig. 8.7 shows that the percolating cluster becomes more dissected with minimum number of parallel paths of current flow.

Hence, the simulation method of the heterogeneous structures based on the impedance measurements that were proposed by C. R. Bowen and D. P. Almond can be implemented by means of the *RC*-EDP. The simulating fractal elements via continuum (continuous) medium give more accurate representation to fractal components than lumped parameters of discrete resistors and capacitors.

8.3 Application of *RC*-EDP Devices in Fractional-Order Dynamics and Control

Fractional-order differential equations that describe fractal processes and objects allow one to design regulators of either integer- or fractional-order dynamical systems to adjust system responses. Various published works on this subject provide theoretical and experimental proofs of the fact that the fractional-order controllers are much more efficient in controlling the fractional-order dynamical processes in comparison to the integer-order ones. Hence, it is obvious that the fractional-order regulators find use in various fields of science and technology.

8.3.1 *Fractional-Order* $\mathrm{PI}^{\lambda}\mathrm{D}^{\delta}$ *Controller Design*

Fractional-order *PID* controllers, shortened as $PI^{\lambda}D^{\delta}$, are a generalization of the integer-order one. This shortened designation of the regulator means the names of the blocks that form the regulator; that is, *P* represents a proportional component, *I* denotes integration, *D* is a differentiating, and λ and δ are the fractional orders of the integrators and differentiators, respectively. Adding two new more parameters makes it difficult to tune $PI^{\lambda}D^{\delta}$ regulators. Hence, one has to carefully look for new design algorithms to design or implement such controllers.

The transfer function of such regulator $C(p)$ in the complex frequency p space is as follows:

$$C(s) = \frac{U(s)}{E(s)} = K_{\mathrm{p}}\left(1 + \frac{1}{T_{\mathrm{i}}s^{\lambda}} + T_{\mathrm{d}}s^{\delta}\right); \quad (\lambda, \delta > 0), \tag{8.6}$$

where K_{p} is the proportionality constant, T_{i} is the integration constant, T_{d} is the differentiation constant, $E(s)$ is the error input signal, and $U(s)$ is the controller response.

The structure of the fractional-order *PID*-regulator (hereinafter designated as PI^{λ} D^{δ}-regulator) is similar to the classical integer-order *PID*-regulator structure since it also is comprised of three blocks connected in parallel: the proportional, the integrating, and the differentiating (see Fig. 8.9).

Assume that $e(t)$ is the input signal and $u(t)$ is the controller response. Then the time-domain differential equation of the $PI^{\lambda}D^{\delta}$ controller follows from (8.6) as follows:

$$u(t) = K_{\mathrm{p}}e(t) + \frac{1}{T_{\mathrm{i}}}D_{\mathrm{t}}^{-\lambda}\, e(t) + T_{\mathrm{d}}D_{\mathrm{t}}^{\delta}\, e(t) \tag{8.7}$$

where the system is assumed to be causal; that is, the system is at rest for $t > 0$.

Obviously, it converges into a classical integer-order *PID* controller at $\lambda = 1$ and $\delta = 1$. A fractional-order lead, PD^{δ}, regulator emerges when $\lambda = 0$ and $T_{\mathrm{i}} = \infty$, while a fractional-order lag controller, PI^{λ}, is obtained when $T_{\mathrm{d}} = 0$. All these types of regulators are particular cases of the fractional-order regulator that is more flexible and provides better adaptability to the dynamical properties of the fractional-order systems. These capabilities of the $PI^{\lambda}D^{\delta}$ regulator are illustrated in Fig. 8.10 that defines both the regions of the integer-order controllers (Fig. 8.10a) and the fractional-order regulators (Fig. 8.10b).

From (8.7), the $PI^{\lambda}D^{\delta}$ regulator possesses five adjustable parameters whereas the classical integer-order *PID* regulator has only three such parameters. It is evident that the increased number of degrees of freedom should stimulate enhancement of the controller's ability to solve different control problems.

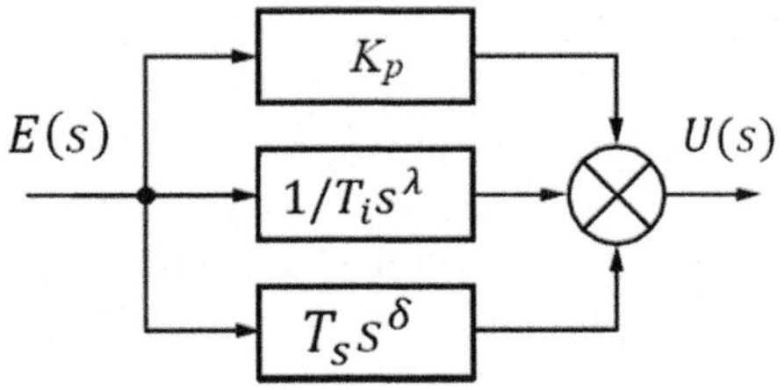

Fig. 8.9 The structure of the fractional-order $PI^{\lambda}D^{\delta}$ regulator

Fig. 8.10 Definition regions of the parameters of (**a**) the integer-order *PID* controllers; (**b**) the fractional-order $PI^{\lambda}D^{\delta}$ controllers

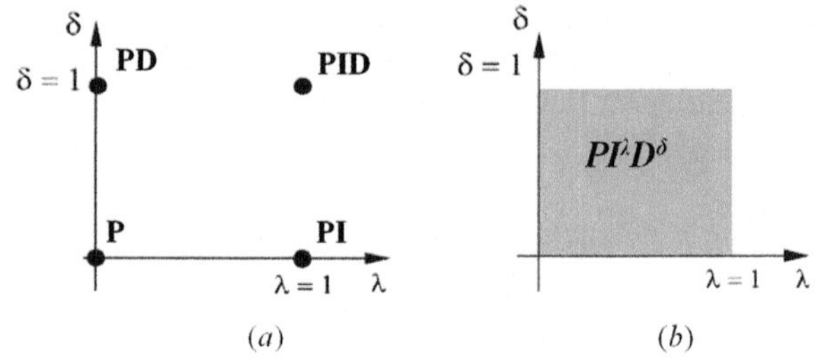

The problem of the $PI^{\lambda}D^{\delta}$ lies in the realization of the lead and the lag components of the controller. Chapter 3 presented three algorithms to approximate the Laplacian fractional-order operators. The special biquadratic structure of *El-Khazali* given by (3.46) and (3.46) can only be used to generate a realizable controller since it approximates both differentiators and integrators.

To simplify the design procedure of a $PI^{\lambda}D^{\delta}$ regulator, we consider the following modified form for (8.6) (see *El-Khazali*, 2013); that is,

$$\bar{C}(s) = \frac{U(s)}{E(s)} = K_{\mathrm{p}}\left(2 + \frac{1}{T_{\mathrm{i}}s^{\lambda}} + T_{\mathrm{i}}s^{\lambda}\right); \quad (T_{\mathrm{i}} = T_{\mathrm{d}}, \quad \text{and} \quad \lambda = \delta > 0) \tag{8.8}$$

$$\bar{C}(s) = K_{\mathrm{c}}\frac{\left(1 + T_{\mathrm{i}}s^{\lambda}\right)^2}{s^{\lambda}}; \quad K_{\mathrm{c}} = \frac{K_{\mathrm{p}}}{T_{\mathrm{i}}}, \tag{8.9}$$

thus reducing the number of parameters to be determined to three. One may think from the first glance that (8.9) is not realizable, but substituting from (3.47a) for $\omega_i = 1$ rad/s yields the following fourth-order transfer function that approximates (8.9); that is,

$$\bar{C}(s) = \frac{K_{\mathrm{c}}\left((a_2 + T_{\mathrm{i}}a_0)s^2 + a_1(1 + T_{\mathrm{i}})s + (a_0 + T_{\mathrm{i}}a_2)\right)^2}{(a_0s^2 + a_1s + a_2)(\,a_2s^2 + a_1s + a_0)}, \tag{8.10}$$

where from (3.47b),

$$\begin{aligned} a_0 &= \lambda^{\lambda} + 3\lambda + 2 \\ a_2 &= \lambda^{\lambda} - 3\lambda + 2 \\ a_1 &= 6\lambda\, \tan\left((2 - \lambda)\pi/4\right) \end{aligned} \quad . \tag{8.11}$$

Therefore, from (8.9) and (8.10), finding K_{c}, T_{i}, and λ completely characterizes the special form of the given $PI^{\lambda}D^{\lambda}$ regulator.

Let φ_{p} be the phase margin of the uncontrolled plant (system), and $\varphi_{\mathrm{c}} \equiv Arg\left(\bar{C}(s)\right)$ be the phase contribution of the controller in (8.10); then for a desired phase margin of the controlled system, $\varnothing_{\mathrm{m}}$, the required controller phase angle, φ_{c}, is calculated from

$$\varphi_c = \varnothing_m - \varphi_p - \pi. \tag{8.12}$$

Since the phase margin of $\text{Arg}(s^\lambda) = \lambda\pi/2$ at $\omega_i = 1$, the contribution of (8.10) at the crossover frequency, ω_{cg}, yields

$$T_i = \left\{ \frac{\tan\left(\frac{\varphi_c}{2}\right) + \tan\left(\frac{2+\lambda}{\pi/4}\right)}{\tan\left(\frac{\varphi_c}{2}\right) - \tan\left(\frac{2+\lambda}{\pi/4}\right)} \right\}; \quad \varphi_c \neq \frac{(2+\mu)}{\pi/8}. \tag{8.13}$$

Equivalently, for a specific integrator time constant, the phase angle, φ_c, can also be found from

$$\varphi_c = 2\tan^{-1}\left\{ \frac{(T_i + 1)\tan\left(\frac{2+\lambda}{\pi/4}\right)}{T_i - 1} \right\}; \quad T_i \neq 1. \tag{8.14}$$

An initial tuning value of the controller gain, K_c, can be found by solving (8.10) at $\omega = \omega_{cp}$ to meet the desired gain margin of the controlled system:

$$K_c = \frac{\frac{g_m}{g_p}\left\{(a_0 - a_2)^2 + a_1^2\right\}}{(a_0 - a_2)^2(1 - T_i)^2 + a_1^2(1 + T_i)^2}; \quad g_p \neq \infty, \quad \text{and} \quad g_p \neq 0, \tag{8.15}$$

where $g_p = \left| G_p\left(\frac{j\omega_{cp}}{\omega_{cp}}\right) \right|$ of the uncontrolled plant, and g_m is the desired gain margin.

Remark 1 If $g_p = 0$, or undefined, one may replace it by a finite value; $g_p = M$, and start tuning K_c to meet the design requirements.

The design algorithm of the proposed $PI^\lambda D^\lambda$ can be summarized as follows:

1. For a desired phase margin, $\varnothing_m$, add a safety margin (few degrees) and use (8.13) to determine the required order, λ, of the $PI^\lambda D^\lambda$.
2. If $\lambda > 1$, determine the number of modules, n, that should be cascaded such that $\frac{\lambda}{n} < 1$, where n is the number of biquadratic modules of (8.10) that can be cascaded.
3. Determine the required regulator phase angle, φ_c, from (8.14).
4. Calculate the integral time constant, T_i, from (8.13).
5. Calculate the controller gain, K_c, using (8.15).
6. Tune K_c and T_i as necessary to meet the design requirements.

Example 2 Consider an open-loop system described by $G_p(s) = \frac{0.25}{s^3 + s^2}$. Suppose it is required to design a $PI^\lambda D^\lambda$ controller to stabilize the unity feedback open-loop system, and to obtain a phase margin $\varnothing_m = 45°$ and a gain margin of $g_m \geq 20$ dB.

Observe that the given system has two poles of the open-loop transfer function at the origin. It is highly unstable and considered here to test the viability of our proposed controller.

The phase angle of the open-loop system is $\theta_p = -205.4°$, at $\omega_{cp} = 0.4752$, $\omega_{cg} = 0$, $g_p = 0$, and a phase margin $\varnothing_{m_o} = -25.4°$.

Since the desired phase margin is 45°, one may add additional safety margin of about 10° to ensure achieving the desired design from the first attempt; thus, choose $\varnothing_m = 55°$. From (8.12), (8.14), and (8.15), the order of the $PI^{\lambda}D^{\lambda}$ controller is equal to $\lambda = 0.8935$, and the expected phase angle contribution of the controller is $\varphi_c = 80.4155°$. From (8.11), $a_o = 5.5848$, $a_1 = 6.3422$, and $a_2 = 0.2238$. Since $\lambda < 1$, a single module of a biquadratic controller centered at $\omega_{cp} = 0.4752$ is described by

$$\left(\frac{s}{\omega_{cp}}\right)^{0.8935} \approx \frac{5.5848\, s^2 + 3.0136\, s + 0.0505}{0.2238\, s^2 + 3.0136\, s + 1.2610}.$$

Solving (8.13) gives $T_i = 5.9739$. Now to find K_c, since $g_p = 0$, one may replace g_m/g_p in (8.15) by a nonzero value, say, $g_m/g_p = 3$. Solving (8.15) gives a controller gain $K_c = 0.0776$. Using these values resulted in a controlled system with $\varnothing_m = 26°$, and a gain margin, $g_m = 17.3$. The phase margin is not met yet. Now, slightly letting $T_i = 10$ and reducing the controller gain, $K_c = 0.0085$, yield the following final form of a fractional-order *PID* controller:

$$\bar{C}(s) = \frac{26.72\,\mathrm{s}^4 + 31.6\,\mathrm{s}^3 + 11.02\,\mathrm{s}^2 + 0.9953\,\mathrm{s} + 0.02652}{1.25\,\mathrm{s}^4 + 17.5\,\mathrm{s}^3 + 16.14\,\mathrm{s}^2 + 3.952\,\mathrm{s} + 0.06371}.$$

Figure 8.11 shows the Bode diagram of both the controlled and the uncontrolled system with $\omega_{cg} = 3.18$ rad/s and $\omega_{cp} = 0.434$ rad/s. Clearly, with minor adjustment to K_c and T_i, the design requirements have been successfully met using a $PI^{0.8935}$ $D^{0.8935}$ controller.

The unit-step response of both the controlled and the uncontrolled system with unity feedback is shown in Fig. 8.12. It is clear that the proposed *PID* controller has stabilized the system. The proposed controller can be realized by a two-port network using Foster or Cauer methods.

The following introduces a Matlab algorithm of a special type of *El-Khazali* $PI^{\lambda}D^{\lambda}$ controller design:

```
% This program is used to design a PID controller using El-Khazali
method
% PI^lambda D^delta
clear all
% define the system to be controlled
   n=0.25;d=[1  1  0  0];
   sys=tf(n,d);
```

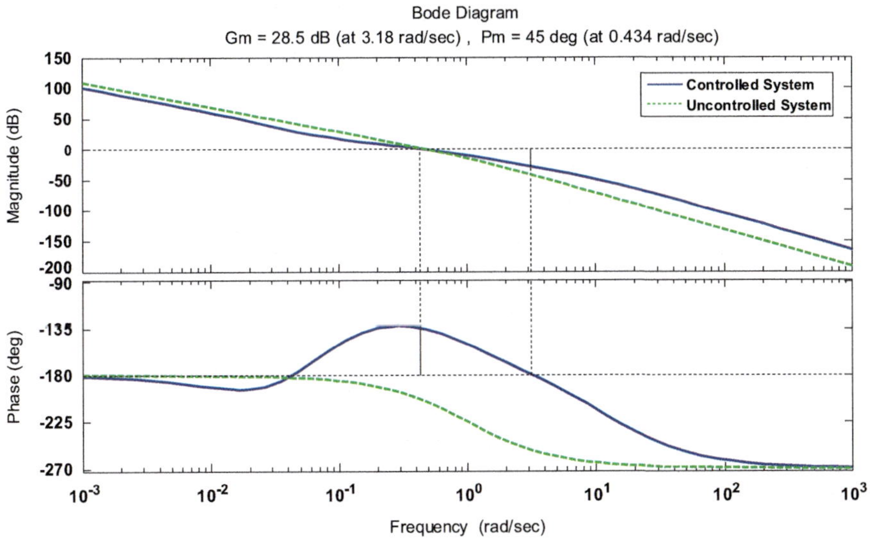

Fig. 8.11 Frequency response of the controlled system using a $PI^{0.7778}D^{0.7778}$ controller

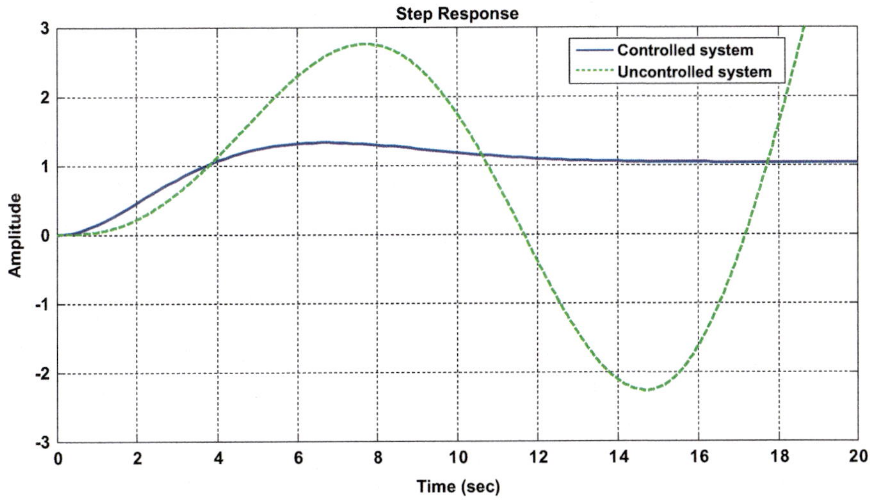

Fig. 8.12 Step response of the controlled and the uncontrolled system

```
% Evaluate the phase and the gain margins
   [Gm,Pm,Wcg,Wcp] = MARGIN(sys)
%margin(sys)
%%
% Assume a desired gain margin of gd dB
   gd= 20;
```

```
% calculate the order of the controller
%  the desired phase margin
     pmd=pi/4;   pmd=180*pmd/pi;
% Determine the angle of the system
      Phs=Pm-180;
% Calculate the required controller phase angle
          phc=(pmd+10-Phs-180);
% change it to Radians
          phc=phc*pi/180;
% Calculate the order of the PID controller
         lam= phc/90;
         %
%  Determine the coefficients of the biquadratic
%  controller. ao, a1, a2, from (8.11)
% or the approximation of the Laplacian operator
%
     ao=lam^lam+3*lam+2;
     a2=lam^lam-3*lam+2;
     a1=(ao-a2)*tan((2-lam)*pi/4);
  %
     N=[ao a1*Wcp a2*Wcp^2];
     D=[a2 a1*Wcp ao*Wcp^2];
%%
% calculate the time constant of the controller
%
     Ti=(tan(phc/2)+tan((2+lam)/(pi/4)))/(tan(phc/2)-tan((2+lam)/
(pi/4)));
%  Calculate the expected phase of the controller
    phcc=(2*atan(((Ti+1)*tan((2+lam)/(pi/4))/(Ti-1))))*180/pi
%%
%  controller gain
if (Gm == 0),
    Gm=60;
%   Calculate the gain
%    set the tuning ration Gm/gd = 60/20 = 3
end
    Kc= (Gm/gd)*((ao-a2)^2+a1^2)/((ao-a2)^2*(1-Ti)^2+a1^2*(1+Ti)^2)
%%
%Tune the Controller gain and the integrator time constant dynamics
%controller TF
Ti=10;Kc=0.0085
  NN=conv((D+Ti*N),(D+Ti*N));DD=conv(N,D);
```

```
Gc=Kc*tf(NN,DD);
 figure(1)
 margin(sys*Gc)
 % closed-loop system response
YC=feedback (sys*Gc,1);
  % Open-loop uncontrolled feedback system
YO=feedback(sys,1);
T=20;
figure(3)
step(YY,T), hold
step(YO,T)
```

8.3.2 Fractional-Order Differentiators and Integrators (Fractional Operators)

Electronic computers today can perform fractional integration and differentiation operations (FID) by means of the fractional calculus mathematics (see Chap. 2). It can also be done with special digital devices. However, there are unresolved problems regarding consideration of the initial conditions when performing these calculations by means of the real-time computer software. Furthermore, computer software implementation of FID in the systems that control the fast-moving objects or chain chemical reactions requires very-high-speed digital processors.

If the *RC*-EDP-based analog devices perform FID operations, then the processing rate will significantly increase and the price of such analog device will be lower than the price of the corresponding computer implementation by more than ten times. That is why design development and production of the cheap analog FID devices could be a good alternative to the expensive high-speed signal processors.

The analog FID devices can be built based on the same circuits as the conventional operation amplifier-based integer-order integrators and differentiators. The difference is that the capacitor is replaced with a two-port fractal impedance (a fractal element).

Figure 8.13 depicts examples of fractional-order analog integrator and differentiator. The fractal element here is designated as Z_F and its fractal impedance conforms to expression (3.2) (valid within a certain frequency range).

Today there are known alternatives of the FID analog devices based on the following fractal impedance implementations described in Chap. 3: electrochemical cells, *RC*-ELP-based circuits, and film *RC*-EDP. It is preferable to use the option of the *RC*-EDP fractal impedance design as fractal elements Z_F that specify the FID order. Such solution provides the following advantages:

- Better mass and dimensions parameters

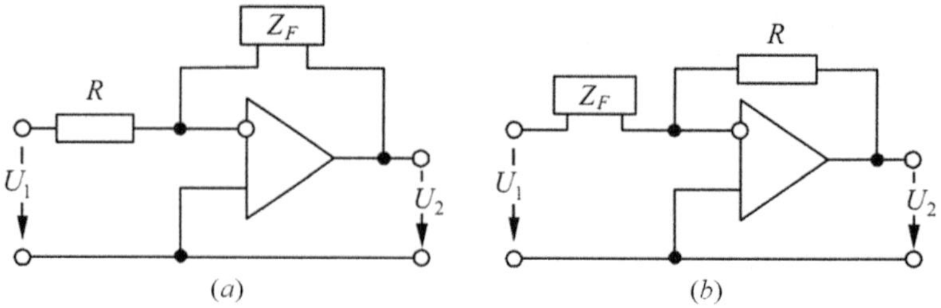

Fig. 8.13 Circuit diagrams of the operation amplifier-based fractional-order integrator (**a**) and differentiator (**b**)

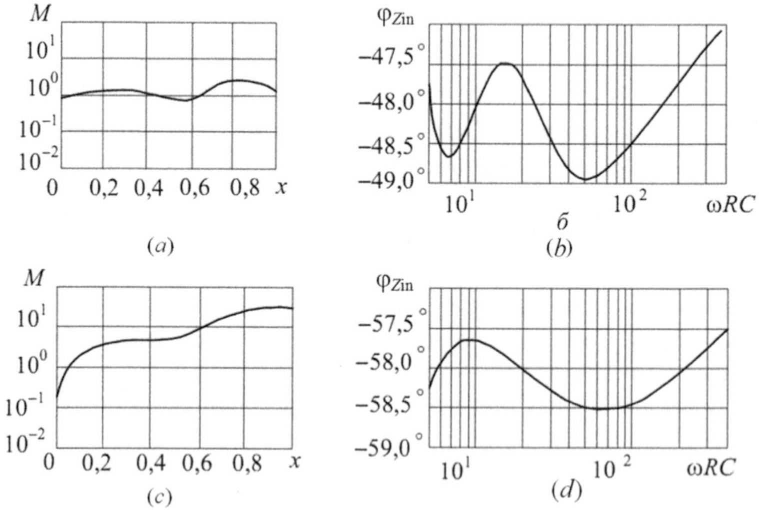

Fig. 8.14 Variation law of the input impedance PFR of *R*-*C*-0-layered *RC*-EDP resistive $\rho_\square$ square layer: (**a**) $\rho_\square$ variation ($f_1(x)$ dependence); (**b**) the input impedance PFR ($\varphi_c = -48.25 \pm 0.75°$) corresponds to $f_1(x)$; (**c**) $\rho_\square$ variation ($f_2(x)$ dependence); (**d**) the input impedance PFR ($\varphi_c = -58 \pm 0.5°$) corresponds to $f_2(x)$

- Compatibility with film elements or integrated circuits of semiconductor production technology
- Implementation of fractal impedances over a wide frequency range
- Possibility of accurate adjustment of the fractal element performance characteristics
- Capability of dynamical adjustment when using the parametric *RC*-EDP

The possibility of varying the *RC*-EDP parameters was discussed in Chap. 7 by applying different physical fields on the *RC*-EDP layers that are made of materials of special electro-physical properties, which are sensitive to these fields. Figure 8.14 reflects the possibility of estimating the PFR of the *RC*-EDP input impedance by specifying a certain variation law of the line resistance along the element's length.

Figure 8.14a,c depicts two cases of variation laws of the relative square layer resistance, $M(x) = \rho_s(x)/\rho_{s0}$, ($\rho_{s0}$ is the square layer resistance at no external action) of the rectangular *R*-*C*-0-layered EDP. Figure 8.14b,d shows the corresponding PFR of the input impedance of the *RC*-EDP when the latter is connected to option 1 (see Table 5.1).

The figure shows that variation of $M(x)$ law by means of the external controlling field is able to cause change in the order of FID. In the present example, this change is from $\alpha = 48.25°/90° = 0.5361$ to $\alpha = 58°/90° = 0.6444$ with errors not exceeding $\pm 1\,\%$ within a two frequency decades. All this is a good illustration of how the parametric *RC*-EDP can be used to implement the FID devices with dynamical adjustment of the operation orders.

8.4 Promising Application Areas of the *RC*-EDP Devices

RC-EDP devices can be used in many different applications. Here we list just few of them.

Hybrid computers: It is well known that analog computers are the most fast to solve differential equations in real-time mode. However, separately used analog or digital computers are not efficient enough to solve the problems that regard controlling the fast-moving objects, optimization and simulation of control systems, making complex simulation systems, etc. This caused relevance of hybrid computers that combine the required solving speed provided by analog computers and the desired solving accuracy provided by digital computers.

Arbitrary fractional-order *RC*-EDP-based integrators and differentiators and special block diagrams of analog functions can be introduced into the analog units of hybrid computers in order to increase the speed, to provide flexibility of choice of solving methods and to increase marketability in the special electronic computer market.

Measurement of physical quantities: The major parts of the processes that take place in technical and natural systems are the spatial-temporal scalar or vector fields of different nature. Measurement and control of the condition of such systems have a dominant role in information management of controlling the modern manufacturing complex processes and technical systems. Professor Yu. K. Evdokimov (Kazan National Research Technical University named after A. N. Tupolev) proposed in 1990s a concept of the distributed measurement environment (DME) along with the corresponding continuum structure detectors to perform measurements of the spatial variation of different natural fields. DME means an artificial material environment that is capable to perform measuring (sensing) and transfer functions. DME were made of the *RC* structures with distributed parameters implemented with electrochemical cells, semiconductor *p*-*n*-junctions, or *R*-*C*-0-layered film EDP.

Certain combinations of electro-physical and physical-chemical parameters of the measurement environment determine its sensitivity to certain measured fields.

For example, application of thermo-, photo-, piezo-, magnetoresistive, semiconductor, and other materials as the components of the measurement environment makes it possible to provide direct or indirect sensitivity of this environment to thermo-physical, hydro- and gas-dynamic, light, force, electromagnetic, and other various fields.

The list of problems that could be solved due to the *RC*-EDP capabilities is not limited with the given examples of these elements' application. The classification diagram given in Fig. 8.15 provides a more exhaustive overview of prospects of application of the *RC*-EDP and their mathematical models.

This classification is based on the performed overview of the *RC*-EDP and their mathematical model application areas and on analysis of the works on fractional calculus application areas in science and technology that have been published from the early 1980s until the present days.

8.5 The Universal Simulation Environment with Distributed *RLCG* Parameter Method

The proposed classification of application areas of the *RC*-EDP and their mathematical models means that every certain application area along with the problem to solve implies the corresponding structure of the *RC*-EDP layers and certain requirements to the elements' performance characteristics that may significantly differ from those regarding other application areas. This may cause development of highly specialized methods for analysis and synthesis of the *RC*-EDP and of the corresponding highly specialized software.

This makes it necessary to develop a basis of universal design for analysis and synthesis of the fractal elements based on multilayered *RC* medium. The latter should take account of the needs to implement certain structures of the *RC*-EDP layers for various areas of application. They also should take into account the available technology of making such elements and adjusting their performance parameters and characteristics.

The uniformity of the fractal element analysis and synthesis methods based on a universal design model should be based on a universal mathematical model that should meet the following requirements:

- To be adaptable to the EDP with different layer structures
- To cover as much types of the heterogeneities that could be used to synthesize the *RC*-EDP desired performance characteristics as possible
- To be applicable both to the parametric and to the nonlinear fractal elements
- To be suitable for analysis and synthesis of the fractal elements within the broad frequency range (from Hz unit fractions to GHz units)

A two-dimensional n-layered *RLCG*-EDP is proposed as such design basis. Therefore, according to the agreed notations, it is called the *generalized*

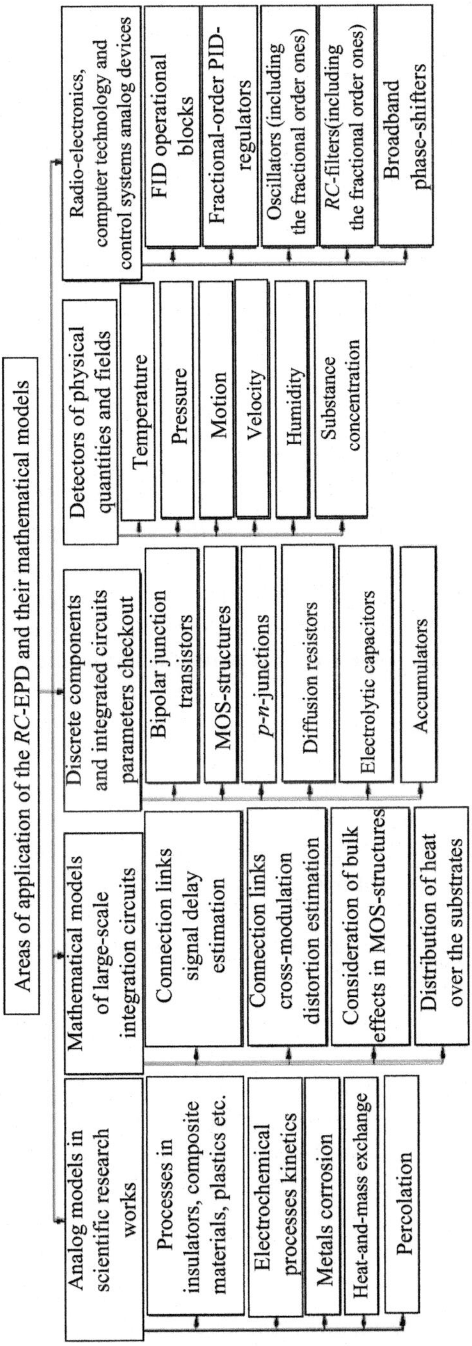

Fig. 8.15 Classification diagram of areas of application of the *RC*-EDP and their mathematical models

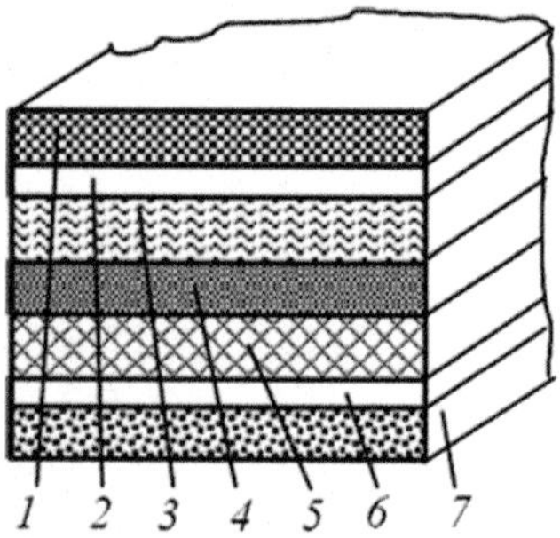

Fig. 8.16 A fragment of the generalized *RCG*-EDP: (*1, 4, 7*) resistive layers R_1, R, R_2, respectively; (*2, 6*) resistive layers G_1, G_2, respectively; (*3, 5*) dielectric layers C_1, C_2, respectively

RLCG-EDP with the corresponding layer structure. Expression "generalized" reflects the fact that such *RLCG*-EDP can be used to create the simulation elements with different layer structures: linear, parametric, and nonlinear. In particular, when the operating frequency range of the *RC*-EDP makes it possible to neglect the inductances of the conducting and resistive layers, the proposed design basis will become a two-dimensional *RCG*-EPD (the *generalized RCG*-EDP) device.

Due to the capabilities of the multilayered *RC*-EDP fabrication technology (see Chap. 6, for instance) and the complexity of the existing simulation objects, this book considers just the generalized *RCG*-EDP ones. A fragment of such element consisting of seven layers with different conductivity is depicted in Fig. 8.16.

The external terminals of the *RC*-EDP with such structure are supposed to be contacting resistive layers R_1, R, and R_2; the resistive layers G_1 and G_2 are supposed to have their own functions only when the specific material of the resistive layers R_1 and R_2 surface resistances tends zero. In the rest of cases, the resistances of the layers G_1 and R_1 and G_2 and R_2 will form the aggregated resistive layers R_1' and R_2', respectively. The latter will possess some resulting specific parameters.

Such generalized *RCG*-EDP can obviously exist only as a mathematical model, which is supposed to be transformed in the course of synthesis of the *RC*-EDP with the desired performance characteristics. Also it can exist in the form of an electrical model that reflects the character of the simulated process or object.

Practice

1. Derive the fractional-order integrator transfer function expression (see Fig. 8.11a) taking into account that Z_F is represented by expression (3.2).
2. Derive the fractional-order differentiator transfer function expression (see Fig. 8.11b) taking into account that Z_F is represented by expression (3.2).

Test Questions

1. Give examples of objects and processes that could be simulated by means of the *RC*-EDP.
2. Illustrate that the aluminum oxide electrolytic capacitor model can be represented in the form of an element with distributed parameters.

3. What elements of the equivalent circuit of the ion-lithium polymer battery input impedance could be replaced with the *RC*-EDP?
4. Give examples of the objects with heterogeneous composite medium.
5. What design variants of the *RC*-EDP can be used to simulate the objects representing a homogeneous composite medium?
6. Give the definition of the fractional-order *PID*-regulator. What is the reason of higher control efficiency of these regulators in comparison to the integer-order *PID*-regulators?
7. Depict the fractional-order integrator and differentiator circuit diagrams.
8. Describe the methods to vary the order of the fractional-order differentiation and integration operations.
9. Describe the possible areas of application of the *RC*-EDP and of their mathematical models in science and technology.
10. Design a $PI^{\lambda}D^{\lambda}$ to stabilize a unity feedback system with an open-loop system described by $G_{\mathrm{p}}(s) = \frac{10}{s^3+s^2+s}$ to a achieve a phase margin $\varnothing_{\mathrm{m}} = 45°$.

Chapter 9
Modeling of *RC*-EDP Circuits

It follows from the previous chapters that the *RC*-EDP is an electrical circuit with distributed parameters. Therefore, the *RC*-EDP analysis problem does not differ from the problem of analyzing distributed parameter circuits. Thus, the goal is to find out the equilibrium equations of the electric circuit, to determine the distribution of currents and voltages along the coordinates of the circuit, and to find the current-voltage relationship at the circuit's terminals.

This chapter considers one of the possible approaches to derive the state equation of such type of electric circuits. We develop a partial differential system of equations for the two-dimensional generalized *RCG*-EDP devices.

Consider Fig. 8.16, and for the sake of simplicity, assume that the two upper resistive layers R_1-G_1 are integral layers with no jumps of conductivity at the interface between them. The same kind of assumption is applied for the lower two resistive layers G_2-R_2.

9.1 General Model of *R*1-*C*1-*R*2-*C*2-*R*3-Layered *RC*-EDP Devices

Consider the layer structure of a typical *RC*-EDP device as depicted in Fig. 9.1. It represents fragment of *R*1-*C*1-*R*2-*C*2-*R*3-layered *RC*-EDP structure. Assume that the design of the given *RC*-EDP device is symmetrical for the resistive layer *R*2 under certain conditions. In addition, it will also be assumed that the higher and the lower parts of the structure interact only via the common resistive layer *R*2.

For simplicity, the following assumptions are considered:

1. The thickness of each layer of the structure is constant.
2. Each layer is considered linear and isotropic according to its electrical properties but it is considered heterogeneous over the structure surface.

A.Kh. Gil'mutdinov et al., *Fractal Elements and their Applications*, Analog Circuits and Signal Processing, DOI 10.1007/978-3-319-45249-4_9

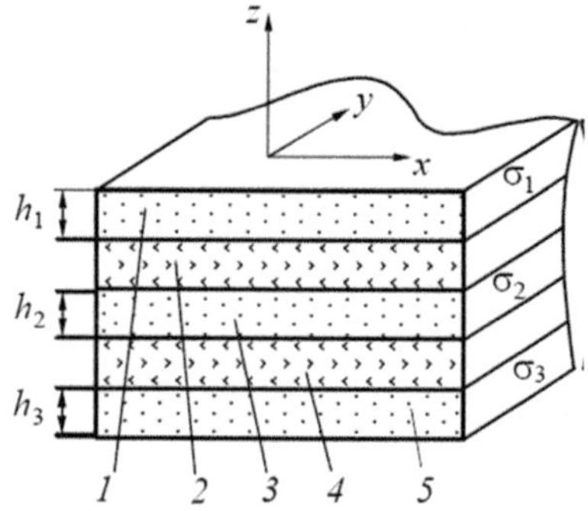

Fig. 9.1 A fragment of the *R*1-*C*1-*R*2-*C*2-*R*3-layered *RC*-EDP: (*1*, *3*, *5*) resistive layers *R*1, *R*2, *R*3, respectively; (*2*, *4*) dielectric layers *C*1, *C*2, respectively

3. The electrical field frequencies are not very high in order to neglect the inductance of the structure.

From the previous assumptions, the dependence of the electric field intensity on the conducting medium and the potential is expressed as follows:

$$\mathbf{E} = -\text{grad}\ \varphi, \tag{9.1}$$

where **E** is the electric field intensity, and the potential, φ, remains unchanged over the thickness of the resistive layer; that is, there is parallel plane where $\varphi = \varphi(x, y)$ of the corresponding electric field.

The heterogeneous electrical properties of the structure cause uncompensated charges within this structure to appear due to an alternating electric field. The corresponding spatial charge density in layer *R*1 can be expressed as follows:

$$\rho_1(x, y) = dq_1/(h_1 dxdy), \tag{9.2}$$

where dq_1 is the elementary electric charge within a volume $dV_1 = h_1 dxdy$ and h_1 is the thickness of the upper resistive layer *R*1.

The value of charge, q_1, can be calculated using the definition of the capacitance of a parallel-plate capacitor formed by the *R*1-*C*1-*R*2 layers as follows:

$$dq_1 = C_1(x, y)[\varphi_1(x, y) - \varphi_2(x, y)], \tag{9.3}$$

where $C_1(x, y)$ is the total electrical capacitance between the layers of *R*1 and *R*2 and $\varphi_1(x, y)$ and $\varphi_2(x, y)$ are the potentials of upper and middle resistive layers *R*1 and *R*2, respectively.

The principle of continuity of the electrical current in the materials implies the following:

$$\text{div}(\sigma\mathbf{E}) + \frac{\partial\rho}{\partial t} = 0, \tag{9.4}$$

where σ is the resistive layer material-specific conductance, and ρ is the space charge density.

Substituting from expressions (9.2) and (9.3) into (9.4) yields the following:

$$\mathrm{div}(\sigma_1 \mathbf{E}) = -\frac{C_{01}(x,y)}{h_1}\frac{\partial}{\partial t}[\varphi_1(x,y) - \varphi_2(x,y)], \tag{9.5}$$

where $C_{01}(x, y)$ is the value of the capacitance per unit area of the dielectric layer surface of $C1$, and σ_1 denotes the material-specific conductance of layer $R1$.

Substitute from (9.1) into (9.5) gives

$$\mathrm{div}\ [\sigma_1 \mathrm{grad}\ \varphi_1(x,y)] = \frac{C_{01}(x,y)}{h_1}\left(\frac{\partial \varphi_1(x,y)}{\partial t} - \frac{\partial \varphi_2(x,y)}{\partial t}\right) \tag{9.6}$$

Use the known relation $\mathrm{div}(a\mathbf{F}) = a(\mathrm{div}\mathbf{F}) + \mathbf{F}(\mathrm{grad}(a))$ to simplify (9.6) further to take the following expression:

$$\sigma_1 \nabla^2 \varphi_1 + (\nabla \varphi_1)(\nabla \sigma_1) = \left(\frac{C_{01}}{h_1}\right)\left(\frac{\partial \varphi_1}{\partial t} - \frac{\partial \varphi_2}{\partial t}\right) \tag{9.7}$$

where ∇^2 and ∇ are the Laplace and Hamilton operators over (x, y) plane.

Denote the distribution of the square resistance over the upper resistive layer surface as $r_1(x, y) = [h_1\sigma_1(x, y)]^{-1}$. Since,

$$\nabla\left(\frac{1}{r_1}\right) = -\frac{1}{r_1^2}\nabla r_1$$

the expression of (9.7) can be rewritten as follows:

$$\nabla^2 \varphi_1(x,y) - \frac{\nabla r_1(x,y)}{r_1(x,y)}\nabla \varphi_1(x,y) = r_1(x,y) C_{01}(x,y)\left(\frac{\partial \varphi_1(x,y)}{\partial t} - \frac{\partial \varphi_2(x,y)}{\partial t}\right) \tag{9.8}$$

Equation (9.8) represents the distribution of the electrical potential over the upper resistive layer $R1$.

The expression of the electrical potential distribution of the lower resistive layer $R3$ is derived in the same manner. The corresponding space-charge density of layer $R3$ will be

$$\rho_3(x,y) = dq_3/(h_3 dxdy), \tag{9.9}$$

where dq_3 is the elementary electric charge within a volume $dV_2 = h_3 dxdy$, and where h_3 is the thickness of upper resistive layer $R3$.

In a similar argument, the value of the electric charge, q_3, can be calculated using the capacitance of the parallel-plate capacitor formed by the layers *R*3-*C*2-*R*2 as follows:

$$dq_3 = -C_2(x,y)[\varphi_3(x,y) - \varphi_2(x,y)], \tag{9.10}$$

where $C_2(x, y)$ is the total electrical capacitance between *R*3-*C*2-*R*2 layers, and $\varphi_3(x, y)$ and $\varphi_2(x, y)$ are the potentials of the resistive layers *R*3 and *R*2, respectively.

Substituting from (9.9) and (9.10) into (9.4) yields the following:

$$\text{div}\ (\sigma_3 \mathbf{E}) = \frac{C_{02}(x,y)}{h_3}\left(\frac{\partial \varphi_3(x,y)}{\partial t} - \frac{\partial \varphi_2(x,y)}{\partial t}\right) \tag{9.11}$$

where $C_{02}(x, y)$ is the value of the capacitance per unit area of the dielectric layer surface of C_2, and where σ_3 is the specific conductance of the *R*3 lower layer material.

Substitute expression (9.1) into (9.11) and perform some transformations:

$$\nabla^2\varphi_3(x,y) - \frac{\nabla r_3(x,y)}{r_3(x,y)}\nabla\varphi_3(x,y) = r_3(x,y)C_{02}(x,y)\left(\frac{\partial \varphi_3(x,y)}{\partial t} - \frac{\partial \varphi_2(x,y)}{\partial t}\right) \tag{9.12}$$

where $r_3(x, y) = [h_3\sigma_3(x, y)]^{-1}$ is the resistance of a square of resistive layer *R*3.

The equation of the potential distribution over the *R*2 layer will be derived based on the following considerations. Expressions (9.5) and (9.11) both describe the full charging currents of the capacitors C_1 and C_2, where both currents flow through the *R*2 layer. Therefore, the current density in *R*2 layer equals to

$$\begin{aligned}\text{Div}\ (\sigma_2 \mathbf{E}) &= -\text{div}\ (\sigma_1 \mathbf{E}) - \text{div}\ (\sigma_3 \mathbf{E}) = \left(\frac{C_{01}(x,y)}{h_1}\right)\left(\frac{\partial \varphi_1(x,y)}{\partial t} - \frac{\partial \varphi_2(x,y)}{\partial t}\right) \\ &+ \left(\frac{C_{02}(x,y)}{h_3}\right)\left(\frac{\partial \varphi_3(x,y)}{\partial t} - \frac{\partial \varphi_2(x,y)}{\partial t}\right)\end{aligned} \tag{9.13}$$

Substituting from (9.1) into (9.13) and performing some transformations yield

$$\begin{aligned}\nabla^2\varphi_2(x,y) &- \frac{\nabla r_2(x,y)}{r_2(x,y)}\nabla\varphi_2(x,y) = \\ &= -\frac{h_2}{h_3} r_2(x,y)C_{02}(x,y)\left(\frac{\partial \varphi_3(x,y)}{\partial t} - \frac{\partial \varphi_2(x,y)}{\partial t}\right) \\ &- \frac{h_2}{h_3} r_2(x,y)C_{01}(x,y)\left(\frac{\partial \varphi_1(x,y)}{\partial t} - \frac{\partial \varphi_2(x,y)}{\partial t}\right)\end{aligned} \tag{9.14}$$

where $r_2(x, y) = [h_2\sigma_2(x, y)]^{-1}$ is the resistance of a square of resistive layer $R2$.

Since the thickness of the layers is assumed constant and is almost the same for all of them (it is true for most of the thin- and thick-film resistors, where their square resistances are specified by the used materials), and since the ratios h_2/h_1 and h_2/h_3 are assumed to be equal to one, then the steady-state value of the expressions (9.8), (9.12), and (9.14) can be rewritten as follows:

$$\begin{aligned}\nabla^2\Phi_1(x,y,s) - \frac{\nabla r_1(x,y)}{r_1(x,y)}\nabla\Phi_1(x,y,s) = \\ = sr_1(x,y)C_{01}(x,y)[\Phi_1(x,y,s) - \Phi_2(x,y,s)]\end{aligned} \tag{9.15}$$

$$\begin{aligned}\nabla^2\Phi_2(x,y,s) - \frac{\nabla r_2(x,y)}{r_2(x,y)}\nabla\Phi_2(x,y,s) = \\ = -sr_2(x,y)\{C_{02}(x,y)[\Phi_3(x,y,s) - \Phi_2(x,y,s)] \\ + C_{01}(x,y)[\Phi_1(x,y,s) - \Phi_2(x,y,s)]\}\end{aligned} \tag{9.16}$$

$$\begin{aligned}\nabla^2\Phi_3(x,y,s) - \frac{\nabla r_3(x,y)}{r_3(x,y)}\nabla\Phi_3(x,y,s) = \\ = sr_3(x,y)C_{02}(x,y)[\Phi_3(x,y,s) - \Phi_2(x,y,s)]\end{aligned} \tag{9.17}$$

where $\Phi_1(x, y, s)$, $\Phi_2(x, y, s)$, and $\Phi_3(x, y, s)$ are the operator form of the potential fields of the resistive layers $R1$, $R2$, and $R3$, respectively, and $s = \sigma + j\omega$ is the Laplace (or complex frequency) operator.

Such multilayered structures with distributed parameters can be analyzed by solving the boundary-value problem for the potentials $\Phi_i(x, y, p)$ in the resistive layers. Certain configurations of the contact groups and certain electrical potentials and currents in these contacts specify the boundary conditions.

As for the complex configuration of the structure, the analytical solution of its boundary-value problem is practically impossible. Therefore, it is highly important to figure out a numerical method to solve (9.15)–(9.17) that satisfy the fabrication structure of the $R1$-$C1$-$R2$-$C2$-$R3$-layered RCG-EDP with the desired performance characteristics.

The acquired system of equations makes it possible to classify the RC-EDP based on the potential distribution over the resistive layers. To clarify the proposed principles of classification, we will consider the $R1$-C-$R2$-layered RC-EDP, which is considered as a particular case of the more general $R1$-$C1$-$R2$-$C2$-$R3$-layered structure.

9.2 Classification of *R*1-*C*-*R*2-Layered *RC*-EDP Based on Potential Distribution in Resistive Layers

A fragment of the *R*1-*C*-*R*2-layered *RC*-EDP is depicted in Fig. 9.2. Such structure can be derived from the structure given in Fig. 9.1 by making the layer's capacitance, *C*2, zero, i.e., by excluding the connection between layers *R*2 and *R*3. This can be done in two ways: reduce the overlapping area to zero or reduce specific capacitance C_{02} between the layers to zero.

Equations (9.15)–(9.17) comprise specific capacitance C_{02}. Let C_{02} be zero in order to transform these equations into the expressions that describe electrical processes in the *R*1-*C*-*R*2-layered *RC*-EDP. The resulting system of differential equations will be

$$\begin{aligned}\nabla^2\Phi_1(x,y,s) &- \frac{\nabla r_1(x,y)}{r_1(x,y)}\nabla\Phi_1(x,y,s) = \\ &= sr_1(x,y)C_{01}(x,y)[\Phi_1(x,y,s)-\Phi_2(x,y,s)]\end{aligned} \tag{9.18}$$

$$\begin{aligned}\nabla^2\Phi_2(x,y,s) &- \frac{\nabla r_2(x,y)}{r_2(x,y)}\nabla\Phi_2(x,y,s) = \\ &= sr_2(x,y)C_{01}(x,y)[\Phi_2(x,y,s)-\Phi_1(x,y,s)]\end{aligned} \tag{9.19}$$

Equations (9.18) and (9.19) show that the potential distribution in the upper (*R*1) and the lower (*R*2) resistive layers is two-dimensional. They also show that the electrical properties of the materials of the upper and the lower resistive layers are heterogeneous over the surface, i.e., $\nabla r_i(x,y) \neq 0$. That is why the *R*1-*C*-*R*2-layered *RC*-EDP will be called the two-dimensional *R*1-*C*-*R*2 EDP with surface-heterogeneous upper and lower resistive layers (shortened as TDHet-TDHet *R*1-*C*-*R*2 EDP).

If one assumes this device as the initial one, then it can be used as a base to form other types of structures of the same class. It can be done by introducing certain limitations of the element's design parameters and electro-physical properties of the used materials.

The classification diagram of the *R*1-*C*-*R*2-layered *RC*-EDP is depicted in Fig. 9.3a. The equations that correspond to this classification that describe the

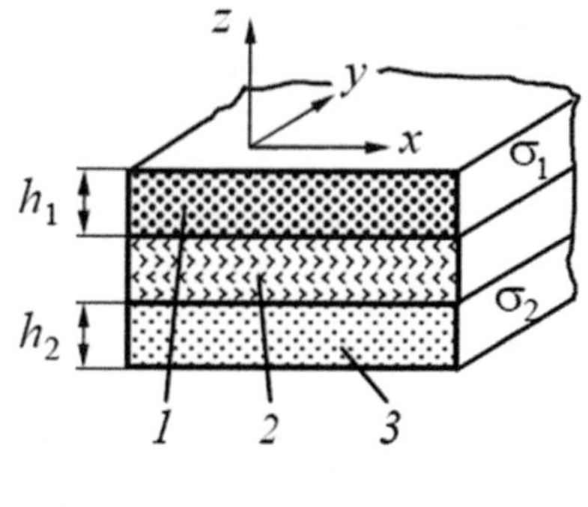

Fig. 9.2 A fragment of the *R*1-*C*-*R*2-layered *RC*-EDP: (*1*) resistive layer *R*1, (*2*) insulator; (*3*) resistive layer *R*2

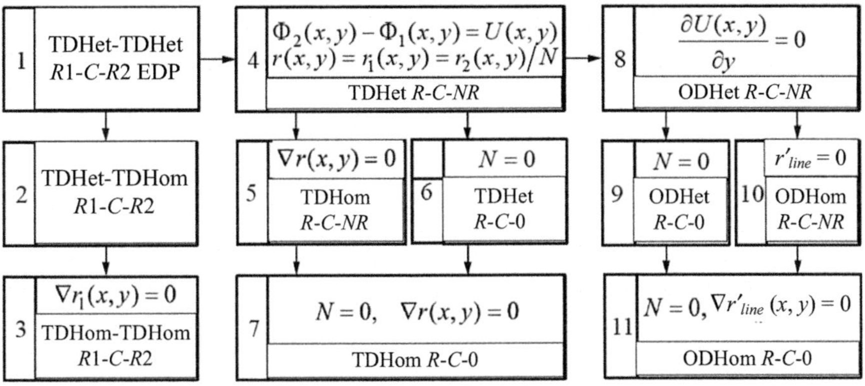

Fig. 9.3 The *R*1-*C*-*R*2-layered *RC*-EDP classification diagram

distribution of electric potentials over the *RC*-EDP resistive layers are given in Table 9.1.

The following designations are listed in Fig. 9.3 and Table 9.1 and defined here for completeness:

- $r_{\text{line}}(x)$ and $c_{\text{line}}(x)$ correspond to the line resistance and line capacitance of the *RC*-EDP.
- r is a constant resistance of a resistive square layer.
- C_0 is a constant capacitance over a unit surface area of the dielectric layer.
- TD denotes a two-dimensional; OD: one-dimensional.
- Hom: means homogeneous; Het: heterogeneous.

The meaning of the shortenings used in the diagram is clarified with the following examples:

1. The laws of variations of the resistive layer square resistances have the following relation: $r_2(x, y) = Nr_1(x, y)$ where $N = \text{const.}$
2. The boundary conditions are specified in the form of voltage $u(x, y) = \varphi_2(x, y) - \varphi_1(x, y)$; that is, the shape of both resistive layers and their electrode (contact pads) location are the same.

Let such kind of *RC*-EDP be called the two-dimensional surface-heterogeneous *R*-*C*-*NR* EDP (shortened as TDHet *R*-*C*-*NR* EDP).

In the same manner as class 1 (Table 9.1), the TDHet-TDHom *R*1-*C*-*R*2 EDP (class 2) is an *R*1-*C*-*R*2 EDP, that is, two-dimensional surface-heterogeneous for the higher resistive layer and two-dimensional surface-homogeneous for the lower resistive layer.

The OD *R*-*C*-*NR* EDP (class 8) is a one-dimensional (the distribution of the electric potential over the resistive layers depends just on one spatial coordinate) length-heterogeneous *R*-*C*-*NR*-layered *RC*-EDP.

Table 9.1 The *R*1-*C*-*R*2-layered *RC*-EDP classification

No	*RC*-EDP name	Equations that describe the distribution of electric potentials over the *RC*-EDP resistive layers
1.	TDHet-TDHet	$\nabla^2\Phi_1(x,y,s) - \dfrac{\nabla r_1(x,y)}{r_1(x,y)}\nabla\Phi_1(x,y,s) = sr_1(x,y)C_0(x,y)[\Phi_1(x,y,s) - \Phi_2(x,y,s)]$
	*R*1-*C*-*R*2	$\nabla^2\Phi_2(x,y,s) - \dfrac{\nabla r_2(x,y)}{r_2(x,y)}\nabla\Phi_2(x,y,s) = sr_2(x,y)C_0(x,y)[\Phi_2(x,y,s) - \Phi_1(x,y,s)]$
2.	TDHet-TDHom	$\nabla^2\Phi_1(x,y,s) - \dfrac{\nabla r_1(x,y)}{r_1(x,y)}\nabla\Phi_1(x,y,s) = sr_1(x,y)C_0(x,y)[\Phi_1(x,y,s) - \Phi_2(x,y,s)]$
	*R*1-*C*-*R*2	$\nabla^2\Phi_2(x,y,s) = sr_2(x,y)C_0(x,y)[\Phi_2(x,y,s) - \Phi_1(x,y,s)]$
3.	TDHom-TDHom	$\nabla^2\Phi_1(x,y,s) = sr_1(x,y)C_0(x,y)[\Phi_1(x,y,s) - \Phi_2(x,y,s)]$
	*R*1-*C*-*R*2	$\nabla^2\Phi_2(x,y,s) = sr_2(x,y)C_0(x,y)[\Phi_2(x,y,s) - \Phi_1(x,y,s)]$
4.	TDHet *R*-*C*-*NR*	$\nabla^2U(x,y,s) - \dfrac{\nabla r(x,y)}{r(x,y)}\nabla U(x,y,s) = s(N+1)rC_0U(x,y,s)$
5.	TDHom *R*-*C*-*NR*	$\nabla^2U(x,y,s) = s(N+1)rC_0U(x,y,s)$
6.	TDHet *R*-*C*-0	$\nabla^2U(x,y,s) - \frac{\nabla r(x,y)}{r(x,y)}\nabla U(x,y,s) = srC_0U(x,y,s)$
7.	TDHom *R*-*C*-0	$\nabla^2U(x,y,s) = srC_0U(x,y,s)$
8.	ODHet *R*-*C*-*NR*	$\dfrac{\partial^2U(x,s)}{\partial x^2} - \dfrac{1}{r_{\text{line}}(x)}\dfrac{\text{d}r_{\text{line}}}{\text{d}x}\dfrac{\partial U(x,s)}{\partial x} = s(N+1)r_{\text{line}}(x)C_{\text{line}}(x)U(x,s)$
9.	ODHom *R*-*C*-*NR*	$\dfrac{\partial^2U(x,s)}{\partial x^2} - \dfrac{1}{r_{\text{line}}(x)}\dfrac{\text{d}r_{\text{line}}}{\text{d}x}\dfrac{\partial U(x,s)}{\partial x} = sr_{\text{line}}(x)C_{\text{line}}(x)U(x,s)$
10.	ODHom *R*-*C*-*NR*	$\dfrac{\partial^2U(x,s)}{\partial x^2} = s(N+1)r_{\text{line}}(x)C_{\text{line}}(x)U(x,s)$
11.	ODHom *R*-*C*-0	$\frac{\partial^2U(x,s)}{\partial x^2} = sr_{\text{line}}(x)C_{\text{line}}(x)U(x,s)$

The ODHom *R*-*C*-0 EDP (class 11) is a one-dimensional length-homogeneous (i.e., line parameters r_{line} and c_{line} are constant) *R*-*C*-0-layered *RC*-EDP (i.e., there is a conducting layer instead of the second resistive one).

It should be noted that the given classification of the *RC*-EDP covers just the possible variants of the *R*1-*C*-*R*2-layered *RC*-EDP with complete layers overlapping. It is evident that designs with incomplete overlapping of the layers can be divided into areas with complete overlapping and purely resistive areas with distributed resistance. The right-hand sides of their equations (see Table 9.1) will be identically equal to zero.

Test Questions

1. Recite the assumptions that simplify the $R1$-$C1$-$R2$-$C2$-$R3$-layered RC-EDP analysis problem. Provide comments regarding the necessity of all the assumptions.
2. Describe the method to figure out the equation of distribution of electrical potential over any resistive layer of the RC-EDP.
3. In correspondence with the classification given in this chapter, describe the RC-EDP designs that were considered in Chap. 4 (Figs. 4.3, 4.5, 4.8–4.11, 4.13, 4.16, 4.19).

Chapter 10
Calculation of One-Dimensional *RC*-EDP Primary Parameters

10.1 Preliminaries

The real *RC*-EDP used to make fractal elements and devices is a structural part with conducting terminals just like all other electrical elements and components, which may or may not be packaged. For this reason, a user's prime interest will be its performance parameters and characteristics that could be detected by measuring the currents and voltages at the terminals of these elements.

It is obvious that the real *RC*-EDP can be represented with various equivalent circuits (models) including, but not limited to, those based on the ideal ODHom, ODHet, and TDHom *RC*-EDP or their mixed models (they comprise the elements with lumped parameters). Hence, the objective of analyzing these circuits is to figure out the methods that would make it possible to switch from the general equations of electrical equilibrium, which describe the distribution of electric potential (see Chap. 9) over the idealized *RC*-EDP resistive layers, to the parameters that characterize the relationships between the currents and voltages at the *RC*-EDP terminals. In the electric circuit theory, these parameters are usually called the primary parameters (also known as the proper or the basic parameters) of the two-port (the multiport) networks.

It is commonly known from electric circuits theory that such performance characteristics of multiport networks can be expressed via several systems, of certainly interrelating parameters (for example, *Z*-, *Y*-, *H*-, *A*-system). The choice of a certain system typically depends on the simplicity of the operation of these parameters and on the expressions acquired as a result of the analysis (these expressions depend on the relation between the elements of the multiport networks).

Practical computer-aided design of analog circuits is commonly conducted based on the nodal method to perform the frequency-domain analysis. This method considers the *y*-parameters of the multiport (usually two-port) networks. That is

A.Kh. Gil'mutdinov et al., *Fractal Elements and their Applications*, Analog Circuits and Signal Processing, DOI 10.1007/978-3-319-45249-4_10

why the multiport *RC*-EDP is described by the *y*-parameter system in this book. From analysis point of view, neither the *RC*-EDP class, nor its structure, nor the number of its terminals matters.

10.2 *Y*-Parameter Calculation of ODHom *R*-*C*-0-Layered EDP Devices

The ODHom *R*-*C*-0-layered EDP devices are sufficiently good example of *RC*-EDP devices because their contact pads completely overlap the resistive layer, their manufacturing technology is industrially developed, and whose specific resistance of their resistive layer is at least hundred times greater than that of the conducting plate.

The ODHom *R*-*C*-0-layered EDP two-port connection circuit is depicted in Fig. 10.1. For the selected class of the *RC*-EDP device, the equation that describes the distribution of the voltage along the *RC*-EDP length will be described as follows (see Table 9.1, class 11):

$$\frac{\partial^2 U(x,p)}{\mathrm{d}x^2} - p\, r_{\mathrm{line}} C_{\mathrm{line}} U(x,p) = 0 \tag{10.1}$$

The general solution of (10.1) will be written in the following form:

$$U(x,p) = \underline{a}_1 e^{\gamma x} + \underline{a}_2 e^{-\gamma x}, \tag{10.2}$$

where $\underline{a}_1$ and $\underline{a}_2$ are complex constants derived from the boundary conditions, and $\gamma = \sqrt{p r_{\mathrm{line}} C_{\mathrm{line}}}$ is the propagation constant of the ODHom *R*-*C*-0-layered EDP.

Since the *y*-parameters refer to the well-known short-circuit parameters of two-port networks, then for completeness, we have

$$y_{ij} = \frac{I_i}{U_j}\Big|_{U_{k \neq j}=0}, \tag{10.3}$$

where $i = \overline{1,2}$, $j = \overline{1,2}$, and $k \neq j$.

The corresponding boundary conditions used to calculate y_{11} and y_{21} parameters are given by

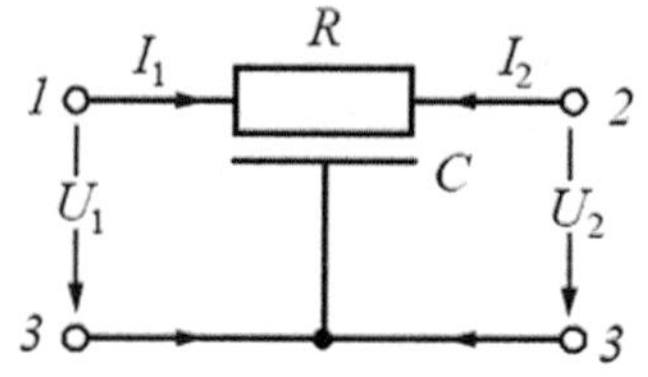

Fig. 10.1 The *RC*-EDP graphical symbol

$$U(x)|_{x=0} = U_1 = 1; \quad U(x)|_{x=l} = U_2 = 0. \tag{10.4}$$

The same kind of boundary conditions to calculate parameters y_{12} and y_{22} will be

$$U(x)|_{x=0} = U_1 = 0; \quad U(x)|_{x=l} = U_2 = 1. \tag{10.5}$$

Substituting from (10.4) into (10.2) gives the following set of algebraic equations:

$$\left.\begin{array}{r} \underline{a}_1 + \underline{a}_2 = 1 \\ \underline{a}_1 e^{\gamma l} + \underline{a}_2 e^{-\gamma l} = 0 \end{array}\right\}$$

which provides an easy calculation of the coefficients:

$$\underline{a}_1 = \frac{e^{-\gamma l}}{e^{\gamma l} - e^{-\gamma l}}; \quad \underline{a}_2 = \frac{e^{\gamma l}}{e^{\gamma l} - e^{-\gamma l}}$$

Substituting these coefficients into (10.2) yields the following value of the electric voltage:

$$U(x,p) = -\frac{\sinh(x - l)}{\sinh \gamma l} \tag{10.6}$$

and the current $I(x, p)$ is derived by means of the differential form of the Ohm law:

$$I(x,p) = -\frac{1}{r_{\text{line}}} \frac{\partial U(x,p)}{\partial x}$$

Then

$$I(x,p) = -\frac{1}{R} \frac{\gamma l}{\sinh \gamma l} \cosh \gamma(x - l) \tag{10.7}$$

Hence, from (10.6) and (10.7), the expressions of parameters y_{11} and y_{21} will be given by

$$y_{11} = \frac{I_1}{U_1}|_{U_2=0} = \frac{\theta}{R} \frac{\cosh \theta}{\sinh \theta}; \; y_{12} = \frac{I_2}{U_1}|_{U_2=0} = -\frac{\theta}{R} \frac{1}{\sinh \theta}$$

where $R = r_{\text{line}} l$ is the RC-EDP overall resistance, and $\theta = \gamma l$ is the intrinsic constant of the ODHom R-C-0-layered RC-EDP.

Formula (10.3) is used to find the expressions of parameters y_{21} and y_{22} in the same manner. Hence, the resulting matrix of the y-parameters of the ODHom R-C-0-layered RC-EDP is given by

$$\begin{bmatrix} y_{11} & y_{12} \\ y_{21} & y_{22} \end{bmatrix} = \frac{\theta}{R} \begin{bmatrix} \coth\theta & -\mathrm{csch}\theta \\ -\mathrm{csch}\theta & \coth\theta \end{bmatrix} \tag{10.8}$$

This expression will further be shown as a fundamental expression of numerical calculations of the y-parameters of the heterogeneous *RC*-EDP.

10.3 *Y*-Parameter Calculation of ODHet *R-C*-0-Layered EDP Devices

Consider the ODHet *R-C*-0-layered EDP devices whose heterogeneity is specified with a law of variation of its width along its length. In this context, the product of their line resistance $r_{\text{line}}(x)$ and line capacitance $c_{\text{line}}(x)$ is constant for any cross section x along the structure's length, i.e., $r_{\text{line}}(x)c_{\text{line}}(x) = \text{const}$. Examples of such *RC*-EDP resistive layer topologies are given in Table 4.1.

According to the *RC*-EDP classification provided in Chap. 9, the distribution of the voltage along the length of the device will be described by the following dynamics (see Table 9.1, class 9):

$$\frac{\partial^2 U(x,p)}{\mathrm{d}x^2} - \frac{1}{r_{\text{line}}(x)} \frac{\mathrm{d}r_{\text{line}}(x,y)}{\mathrm{d}x} \frac{\partial U(x,p)}{\partial x} = p\, r_{\text{line}} C_{\text{line}} U(x,p)$$

which, for simplicity and for further analysis, can be expressed as follows:

$$U'' - \left(\frac{r'}{r}\right) U' - prc\, U = 0 \tag{10.9}$$

where $r = r_{\text{line}}(x)$, $c = c_{\text{line}}(x)$, and the prime symbols signify differential operation with respect to x.

Expression (10.9) is a diffusion equation that has no general closed-form solution. It can be transformed into an equation with known solution in order to acquire an analytical solution.

Let $U = \sqrt{r}f$; then (10.9) can be transformed to the following canonical form as follows:

$$f'' - \eta^2(x) f = 0. \tag{10.10}$$

Differentiating U twice yields

$$U' = f'\sqrt{r} + f\frac{r'}{2r}\sqrt{r}\,; \tag{10.11}$$

$$U'' = f''\sqrt{r} + f'\frac{r'}{r}\sqrt{r} + f\left[\frac{r''}{2r}\sqrt{r} - \left(\frac{r'}{r}\right)^2\sqrt{r}\right] \tag{10.12}$$

Substitution from (10.11) and (10.12) into (10.9) yields the canonical form of (10.10) under the following condition:

$$\eta^2(x) = prc - \frac{r''}{2r} + 3\left(\frac{r'}{2r}\right)^2. \tag{10.13}$$

Analyzing the dimensions in (10.13) that involve $\eta^2(x)$ makes sense when the following equation holds:

$$-\frac{r''}{2r} + 3\left(\frac{r'}{2r}\right)^2 = \frac{k^2}{l^2}, \tag{10.14}$$

where k^2 is a constant that can be both positive and negative, and l is the length of the RC-EDP device.

Thus, considering the aforementioned transformations and its limitations, the resulting solution of (10.9) can be expressed in the following way:

$$U(x,p) = \sqrt{r}\left[\underline{a}_1 e^{\gamma x} + \underline{a}_2 e^{-\gamma x}\right], \tag{10.15}$$

where $\gamma = \sqrt{prc + (k/l)^2}$, $\underline{a}_1$, and $\underline{a}_2$ are the complex constants that depend on the boundary conditions.

It can be shown that the boundary conditions $U(x_1) = 1$ and $U(x_2) = 0$ (x_1 and x_2 are the coordinates of the beginning and the end of the structure) that are used to calculate the two-port parameters y_{11} and y_{21} from the voltage-current relationship of the resistive layer:

$$U(p,x) = \sqrt{\frac{r_{\text{line}}(x)}{r_{\text{line}}(x_1)}\frac{\sinh(x_2 - x)}{\sinh \gamma l}}$$

$$I(x,p) = \frac{1}{\sinh(x)}\frac{\gamma}{\sqrt{r_{\text{line}}(x) r_{\text{line}}(x)}}\left[\cosh\gamma(x_2 - x) - \beta(x)\frac{\sinh(x_2 - x)}{\sinh \gamma l}\right]$$

where $\beta(x) = \frac{r'_{\text{line}}(x)}{2r_{\text{line}}(x)} l$.

The boundary conditions $U(x_1) = 0$ and $U(x_2) = 1$ that are used to calculate the two-port parameters y_{12} and y_{22}, which follows from the voltage-current relationship of the resistive layer, are obtained as follows:

$$U(p,x) = \sqrt{\frac{r_{\text{line}}(x)}{r_{\text{line}}(x_2)}} \frac{\sinh(x-x_1)}{\sinh \gamma l}$$

$$I(p,x) = \frac{1}{l}\frac{\gamma}{\sqrt{r_{\text{line}}(x) r_{\text{line}}(x)}}\left[\cosh \gamma(x-x_1)\frac{\gamma l}{\sinh \gamma l} + \beta(x)\right]$$

Consequently, the four *Y*-parameter matrix of the ODHet *R*-*C*-0-layered EDP is described as follows:

$$\begin{bmatrix} y_{11} & y_{12} \\ y_{21} & y_{22} \end{bmatrix} = \frac{1}{r_{\text{line}}(x)\, l}\begin{bmatrix} \theta_{\text{het}}\coth\theta_{\text{het}} - \beta(x_1) & -\dfrac{\alpha}{\sinh\theta_{\text{het}}} \\ -\dfrac{\alpha}{\sinh\theta_{\text{het}}} & \alpha(\theta_{\text{het}}\coth\theta_{\text{het}} - \beta(x_1)\,) \end{bmatrix} \tag{10.16}$$

where $\alpha = [r_{\text{line}}(x_1)/r_{\text{line}}(x_2)]$, and $\theta_{\text{het}} = \gamma l = \sqrt{p r_{\text{line}} c_{\text{line}} l^2 + k^2}$ is the characteristic constant of the ODHet *RC*-EDP.

It should be noted that this *y*-parameters matrix corresponds only to one certain class of the one-dimensional heterogeneous *RC*-EDP, where its line resistance complies with condition (10.14).

Let us figure out a probable class of distributions that would comply with condition (10.14). To this effect, consider the following substitution:

$$r'/r = 2\nu \tag{10.17}$$

to transform expression (10.14) into a first-order nonlinear differential equation as follows:

$$\nu' = \nu^2 - (k/l)^2. \tag{10.18}$$

Its solution for different values of *k* is given by

$$v(x) = \frac{\alpha}{l}\,\tan\left(\frac{\alpha}{l}(x+m)\right);\quad k^2 = -alp^2 \; < 0 \tag{10.19a}$$

$$v(x) = -\frac{1}{x-m};\; k^2 = 0 \tag{10.19b}$$

$$v(x) = -\frac{k}{l}\tanh\left(\frac{k}{l}(x+m)\right);\; k^2 > 0 \quad \text{and} \quad v^2 < \left(\frac{k}{l}\right)^2 \tag{10.19c}$$

$$v(x) = -\frac{k}{l}\coth\frac{k}{l}(x+m);\; k^2 > 0 \quad \text{and} \quad v^2 > \left(\frac{k}{l}\right)^2 \tag{10.19d}$$

Substituting (10.19a–10.19d) into (10.17) and solving for $r_{\text{line}}(x)$ result in a class of distributions the line resistances which vary in accordance with the following laws:

Trigonometric law:

$$r_{\text{line}}(x) = \frac{r_0}{\cos^2\left[\frac{k}{l}(x+m)\right]}, \tag{10.20a}$$

where $k^2 = \alpha^2 < 0$; α, m—constants; $r_0 = r_{\text{line}}(x_1)$;

Parabolic law:

$$r_{\text{line}}(x) = \frac{r_0}{(x+m)^2}; \tag{10.20b}$$

Hyperbolic law:

$$r_{\text{line}}(x) = \frac{r_o}{\coth^2\left[\frac{k}{l}(x+m)\right]} \tag{10.20c}$$

$$r_{\text{line}}(x) = \frac{r_o}{\sinh^2\left[\frac{k}{l}(x+m)\right]}; \tag{10.20d}$$

Exponential law:

$$r_{\text{line}}(x) = r_o \pm (k/l)x. \tag{10.20e}$$

Such variety of options of distribution laws and their parameters (constants k, l, m, x_1) provide additional degrees of freedom that could be used to synthesize the *RC*-EDP with desired performance characteristics.

However, the general ODHet *R-C*-0-layered *RC*-EDP given by (10.9) can be transformed into the canonical form that possesses the analytical solution only for a limited number of the *RC*-EDP width variation laws. The following laws can be added to the considered laws with known analytical solutions:

Linear law:

$$r_{\text{line}}(x) = \frac{r_0}{(1-kx)}, \tag{10.20f}$$

Bessel law:

$$r_{\text{line}}(x) = r_0 x^k, \tag{10.20g}$$

where k is constant.

10.4 *Y*-Parameter Calculation of ODHom *R*1-*G*1-*C*1-*R*-*C*2-*G*2-*R*2-Layered EDP Devices

Chapter 9 has represented differential equations (9.15)–(9.17) that describe the generalized *RCG*-EDP resistive layers' potential distribution. These equations are valid only for the *R*1-*C*1-*R*-*C*2-*R*2-layered EDP and can be used to calculate the *y*-parameters only of the elements structured in this way.

Systems with distributed parameters where quasi-*TEM*-wave propagates are thus referred to as one-dimensional ones. Such systems are often simulated with electrical models in the form of ladder-connected elements. Let us use this approach to figure out the *y*-parameter expressions of the ODHom *R*1-*G*1-*C*1-*R*-*C*2-*G*2-*R*2-layered EDP.

10.4.1 *Electrical Models of One-Dimensional Systems with Distributed Parameters*

Let there be the following integral characteristics of a distributed system: voltage $u(x, t)$ and current $i(x, t)$. Let the system's electrical model be a ladder connection of elements described with S and P operators (Fig. 10.2). These operators transform u and i variables along the system.

As long as the transverse (perpendicular to x direction of the wave propagation) resistance of the system is finite, the voltage wave causes leakage currents $\mathrm{d}i = -P[(u + \mathrm{d}u)]\mathrm{d}x$ that contribute to the current i wave. In its turn, the current wave generates voltage drops $\mathrm{d}u = -S \cdot i\mathrm{d}x$ due to the finite longitudinal resistance and causes thus the voltage u wave.

Therefore, the voltage-current relationships are described by the following equations:

$$-\frac{\partial i(x,t)}{\partial x} = P[u(x,t) + \mathrm{d}u(x,t)] \approx Pu(x,t) \tag{10.21}$$

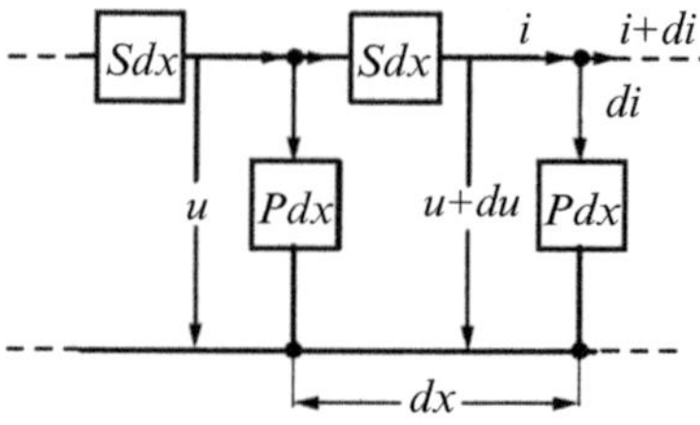

Fig. 10.2 Electrical model of a one-dimensional system with distributed parameters

$$-\frac{\partial u(x,t)}{\partial x} = Si(x,t) \tag{10.22}$$

where the value of $|du| << |u|$ determines the second-vanishing-order correction to di.

In case of a linear and homogeneous medium, the Laplace transform of these equations results in the following:

$$-\frac{\mathrm{d}\dot{I}\,(s,x)}{\mathrm{d}x} = Y_o(s)\dot{U}\,(s,x) \tag{10.23}$$

$$-\frac{\mathrm{d}\dot{U}\,(s,x)}{\mathrm{d}x} = Z_o(s)\dot{I}\,(s,x) \tag{10.24}$$

where $\dot{I}\,(s,x)$ and $\dot{U}\,(s,x)$ are the complex magnitudes (symbolic designations) of the current and the voltage, respectively, that are functions of the distance x, and $Z_0(s)$ and $Y_0(s)$ are the Laplace-transformed linear operators S and P that are the linear longitudinal impedance and transversal immittance of the system's ladder model.

Solving the system of equations (10.23) and (10.24) for $\dot{U}\,(s,x)$ results in

$$\frac{\mathrm{d}^2\dot{U}\,(s,x)}{\mathrm{d}x^2} = Z_o(s)Y_o(s)\dot{U}\,(s,x) \tag{10.25}$$

Equation (10.25) is a second-order linear differential equation. The physical interpretation of the solution of this equation is an electric oscillator where it has a fundamental representation as $\mathring{U}(p,x) = \mathring{A}e^{\gamma x}$.

Substituting this expression into (10.25) and performing the required transformations yield

$$\gamma^2 = Z_0(s)Y_0(s).$$

Hence, the general solution of the studied system comprises two oscillations: $\mathring{U}_1(p,x) = \underline{A}_1 e^{\gamma 1x}$ and $\mathring{U}_2(s,x) = \underline{A}_2 e^{\gamma 2x}$, where $\gamma_1 = \sqrt{Z_o(s)Y_o(s)}$ and $\gamma_2 = -Z_0(s)$ $Y_0(s)$. The complex number $Z_0(s)Y_0(s)$ is called the longitudinal *propagation constant* of the wave. Then the general solution of (10.25) can expressed as follows:

$$\mathring{U}(s,x) = \underline{A}_1 e^{\gamma 1x} + \underline{A}_2 e^{\gamma 2x}. \tag{10.26}$$

Complex numbers are the constants of integration. They depend on the boundary conditions (set voltage and current at the beginning and at the end of the line).

Similarly, the expression for current $\dot{I}(s,x)$ is derived from (10.24) as follows:

$$\dot{I}(s,x) = \frac{1}{Z_o(s)}\frac{d\mathring{U}(s,\ x)}{dx} = \frac{\underline{A}_1 e^{\gamma 1x} + \underline{A}_2 e^{\gamma 2x}.}{Z_o(s)/\gamma} \tag{10.27}$$

where the ratio $Z_0(s)/\gamma$ has the resistance dimension and is called *the wave* or *the characteristic impedance*:

$$Z_{\text{char}}(s) = \sqrt{Z_o(s)/Y_o(s)} \tag{10.28}$$

One way to build the electrical model of the system is to decompose $Z_0(s)$ and $Y_0(s)$ into connection of passive R, L, and C components that are expressed via the medium electro-physical parameters. The other way to do it is to use the Maxwell equations after they are transformed to include the integral variables $u(x, t)$ and $i(x, t)$.

This section develops the given principle of building the electrical models of one-dimensional systems with distributed parameters for the *R*1-*G*1-*C*1-*R*-*C*2-*G*2-*R*2-layered EDP.

10.4.2 Modeling of ODHom R1-G1-C1-R-C2-G2-R2-Layered EDP Devices

The general view of the longitudinal section of the ODHom *R*1-*G*1-*C*1-*R*-*C*2-*G*2-*R*2-layered EDP and its graphical symbol is depicted in Fig. 10.3

The following parameters of the different layers will be set to simplify the analysis and for reasons that will be clear later. Notice that *Nr* denotes line resistance of layer *R*1, g_1 is line transversal conductivity of layer *G*1, *c* is line capacitance between layers *R*1 and *R*, *r* is line resistance of layer *R*, *Mc* is line capacitance between layers *R* and *R*2, g_2 is line transversal conductivity of layer *G*2, and *Kr* is line resistance of layer *R*2. Here *K*, *M*, and *N* are the dimensionless constants.

Therefore, from the previous designation, the area of the ODHom *R*1-*G*1-*C*1-*R*-*C*2-*G*2-*R*2-layered EDP can be replaced with the equivalent circuit given in

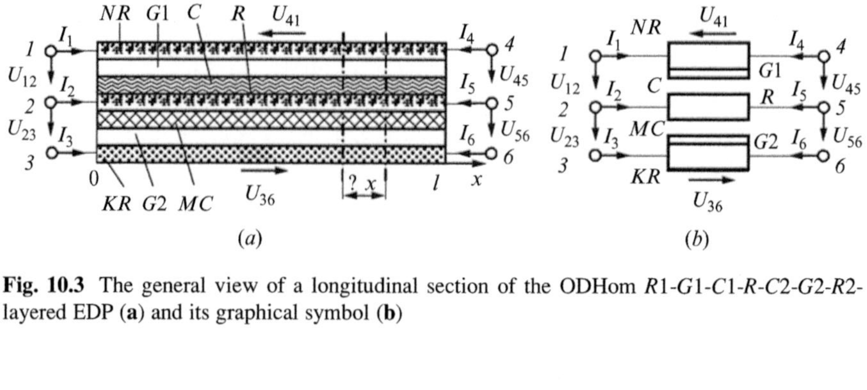

Fig. 10.3 The general view of a longitudinal section of the ODHom *R*1-*G*1-*C*1-*R*-*C*2-*G*2-*R*2-layered EDP (**a**) and its graphical symbol (**b**)

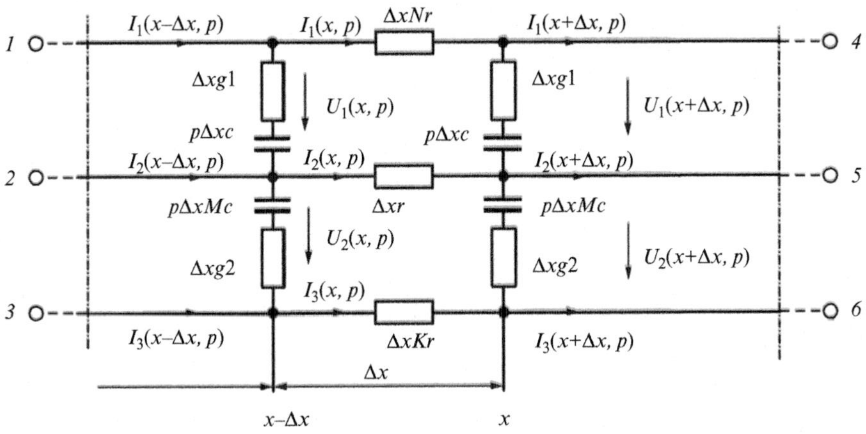

Fig. 10.4 The equivalent circuit of the ODHom *NR-G1-C-R-MC-G2-KR*-layered *RC*-EDP

Fig. 10.4. The currents and voltages in its nodes are represented in the operator form.

The currents and voltages will be expressed hereinafter in a simplified form when deriving the line equations (for example $U_1 \equiv U_1(x, s)$). Then, according to the assumptions given in Fig. 10.4, the state equations of the equivalent circuit will be as follows:

$$\frac{dU_1}{dx} = -(NI_1 - I_2)r; \tag{10.29}$$

$$\frac{dU_2}{dx} = -(I_2 - KI_3)r; \tag{10.30}$$

$$\frac{dI_1}{dx} = -y_1U_1; \tag{10.31}$$

$$\frac{dI_2}{dx} = y_1U_1 - y_2U_2; \tag{10.32}$$

$$\frac{dI_3}{dx} = y_2U_2, \tag{10.33}$$

where $y_1 = \frac{scg_1}{sc+g_1}$; $y_2 = \frac{sMcg_2}{sMc+g_2}$.

It follows from (10.31)–(10.33) that

$$\frac{dI_1}{dx} + \frac{dI_2}{dx} + \frac{dI_3}{dx} = 0. \tag{10.34}$$

differentiate (10.29) with respect to x and using (10.31)–(10.32) yields

$$\frac{d^2U_1}{dx^2} = -\left(N\frac{dI_1}{dx} - \frac{dI_2}{dx}\right)r = (N+1)y_1rU_1 + y_2rU_2.$$

Denote $N_1 = N + 1$, $T_1 = y_1 \cdot r$, and $T_2 = y_2 \cdot r$. Then the last expression can be represented in the following way:

$$\frac{d^2U_1}{dx^2} - N_1T_1U_1 + T_2U_2 = 0. \tag{10.35}$$

Transform (10.30) in the same manner to get the following:

$$\frac{d^2U_2}{dx^2} = -\left(\frac{dI_2}{dx} - K\frac{dI_3}{dx}\right)r = -y_1rU_1 + (K+1)y_2rU_2.$$

Denote $K + 1 = K_1$ and turn the equation to

$$\frac{d^2U_2}{dx^2} + T_1U_1 - T_2K_1U_2 = 0. \tag{10.36}$$

To calculate voltages U_1 and U_2, one needs to solve the system of homogeneous second-order differential equations (10.35) and (10.36). According to the reasoning made in Sect. 10.3 we assume that the general view of the solutions can be expressed in the following manner:

$$U_1(x) = A \cdot e^{sx}; \quad U_2(x) = B \cdot e^{sx}.$$

Substituting these forms of solutions into the studied system of differential equations (10.35) and (10.36), one may use simple transformations to represent the system into a homogeneous system of linear algebraic equations as follows:

$$\begin{cases} (p^2 - N_1T_1)A + T_2B &= 0; \\ AT_1 + (p^2 - T_2K_1)B &= 0, \end{cases} \tag{10.37}$$

which makes it possible to calculate coefficients A and B.

Use the condition for a nontrivial (nonzero) solution of a homogeneous system of algebraic equations: *if the number of equations of a homogeneous system equals the number of the unknowns, then the necessary and sufficient condition for a nonzero (nontrivial) solution is zero value of the determinant made out of its coefficients.*

In our case, this condition is expressed as follows:

$$\begin{vmatrix} s^2 - N_1T_1 & T_2 \\ T_1 & s^2 - T_2K_1 \end{vmatrix} = s^4 - (N_1T_1 + K_1T_2)s^2 + (N_1K_1 - 1)T_1T_2 = 0$$

The acquired equation is called *characteristic equation* and possesses four roots: $s_1 = \gamma_1$, $s_2 = \gamma_2$, $s_3 = -\gamma_1$, and $s_4 = -\gamma_2$, where

$$\gamma_1 = \sqrt{\frac{1}{2}\left[(N_1T_1 + K_1T_2) + \sqrt{(N_1T_1 + K_1T_2)^2 - 4(N_1K_1 - 1)T_1T_2}\right]};$$

$$\gamma_2 = \sqrt{\frac{1}{2}\left[(N_1T_1 + K_1T_2) - \sqrt{(N_1T_1 + K_1T_2)^2 - 4(N_1K_1 - 1)T_1T_2}\right]}.$$

These roots correspond to four sine oscillations in the distributed medium with two propagation constants γ_1 and γ_2.

Solving the equations of system (10.37) for the coefficients A and B defines the oscillatory behavior of the system and yields

$$B_1 = \frac{N_1T_1 - p_1}{T_2}A_1; \quad B_2 = \frac{N_1T_1 - p_2}{T_2}A_2; \quad B_3 = \frac{N_1T_1 - p_1}{T_2}A_3;$$
$$B_4 = \frac{N_1T_1 - p_2}{T_2}A_4.$$

The general solutions of differential equations (10.35) and (10.36) for U_1 and U_2 will be

$$U_1(x) = A_1 \cdot e^{\gamma 1x} + A_2 \cdot e^{\gamma 2x} + A_3 \cdot e^{-\gamma 1x} + A_4 \cdot e^{-\gamma 2x}; \quad (10.38)$$
$$U_2(x) = B_1 \cdot e^{\gamma 1x} + B_2 \cdot e^{\gamma 2x} + B_3 \cdot e^{-\gamma 1x} + B_4 \cdot e^{-\gamma 2x}, \quad (10.39)$$

where constants A_i and B_i are derived from the boundary conditions.

Solve the following system of equations to find currents I_1, I_2, and I_3:

$$\begin{cases} \frac{dU_1}{dx} & = (-NI_1 + I_2)r; \quad (a) \\ \frac{dU_2}{dx} & = (-I_2 + KI_3)r; \quad (b) \\ I_2 & = I_1 - I_3 + C. \quad (c) \end{cases} \quad (10.40)$$

Equation (10.40c) follows from (10.34), and C is the constant of integration. Hence, substituting the expression of the current I_2 from (10.40c) into (10.40a) and (10.40b) gives the following:

$$\begin{cases} \dfrac{dU_1}{dx} &= (-N_1 I_1 - I_3 + C)r; \\ \dfrac{dU_2}{dx} &= (I_1 + K_1 I_3 - C)r. \end{cases}$$

Similarly, the expressions of the currents will be

$$I_1 = -\frac{1}{(N_1K_1 - 1)r}\left(K_1\frac{dU_1}{dx} + \frac{dU_2}{dx}\right) + \frac{KC}{(N_1K_1 - 1)}; \tag{10.41}$$

$$I_2 = \frac{1}{(N_1K_1 - 1)r}\left(K_1\frac{dU_1}{dx} - N\frac{dU_2}{dx}\right) - \frac{NKC}{(N_1K_1 - 1)}; \tag{10.42}$$

$$I_3 = \frac{1}{(N_1K_1 - 1)r}\left(\frac{dU_1}{dx} + N_1\frac{dU_2}{dx}\right) + \frac{NC}{(N_1K_1 - 1)}. \tag{10.43}$$

Therefore, all the expressions of currents and voltages that describe the electrical processes within a Δx-long element of the line have been found.

10.4.3 Derivation of Analytical Expressions of the Y-Parameters

To figure out the expressions of the currents and voltages that correspond to the poles of the ODHom *R*1-*G*1-*C*1-*R*-*C*2-*G*2-*R*2-layered EDP (see Fig. 10.2), the following expressions describing the voltages and currents at a length element of this *RC*-EDP will be used:

$$U_{12} = U_1(x)\big|_{x=0}; \tag{10.44}$$

$$U_{23} = U_2(x)\big|_{x=0}; \tag{10.45}$$

$$U_{41} = -\int_0^l I_1(x)Nr\,dx; \tag{10.46}$$

$$U_{25} = -\int_0^l I_2(x)Nr\,dx; \tag{10.47}$$

$$U_{36} = \int_0^l I_3(x)Kr\,dx; \tag{10.48}$$

$$U_{45} = U_1(x)\big|_{x=l}; \tag{10.49}$$

$$U_{56} = U_2(x)\big|_{x=l}. \tag{10.50}$$

Here l stands for the length of the RC-EDP.

Integrating the expressions of (10.46)–(10.48) results in the following:

$$U_{41} = -\frac{N}{N_1K_1 - 1}[(K_1U_{12} + U_{23}) - (K_1U_{45} + U_{56})] - R\frac{NKC}{N_1K_1 - 1}; \tag{10.51}$$

$$U_{25} = \frac{1}{N_1K_1 - 1}[(KU_{12} - NU_{23}) - (KU_{45} - NU_{56})] - R\frac{NKC}{N_1K_1 - 1}; \tag{10.52}$$

$$U_{36} = \frac{K}{N_1K_1 - 1}[(U_{12} + N_1U_{23}) - (U_{45} + N_1U_{56})] - R\frac{NKC}{N_1K_1 - 1}, \tag{10.53}$$

where R is the overall resistance of the RC-EDP layer R (see Fig. 10.2).

Unlike the currents that are specified with expressions (10.41)–(10.43), the currents of the RC-EDP poles will be denoted with subscripts "E" to denote an "Element." It is evident that these currents can be found based on the following conditions:

$$I_{1\mathrm{E}} = I_1(x)\big|_{x=0}; \tag{10.54}$$

$$I_{2\mathrm{E}} = I_2(x)\big|_{x=0}; \tag{10.55}$$

$$I_{3\mathrm{E}} = I_3(x)\big|_{x=0}; \tag{10.56}$$

$$I_{4\mathrm{E}} = -I_1(x)\big|_{x=l}; \tag{10.57}$$

$$I_{5\mathrm{E}} = -I_2(x)\big|_{x=l}; \tag{10.58}$$

$$I_{6\mathrm{E}} = -I_3(x)\big|_{x=l}. \tag{10.59}$$

The boundary conditions that correspond to the y-parameter measurement modes will be used to determine the unknown constants A_i and B_i using the general solutions of the differential equations (10.38)–(10.39). To this effect, connect the RC-EDP according to the circuit diagram depicted in Fig. 10.5.

Then the following conditions will be used to find out the expressions of the conductivity matrix ($y_{11} \ldots y_{61}$) of the generalized finite distributed element (GFDE):

$$U_1 = 1, \quad U_{i \neq 1} = 0, \quad i = \overline{2,6}.$$

The corresponding voltages at the RC-EDP terminals (see Fig. 10.4) will be

$$U_{12} = 1; \quad U_{41} = -1; \quad \text{other } U_{ij} = 0. \tag{10.60}$$

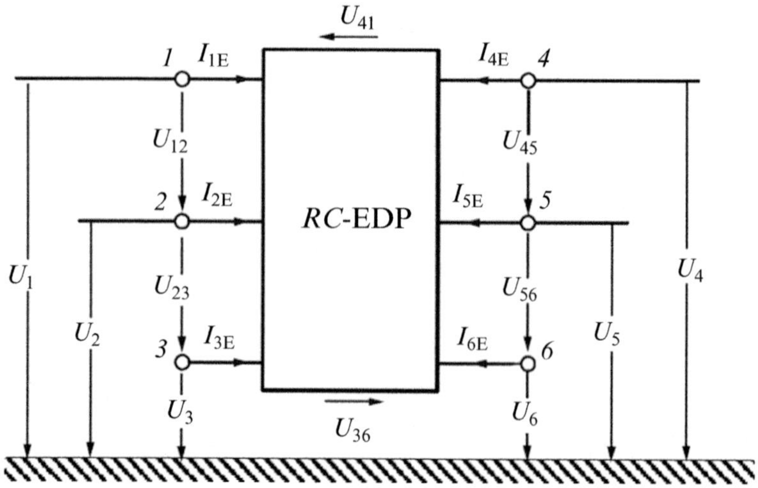

Fig. 10.5 The *RC*-EDP connection circuit aimed at estimation of the six-terminal *y*-parameters

In accordance with the specified boundary conditions, use the following system of equations taken from (10.44)–(10.50) to calculate the unknown A_i in (10.38)–(10.50):

$$\begin{cases} U_{12} & A_1 + A_2 + A_3 + A_4 = 1; \\ U_{23} & k_1A_1 + k_2A_2 + k_1A_3 + k_2A_4 = 0; \\ U_{45} & A_1 \cdot e^{\gamma_1 l} + A_2 \cdot e^{\gamma_2 l} + A_3 \cdot e^{-\gamma_1 l} + A_4 \cdot e^{-\gamma_2 l} = 0; \\ U_{56} & k_1A_1e^{\gamma_1 l} + k_2A_2e^{\gamma_2 l} + k_1A_3e^{-\gamma_1 l} + k_2A_4e^{-\gamma_2 l} = 0, \end{cases} \tag{10.61}$$

where $k_1 = \frac{N_1T_1 - P_1}{T_2}$, $k_2 = \frac{N_1T_1 - P_2}{T_2}$.

Solving the system of equations (10.61) gives the values of coefficients A_i:

$$A_1 = \frac{k_2}{2(k_1 - k_2)\mathrm{sh}\theta_1} e^{-\gamma_1 l}; \tag{10.62a}$$

$$A_2 = \frac{k_1}{2(k_1 - k_2)\mathrm{sh}\theta_2} e^{-\gamma_2 l}; \tag{10.62b}$$

$$A_3 = \frac{k_2}{2(k_1 - k_2)\mathrm{sh}\theta_1} e^{\gamma_1 l}; \tag{10.62c}$$

$$A_4 = \frac{k_1}{2(k_1 - k_2)\mathrm{sh}\theta_2} e^{\gamma_2 l}. \tag{10.62d}$$

where $\theta_1 = \gamma_1 l$, and $\theta_2 = \gamma_2 l$.

In order to determine the expressions of currents $I_{j\mathrm{E}}$ through the *RC*-EDP terminals ((10.54)–(10.59)), we should find the values of the currents (10.41)–(10.43), integration constant *C*, and the derivatives $\mathrm{d}U_1/\mathrm{d}x$ and $\mathrm{d}U_2/\mathrm{d}x$.

The integration constant can be found, for example, using (10.51) under the condition (10.60). The result is

$$C = \frac{1}{NR.}$$

Moreover, the expressions of the derivatives are acquired via (10.38)–(10.39). Then,

$$\frac{dU_1}{dx} = \gamma_1 A_1 e^{\gamma_1 x} + \gamma_2 A_2 e^{\gamma_2 x} - \gamma_1 A_3 e^{-\gamma_1 x} - \gamma_2 A_4 e^{-\gamma_2 x}.$$

Hence:

$$\left.\frac{dU_1}{dx}\right|_{x=0} = \gamma_1(A_1 - A_3) + \gamma_2(A_2 - A_4);$$

$$\left.\frac{dU_1}{dx}\right|_{x=l} = \gamma_1 A_1 e^{\gamma_1 l} + \gamma_2 A_2 e^{\gamma_2 l} - \gamma_1 A_3 e^{-\gamma_1 l} - \gamma_2 A_4 e^{-\gamma_2 l}.$$

Replacing the coefficients, A_i, with their expressions (10.62a–10.62d) to acquire the following final formulae:

$$\frac{dU_1}{dx}\Big|_{x=0} = \frac{1}{(k_1 - k_2)}\left(\frac{\theta_1 k_2}{\tanh\theta_1} - \frac{\theta_2 k_1}{\tanh\theta_2}\right) \quad (10.63)$$

$$\frac{dU_1}{dx}\Big|_{x=l} = \frac{1}{(k_1 - k_2)}\left(\frac{\theta_1 k_2}{\sinh\theta_1} - \frac{\theta_2 k_1}{\sinh\theta_2}\right) \quad (10.64)$$

Another derivative is found in the same manner:

$$\frac{dU_2}{dx} = -(k_1\gamma_1 A_1 e^{\gamma_1 x} + k_2\gamma_2 A_2 e^{\gamma_2 x} - k_1\gamma_1 A_3 e^{-\gamma_1 x} - k_2\gamma_2 A_4 e^{-\gamma_2 x})$$

$$\frac{dU_2}{dx}\Big|_{x=0} = \frac{k_1 k_2}{(k_1 - k_2)l}\left(\frac{\theta_1}{\tanh\theta_1} - \frac{\theta_2}{\tanh\theta_2}\right) \quad (10.65)$$

$$\frac{dU_2}{dx}\Big|_{x=l} = \frac{k_1 k_2}{(k_1 - k_2)}\left(\frac{\theta_1}{\sinh\theta_1} - \frac{\theta_2}{\sinh\theta_2}\right) \quad (10.66)$$

Substituting from (10.63)–(10.66) into (10.54)–(10.59) and considering the conditions of (10.41)–(10.43) allow one to find out the expressions of the currents through all the terminals of the *RC*-EDP.

Whence the *y*-parameter matrix first column elements are calculated from the expression $y_{i1} = I_i/U_1$ at $U_1 = 1$, they are the same as the expressions of the currents.

Hence, as a result of some transformations, the following analytical expressions of the *Y*-matrix coefficients are given by

$$y_{11} = D[k_1 k_2(\theta_{t1} - \theta_{t2}) - K_1(k_2\theta_{t1} - k_1\theta_{t2})] + K \cdot G; \tag{10.67}$$

$$y_{21} = D[K(k_2\theta_{t1} - k_1\theta_{t2}) + Nk_1 k_2(\theta_{t1} - \theta_{t2})] + K \cdot N \cdot G; \tag{10.68}$$

$$y_{31} = D[(k_2\theta_{t1} - k_1\theta_{t2}) - Nk_1 k_2(\theta_{t1} - \theta_{t2})] + N \cdot G; \tag{10.69}$$

$$y_{41} = D[K_1(k_2\theta_{s1} - k_1\theta_{s2}) - k_1 k_2(\theta_{s1} - \theta_{s2})] - K \cdot G; \tag{10.70}$$

$$y_{51} = -D[K(k_2\theta_{s1} - k_1\theta_{s2}) + Nk_1 k_2(\theta_{s1} - \theta_{s2})] - K \cdot N \cdot G; \tag{10.71}$$

$$y_{61} = -D[K(k_2\theta_{s1} - k_1\theta_{s2}) - N_1 k_1 k_2(\theta_{s1} - \theta_{s2})] - N \cdot G, \tag{10.72}$$

where

$$D = \frac{1}{(N_1 K_1 - 1)(k_1 - k_2) r.l}, \quad \theta_{si} = \frac{\theta_i}{\sinh \theta_i}, \quad \theta_{ti} = \frac{\theta_i}{\tanh \theta_i},$$

$$G = \frac{1}{N(N_1 K_1 - 1)}$$

The expressions of the other y-parameters can be obtained by performing the given above transformations under the corresponding boundary conditions for each terminal of the *RC*-EDP.

Practice

1. Derive the expressions of y_{21} and y_{22} for the ODHom *R-C*-0-layered EDP.
2. Prove condition (10.13).
3. Figure out the expressions of the voltage and the current through the resistive layer of the ODHet *R-C*-0-layered EDP under the following boundary conditions: $U(x_1) = 0$; $U(x_2) = 1$.
4. Derive the expressions of y_{21} and y_{22} for the ODHet *R-C*-0-layered EDP under the parabolic law of variation of the width (10.20b).
5. Derive the expression of y_{11} for the ODHom 0-*C*1-*R*-*C*2-*G*2-*R*2-layered EDP.

Test Questions

1. Why is it reasonable to describe the *RC*-EDP with the y-parameters when analyzing the devices comprising such EDP?
2. Write the y-parameter matrix of the ODHom *R-C*-0-layered EDP.
3. What laws of distribution of line resistance make it possible to derive the analytical expressions of the y-parameters of the ODHet *R-C*-0-layered EDP?
4. Is it possible to figure out an analytical expression of the y-parameters of the ODHet *R-C-NR*-layered EDP? Why?
5. Formulate the concept of electrical models of the one-dimensional systems with distributed parameters.

6. Is it possible to create electrical models of the two-dimensional systems with distributed parameters? What is the probable view of such a model?
7. Formulate the method to derive the equations of the ODHom $R1$-$G1$-$C1$-R-$C2$-$G2$-$R2$-layered EDP.
8. Depict the six-pole parameter y_{61} measurement circuit.

Chapter 11
Numerical Methods of *R*-*C*-0-EDP Fractal Elements

11.1 Preliminaries

There are a large number of numerical methods to solve the mathematical differential equations like the ones given by (9.15)–(9.17) that describe electric potential distribution of the *RCG*-EDP resistive layers. These methods include finite differences method, variable separation method (Fourier method), finite element method, and others. One has to study the object or the problem to solve and specify applicability of these methods.

The fundamental idea of *variable separation method* is to find a nontrivial (i.e., not identically zero) solution of equations like (9.15)–(9.17) that are subject to certain boundary conditions in the form of a product of two single-variable functions.

Assume that one wishes to solve an equation that describes the TDHom *RC*-EDP potential distribution (see Table 9.1, option 7). In the Cartesian coordinate system, under constant value of $j\omega rC_0 = j\omega_{\text{norm}}$, that equation allows a separation of variables due to the following. The assumed solution

$$U(x, y) = X(x)Y(y) \tag{11.1}$$

can be substituted into the studied equation. The forthcoming transformations easily result in

$$\frac{X''}{X} + \frac{Y''}{Y} - j\omega_{\text{norm}} = 0 \tag{11.2}$$

Hence, the separated equations will be

A.Kh. Gil'mutdinov et al., *Fractal Elements and their Applications*, Analog Circuits and Signal Processing, DOI 10.1007/978-3-319-45249-4_11

$$X'' = -\alpha^2 X \tag{11.3}$$

and

$$Y'' = \left(\alpha^2 + j\omega_{\mathrm{norm}}\right)Y, \tag{11.4}$$

where α^2 is a separation constant.

The following expressions are the solutions of (11.3) and (11.4), respectively:

$$X = A_1 e^{j\alpha x} + A_2 e^{-j\alpha x}; \tag{11.5}$$

$$Y = B_1 e^{\sqrt{\alpha^2 + j\omega_{\mathrm{norm}}}\, y} + B_2 e^{-\sqrt{\alpha^2 + j\omega_{\mathrm{norm}}}\, y} \tag{11.6}$$

The general solution can be represented as a linear combination of the product of the separated solutions (11.5) and (11.6) for all possible values of the separation constant in the following manner:

$$U(x,y) = \sum_{n=0}^{\infty} X_n(x) Y_n(y), \tag{11.7}$$

where $X_n(x)$ and $Y_n(y)$ are the separate solutions at $\alpha = \alpha_n$.

However, when analyzing the *RC*-EDP, this method can be recommended basically in cases of relatively simple boundary conditions of the *R*-*C*-0-layered *RC*-EDP.

The general idea of *the finite differences method* (FDM) depends on the following procedure: the range of the continuous variation of the argument is replaced with a discrete (finite) set of points called the grid. The continuous-argument functions are replaced with the discrete-argument functions defined at the grid nodes. The derivatives comprised in the differential equation are replaced with the corresponding difference expressions (the finite differences).

For example, the equivalent of the second derivative at point (x_i, y_i) will be the following finite difference:

$$\left.\frac{\partial^2 U}{\partial x^2}\right|_{x_i, y_i} \approx \frac{U(x_i + h, y_i) - 2U(x_i, y_i) + U(x_i - h, y_i)}{h^2},$$

where h is the grid step. The derivative with respect to y is estimated in the same manner.

The whole of the finite differences at a set of the grid points is a system of algebraic equations. The initial and boundary conditions are also replaced by difference initial and boundary conditions for the approximation function.

The solution of the algebraic system of equations subject to the initial and boundary conditions is an approximate distribution of an electric potential over the resistive layer. This distribution is further used to calculate the *RC*-EDP *y*-parameters.

The method of finite-difference approximation of (9.15)–(9.17)-type differential equations has a fine electrical interpretation; that is, it is possible to apply this method to analyze complex electrical circuits to simulate their electric fields. For example, it can be applied to those equivalent circuits that describe part of the ODHom *RC*-EDP as depicted in Fig. 10.4.

However, the precise solution by means of such kind of analog circuit requires significant amounts of *RC*-EDPs. Furthermore, the FDM capability to solve problems with heterogeneous and anisotropic distribution of parameters is poor.

The finite element method (FEM), just like the FDM, is a grid method. The two methods differ in the way of acquiring an approximate solution. In the case of an FDM, the equation and the boundary conditions are approximated, while for FEM, it approximates the desired solution. In addition, unlike FDM, FEM provides both a method to find the approximate solution in the grid nodes and a procedure to calculate the approximate solution at any point of any area not included in the grid nodes (the interpolation way).

The general idea of FEM is that any continuous quantity can be approximated with a discrete model based on numerous piecewise-continuous functions defined on a finite number of subareas. The piecewise-continuous functions are defined with values of the continuous quantity at a finite number of points of the studied area.

The basic stages of applying FEM to solve problems are listed as follows:

1. The physical area is divided into subareas or finite elements (FinE). The dimension of an element depends on the dimension of the problem. The shape of the FinE should provide proper approximation of the boundary conditions.
2. The dependent variable (one or several) is approximated with a special function within each FinE and, consequently, on the whole of the area (usually it is performed with a polynomial like $U(x, y) = a_0 + a_1x + a_2y + a_3xy$). The parameters of these approximations (the values of the continuous quantity at every node of the FinE) become the desired parameters of the problem.
3. Substituting the approximations into the constituting equations results in a system of algebraic equations with the unknown parameters. Solving this system provides an approximate distribution of the continuous quantity over its defined domain.

This concept of FEM is illustrated by Fig. 11.1. The whole of the modeled area is divided, for instance, into triangle elements (as shown in Fig. 11.1a) that are small enough to assume the linear dependence of the potential and the coordinates within each triangle. One triangle element with vertexes *1*, *2*, and *3* is shown in Fig. 11.1b. It means that the initial smooth surface of the real potential distribution is replaced with a faceted surface, which performs piecewise-planar approximation of the ideal solution.

The potentials of the corresponding vertexes of the neighboring elements should be set in order to avoid discontinuities at joints of these elements. For example, the

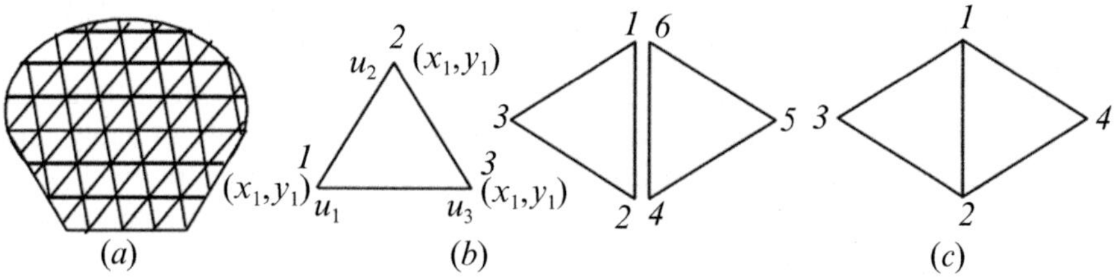

Fig. 11.1 Illustration of the finite element method: (**a**) Fragmentation, (**b**) description of one element, (**c**) assembly of elements

potentials of vertexes *1* and *6* as well as *2* and *4* of the two neighboring elements should be the same (see Fig. 11.1b). Then the elements are united and the vertexes are renumbered (see Fig. 11.1c). The local stiffness matrix stipulates the relation between the coordinates of the neighboring element vertices and the approximating polynomial coefficients. After that, the local matrices are continuously united into one global matrix.

It should be noted that in case of bilinear approximating function at every FinE, FEM possesses the same electrical interpretation as FDM. Therefore, in this sense, analog simulation of the solution of the electrodynamical problem using FEM has no particular difference from the same thing done with FDM.

However, FEM makes it possible to set the type of the approximating function within every FinE and it is an advantage when solving problems using this method in comparison to FDM. It will be shown further that this feature leads to create highly efficient method of analog simulation.

11.2 Finite Distributed Element Method

The problem of applying the finite element method to estimate the potential distribution over the *R*-*C*-0-layered *RC*-EDP resistive layer was set up and analytically studied in a book written by one of the authors ([1]A. Kh. Gil'mutdinov. Resistive-capacitance elements with distributed parameters: Analysis, synthesis and application (Rezistivno-emkostnye elementy s raspredelennymi parametrami: Analiz, sintez i primenenie) (In Russian). Kazan state technical university publishing. Kazan. 2005. 350 p.).

[1]A.Kh. Gil'mutdinov, A.A. Goppe. Analysis of the *RC*-elements with surface-distributed parameters by means of the finite distributed elements method (Analiz *RC*-elementov s poverkhnostno-raspredelennymi parametrami metodom konechnykh raspredelennykh elementov) (In Russian) // Proceedings of the scientific-and-technical conference of the Kazan Aviation Institute folowing the results of 1992–1993 (Kazan, April 14–15, 1994). Kazan Aviation Institute, Kazan. 1994. P.218.

The proposed studies show that the precision of calculating the *RC*-EDP *y*-parameters using FEM with the bilinear functions for the FinE is frequency dependent, all things being equal. The method can be applied only for $\omega_{\text{norm}} \leq 1$. The approximation function expressed in (11.8) was proposed to get rid of this limitation. This function is equal to the known solution of the ODHom *R*-*C*-0-layered *RC*-EDP for every FinE (see expression (10.6)):

$$U(x_i) = -\frac{\sinh \gamma(x_i - x)}{\sinh \gamma h_i} \tag{11.8}$$

where γ is the propagation constant, and h_i is the grid step.

In case of such kind of approximation, the precision of the solution already does not depend on frequency. The proposed FEM procedure called "hybrid" has been used as the basis of the finite distributed element method (FDEM).

Let us illustrate the general idea of FDEM and realize its efficiency in comparison to the *RC*-EDP-based simulation. To do that, consider an ODHom *R*-*C*-0-layered *RC*-EDP, which has a structure divided into finite elements and is depicted in Fig. 11.2a. Applying the electrical interpretation of FEM (FDM) implies that the electrical equivalent circuit of the studied structure will be like the one depicted in Fig. 11.2b.

Each elementary *RC* circuit simulates the *RC*-EDP length potential distribution that approximates the ideal solution with bilinear functions for each FinE. The number of the circuits depends on the required precision of the solution.

Assume that we know the analytical expression of the ODHom *R*-*C*-0-layered *RC*-EDP parameter y_{22} (11.8). Now estimate the precision of solution when calculating this parameter by simulating the ODHom *R*-*C*-0-layered *RC*-EDP equivalent

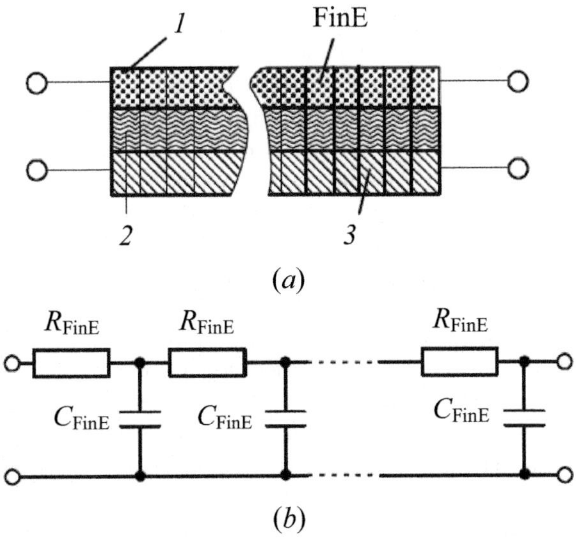

Fig. 11.2 Illustration of FEM when FinE are replaced with their electrical equivalence under bilinear approximating function: (**a**) the ODHom *R*-*C*-0-layered *RC*-EDP divided into FinE, (**b**) equivalent circuit: (*1*) resistive layer; (*2*) insulator; (*3*) ideal conductor

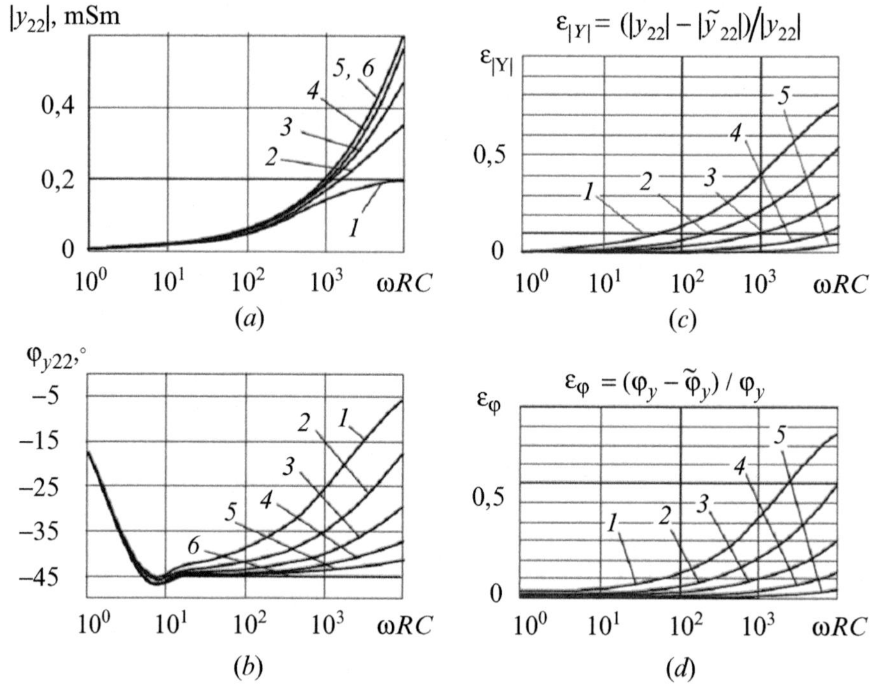

Fig. 11.3 Simulation results of the ODHom *R-C*-0-layered EDP using a multistaged *RC* circuit: (**a**) y_{22} parameter MFR; (**b**) y_{22} parameter PFR; (**c, d**) graphs of absolute value of y_{22} and phase calculation accuracy with respect to the number of elementary circuits and frequency; (*1*) $n = 32$; (*2*) $n = 64$; (*3*) $n = 128$; (*4*) $n = 256$; (*5*) $n = 512$; (*6*) precise analytical expression

circuit with *OrCAD* software at different number of FinE (number of elementary circuits).

The results of simulation in the form of parameter y_{22} MFR and PFR compared to the corresponding MFR and PFR calculated via formula $y_{22} = \gamma \cdot \coth \gamma$ are given in Fig. 11.3a,b. Notice that Fig. 11.3c,d depicts the graphs of dependence of the relative accuracy of this parameter FEM based on frequency and on the number of elementary *RC* circuits that constitute the ODHom *R-C*-0-layered EDP equivalent circuit.

The pictures show that the absolute value of the relative accuracies of y_{22} and phase calculation possess the same character, as expected, and grow with increasing frequency. For $n = 32$, the range of frequencies, where the calculation accuracy of the *y*-parameter does not exceed 10 %, is about two decades; however, for $n = 64$ this range is two and a half decades. To obtain calculation accuracy of not less than 1 % within three decades, at least 512 elementary circuits will be needed.

It follows from this small investigation that it takes significant computational costs to replace the ODHom *R-C*-0-layered EDP with an *RC*-EDP-based equivalent circuit (classical FEM) with the aim to estimate the primary parameters of the ODHom or ODHet *RC*-EDP with appropriate accuracy (for example not less than

1 %) within a wide frequency range. For example, it took a 1.7 GHz pentium-5-based personal computer with 1.5 Gb RAM 1.67 s to calculate the ODHom R-C-0-layered EDP equivalent circuit y_{22} parameter.

On another hand, the "hybrid" version of FEM can also be represented in the form of its electrical interpretation. As long as the approximation function here is the ODHom R-C-0-layered EDP equation solution, each FinE can be replaced with its physical equivalent in the form of the ODHom R-C-0-layered EDP segment. The latter will be hereinafter called the R-C-0-layered one-dimensional homogeneous finite distributed element (ODHom FDE). Then the computational scheme of FDEM in the context of the ODHom R-C-0-layered EDP (see Fig. 11.2a) can be represented in the form of serially connected ODHom R-C-0-layered EDP segments.

It is evident that it makes no sense to divide the ODHom EDP into pieces as long as the y-matrix of the i-th FinE in our example differs from the corresponding ODHom R-C-0-layered EDP only with the propagation constant as follows:

$$[Y_i]_{R-c-0} = \frac{\theta_i}{R_i \sinh\theta_i} \begin{bmatrix} \cosh\theta_i & -1 \\ -1 & \cosh\theta_i \end{bmatrix} \tag{11.9}$$

where $\theta_i = \sqrt{\omega R_i C_i}$ is the i-th FinE propagation constant, and $R_i C_i$ is the i-th FinE total resistance and capacitance, respectively.

Therefore, the ODHom EDP can be represented with one ODHom FDE. The corresponding calculation of any y-parameter according to formula (11.8) under the same computational capabilities lasts for tenths of a microsecond.

It is obvious that FDEM is worth using to analyze the ODHet, TDHom, and TDHet RC-EDP with various static and dynamic heterogeneities. Hence, FDEM (that is an electrical equivalence of the FEM "hybrid" scheme) has been obviously shown to be significantly more efficient in comparison to the classical FEM (or FDM). Furthermore, it will be shown that FDEM will make it possible to calculate the y-parameters of the ODHet, TDHom, and TDHet RC-EDP with several resistive layers (for example, the $R1$-$G1$-$C1$-R-$C2$-$G2$-$R2$ EDP). The known information sources such as special literature have no references that would describe the classical methods of solving the partial differential equations applied to solve such kind of problems.

11.3 Application of Finite Distributed Element Method to Calculate the ODHet *RC*-EDP Primary Parameters

11.3.1 FDEM-Based Calculation of the Y*-Parameters of the ODHet* RC*-EDP of Heterogeneous Width*

Assume that it is required to calculate the y-parameters of the ODHet R-C-0-layered EDP device. Let the distribution of the electric potential over the resistive layer is provided and varies along its width according to the law, $b(x)$ (under constant values of the following specific parameters of the square-resistive and the dielectric layers: $\rho_\square = \text{const}$, $C_{\text{spec}} = \text{const}$).

Without loss of generality, one may assume $b(x)$ a continuous function with the parameters that do not break the structure design or the conditions of its physical realizability. Assume that this function has no analytical solution of the equation that describes the potential distribution over the resistive layer (Fig. 11.4a).

Let us introduce an *algorithm to calculate the y-parameters* of the ODHet R-C-0-layered EDP.

1. According to the proposed FDEM, let the ODHet R-C-0-layered EDP be divided into FinE Δx_i of length n, where n—the number of FinE ($i = 1,\ldots, n$), so that every FinE could be replaced by an ODHom FDE with the same layer structure. Then the ODHet R-C-0-layered EDP equivalent circuit will look like the structure depicted in Fig. 11.4b. The FDE conductivity matrices are defined by the expression (11.9).
2. Express R_i and C_i via the specific parameters of the layers as follows:

$$R_i = \rho_\square \cdot \Delta x_i / b_i; \quad C_i = C_{\text{spec}} b_i \Delta x_i, \tag{11.10}$$

   where b_i is the i-th FinE width that can be specified at the beginning and at the end of FinE in the form of their mean value.
3. Substitute expressions (11.10) into (11.9) to get the width-heterogeneous R-C-0-EDP i-th FDE y-matrix as follows:

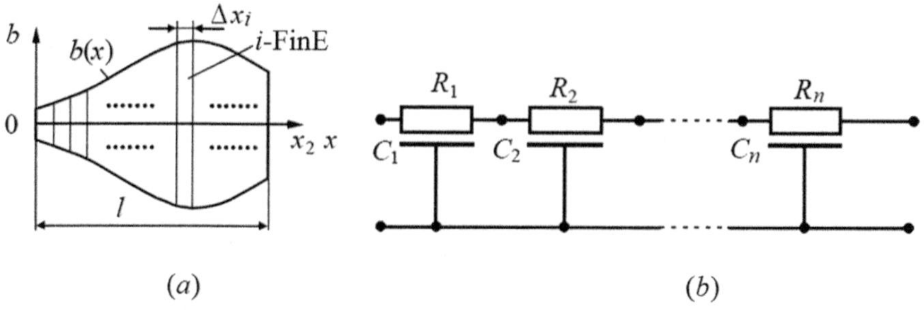

Fig. 11.4 ODHet R-C-0-layered EDP: (**a**) layer geometry; (**b**) equivalent circuit

$$[Y_i]_{b=b(x)} = \frac{\theta_i b_i}{\rho_\square \, \Delta x_i} \begin{bmatrix} \coth\theta_i & -\mathrm{csch}\theta_i \\ -\mathrm{csch}\theta_i & \coth\theta_i \end{bmatrix} \tag{11.11}$$

where $\theta_i = \Delta x_i \sqrt{j\omega\rho_\square \, C_{\text{spec}}}$.

4. According to the theory of electric circuits, it is recommended to multiply the A-matrices of the ODHom FDE of the compound two-port (four-terminal) network (see Fig. 11.4b) to obtain the aggregate parameter matrix. Therefore, it makes sense to transform the y-matrix of the i-th ODHom FDE (11.11) into an A-matrix by means of the known formulae of transformation of the two-port (four-terminal) matrices. In our case, the A-matrix of the i-th ODHom FDE will be

$$[A_i]_{b=b(x)} = \begin{bmatrix} \cosh\theta_i & \dfrac{\rho_\square \Delta x_i}{\theta_i b_i} \sinh\theta_i \\ \dfrac{\theta_i b_i}{\rho_\square \Delta x_i} \sinh\theta_i & \cosh\theta_i \end{bmatrix} \tag{11.12}$$

5. After multiplying all of the FDE A-matrices, where their coefficients vary according to the width variation law of the RC-EDP, the initial parameter formulae of the two-port circuit are applied to use again the y-parameters instead of the a-parameters in order to figure out the ODHet R-C-0 EDP y-matrix.

11.3.2 Y-Parameter Calculation of the Complementary ODHet RC-*EDP Devices*

A particular case of the width-heterogeneous RC-EDP is the so-called complementary RC-EDP. It has been patented first by the authors of the present book.

The one-dimensional complementary (ODC) RC-EDP has been described in Sect. 4.6 as a couple of complementary ODHet RC-EDP acquired from the initial rectangular ODHom RC-EDP by slicing it along its x-axis into two pieces (longitudinal section) according to a certain law. If the initial ODHom RC-EDP is an R-C-0-layered one, then the corresponding ODC R-C-0-layered RC-EDP is a six-terminal circuit. Its graphical symbol is given in Fig. 11.5.

Since the ODC RC-EDP is compound and the parallel connection of its element ODHet RC-EDP produces the initial ODHom ODC R-C-0-layered RC-EDP, then, if neglecting the gap width, the initial RC-EDP y-parameter indefinite matrix will be equal to the sum of indefinite matrices of the y-parameters of the initial RC-EDP elements as follows:

$$[Y] = [Y]' + [Y]'', \tag{11.13}$$

where $[Y]'$ and $[Y]''$ are the indefinite matrices of the y-parameters of the lower and the upper heterogeneous elements, respectively (see Fig. 11.5).

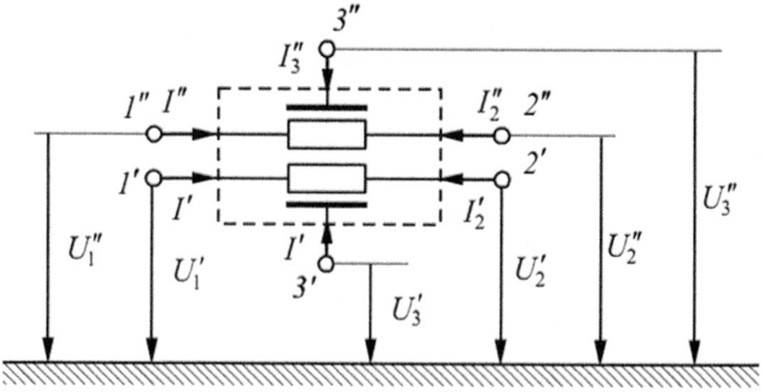

Fig. 11.5 The six-terminal equivalent circuit of the ODC *R*-*C*-0-layered *RC*-EDP

To reduce the computational costs, the indefinite matrix properties can be used to calculate only the four basic *y*-parameters: y_{ij}, where $i, j = 1, 2$. The rest of these parameters are calculated according to the following expressions:

$$y_{i3} = -\sum\nolimits_{j=1}^{2} y_{ij}, \quad \text{where } i = 1,2;$$
$$y_{3j} = -\sum\nolimits_{i=1}^{2} y_{ij}, \quad \text{where } j = 1,2;$$
$$y_{33} = \sum\nolimits_{i=1}^{2} \sum\nolimits_{j=1}^{2} y_{ij}, \quad \text{where } i = 1,2, \quad j = 1,2.$$

Then expression (11.13) can be converted as follows:

$$\begin{bmatrix} y_{11} & y_{12} \\ y_{21} & y_{22} \end{bmatrix} = \begin{bmatrix} y'_{11} & y''_{11} & y'_{12} & y''_{12} \\ y'_{21} & y''_{21} & y'_{22} & y''_{22} \end{bmatrix}. \tag{11.14}$$

The property of the complementary *RC*-EDP makes it possible to perform FDEM-based calculation of the *y*-parameters of just one part of its component and to calculate the second components of the *y*-parameters using (11.13).

The relation between the currents and voltages at the terminals of the six-terminal ODC *R*-*C*-0-layered EDP given in Fig. 11.5 is specified by the following systems of equation and written in a matrix form as follows:

$$\begin{bmatrix} y'_{11} & y'_{12} & y'_{13} & 0 & 0 & 0 \\ y'_{21} & y'_{22} & y'_{23} & 0 & 0 & 0 \\ y'_{31} & y'_{32} & y'_{33} & 0 & 0 & 0 \\ 0 & 0 & 0 & y''_{11} & y''_{12} & y''_{13} \\ 0 & 0 & 0 & y''_{21} & y''_{22} & y''_{23} \\ 0 & 0 & 0 & y''_{31} & y''_{32} & y''_{33} \end{bmatrix} \cdot \begin{bmatrix} U'_1 \\ U'_2 \\ U'_3 \\ U''_1 \\ U''_2 \\ U''_3 \end{bmatrix} = \begin{bmatrix} I'_1 \\ I'_2 \\ I'_3 \\ I''_1 \\ I''_2 \\ I''_3 \end{bmatrix}. \tag{11.15}$$

It is obvious that knowledge of the six-terminal ODC *R*-*C*-0-layered EDP *y*-parameters allows one to determine the circuit functions of these devices.

11.3.3 Y-Parameter Calculation of the Parametric ODHet RC-EDP Devices

Chapter 7 has provided the fundamentals of the parametric *RC*-EDP. Let us consider a parametric *R*-*C*-0-layered EDP device with constant width b_0. The variation law $\rho_\square(x, t)$ of the resistive layer square resistance can be managed by means of a time-variable external field.

The resistive layer square resistance is calculated from $\rho_\square = \rho/d$, where ρ is the specific volume resistance of the resistive layer, and d is the resistive layer thickness. Therefore, $\rho_\square$ can be specified in two ways: by changing the layer material electro-physical characteristic performance (for example, in case of photo-, temperature-, or magneto-sensitive resistive materials) or by changing the space-charge region thickness (in MOS structures, for instance).

To perform further analysis, let us choose an arbitrary time moment and consider the problem of estimating the *Y*-matrix for ODHet *RC*-EDP devices where its heterogeneity is determined with the $\rho_\square(x)$ law. Then, the product of the line parameters $r(x)c(x) \neq \text{const}$. It follows from Chap. 10 that there has been found no analytical solution for such differential equations that describe the potential distribution over the *RC*-EDP resistive layer. For this reason, FDEM will be used to calculate the primary parameters of the ODHom FDE the ODHet *RC*-EDP divided into FinE elements (see Fig. 11.6).

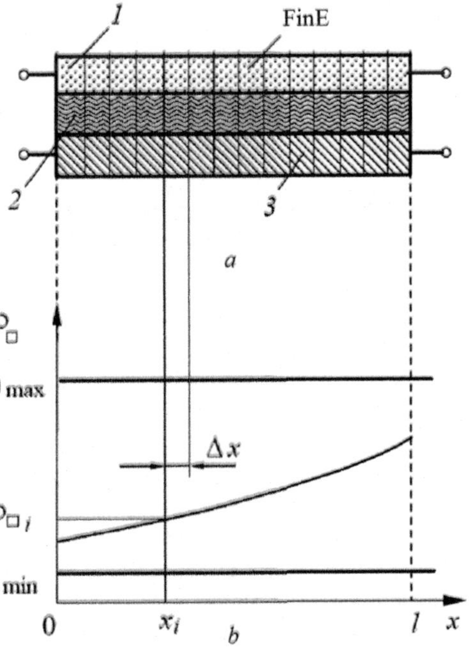

Fig. 11.6 Illustration of the parametric *R*-*C*-0-layered EDP *y*-parameter calculation: (**a**) the *RC*-EDP structure divided into FinE; (**b**) $\rho_\square(x)$ plot: (*1*) resistive layer; (*2*) insulator; (*3*) ideal conductor

Let the variation range of the resistive-layer resistance be limited to the maximum and minimum square resistance values $\rho_{\square\min}$ and $\rho_{\square\max}$. The current value of $\rho_{\square i}$ of any FinE can be within this range. The parameters of ODHom FDE that substitute each FinE (Fig. 11.6a) can be expressed as follows:

$$R_i = \frac{\Delta x_i}{b_o}\ ; \quad C_i = C_{\text{spec}}\, b_o \Delta x_i \tag{11.16}$$

The i-th ODHom FDE propagation constant, θ_i, with a distribution law, $\rho_\square(x)$, is expressed as follows:

$$\theta_i = \sqrt{j\omega\, \rho_{\square\,i}\, C_{\text{spec}}\, \Delta x_i}\ ; \quad i = 1,\ \ldots,\ n \tag{11.17}$$

Then the Y-matrix of the i-th ODHom FDE of the R-C-0-layered EDP that possesses square resistance heterogeneity of the resistive layer will be as follows:

$$[Y_i]_{\rho_\square = \rho_\square(x)} = \frac{\theta_i b_o}{\rho_{\square i} \Delta x_i} \begin{bmatrix} \coth\theta_i & -\text{csch}\theta_i \\ -\text{csch}\theta_i & \coth\theta_i \end{bmatrix} \tag{11.18}$$

The calculation of the y-parameters of the ODHom RC-EDP that possesses heterogeneous square resistance of its resistive layer requires executing steps 4 and 5 of the y-parameter estimation algorithm of the width-heterogeneous R-C-0-layered EDP (see Sect. 11.3.1).

The variation of the external field in time that affects the RC-EDP causes a change in the $\rho_\square(x)$ law. Thus, this procedure should be performed many times in order to accurately analyze the influence of the external action on the RC-EDP performance characteristics.

11.4 Application of Finite Distributed Element Method to Calculate the TDHom *RC*-EDP *Y*-Parameters

Consider a TDHom R-C-0-layered EDP device. The view of its resistive and contact pad layers is depicted in Fig 11.7. According to the FEM theory, the whole surface of the RC-EDP is divided into a grid $(N_1 - 1)(N_2 - 1)$ rectangular finite element. Here N_1 and N_2 denote the number of the grid nodes along the length l and width b of the element, respectively. Generally speaking, the shape of the finite elements cannot only be rectangular. It is chosen here for convenience to describe the RC-EDP borders.

The structure of each FinE is depicted in Fig. 11.8a. The FinE are replaced by TDHom FDE element that has the circuit diagram shown in Fig. 11.8b. The equivalent circuit diagram is comprised of four ODHom R-C-0-layered EDP. It is described by the Y-matrices that differ depending on the direction of the FinE facets as follows:

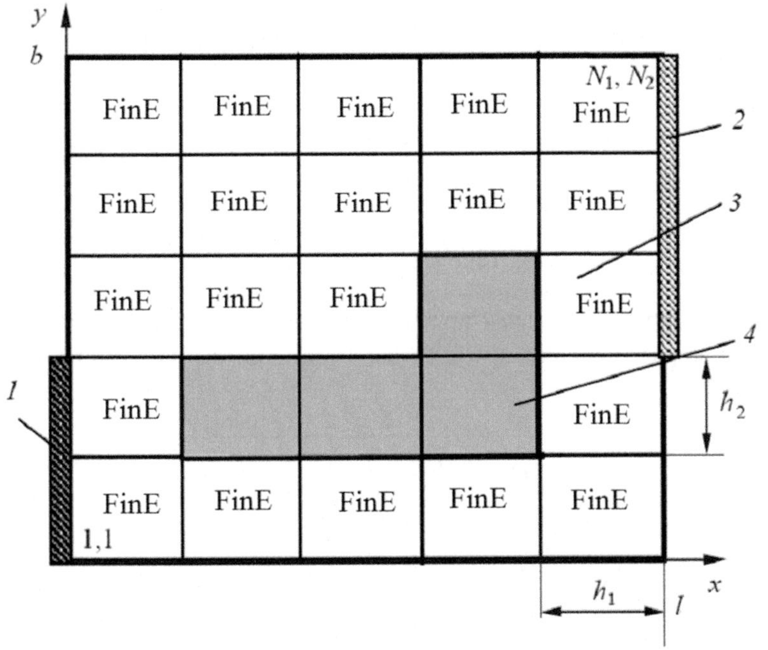

Fig. 11.7 FDEM illustrated by means of the TDHom *R*-*C*-0-layered EDP: (*1*, *2*) contact pads of the resistive layer, (*3*) resistive layer, (*4*) cutout in the resistive layer, FinE: rectangular finite element

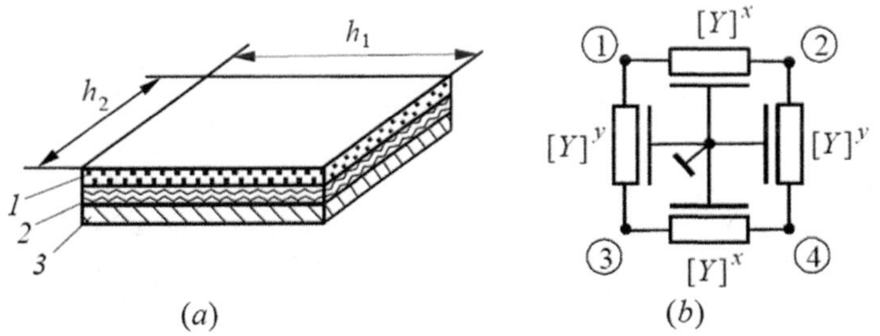

Fig. 11.8 The structure of the FinE of the TDHom *R*-*C*-0-layered EDP (**a**) and its equivalent circuit in the form of a TDHom FDE (**b**): (*1*) resistive layer, (*2*) insulator, (*3*) ideal conductor

$$[Y]^x = \frac{\theta_x}{R_x \sinh\theta_x} \begin{vmatrix} \cosh\theta_x & -1 \\ -1 & \cosh\theta_x \end{vmatrix} \tag{11.19}$$

$$[Y]^y = \frac{\theta_y}{R_x \sinh\theta_y} \begin{vmatrix} \cosh\theta_y & -1 \\ -1 & \cosh\theta_y \end{vmatrix} \tag{11.20}$$

Let us calculate parameters R_x, R_y, θ_x, and θ_y assuming that the resistive layer square resistance, $\rho_\square$, and the insulator layer-specific capacitance, C_{spec}, are known.

The resistance, R_x, will be determined by short-circuiting the nodes 1–3 and 2–4, and R_y will be determined by short-circuiting the nodes 1–2 and 3–4 (see Fig. 11.8b). Then

$$R_x = 2\rho_\square \frac{h_1}{h_2}; \quad R_y = 2\rho_\square \frac{h_2}{h_1} \tag{11.21}$$

The FinE capacitance does not depend on the direction of the FinE facets and it equals to $C_{\mathrm{FinE}} = C_{\mathrm{spec}} h_1 h_2$. Take into account the fact that this capacitance is determined by the capacitances of four ODHom *RC*-EDP facets, and each ODHom *RC*-EDP capacitance in the equivalent circuit equals to $C = C_{\mathrm{FinE}}/4$. Then the propagation constants θ_x, θ_y that characterize the FinE can be described as follows:

$$\theta_x = \sqrt{j\omega 2\rho_\square \frac{h_1}{h_2} C_{\mathrm{spec}} \frac{h_1 h_2}{4}} = \frac{h_1}{\sqrt{2}} \sqrt{j\omega\rho_\square C_{\mathrm{spec}}} = \frac{h_1}{\sqrt{2}} \theta \tag{11.22}$$

$$\theta_y = \sqrt{j\omega 2\rho_\square \frac{h_2}{h_1} C_{\mathrm{spec}} \frac{h_1 h_2}{4}} = \frac{h_2}{\sqrt{2}} \sqrt{j\omega\rho_\square C_{\mathrm{spec}}} = \frac{h_2}{\sqrt{2}} \theta \tag{11.23}$$

where $\theta = \sqrt{j\omega\rho_\square C_{\mathrm{spec}}}$.

Substituting from (11.21)–(11.23) into (11.19)–(11.20) yields

$$[Y]^x = \frac{\theta h_2}{2\sqrt{2}\rho_\square \sinh\theta_x} \begin{vmatrix} \cosh\theta_x & -1 \\ -1 & \cosh\theta_x \end{vmatrix} \tag{11.24}$$

$$[Y]^y = \frac{\theta h_1}{2\sqrt{2}\,\rho_\square \sinh\theta_y} \begin{vmatrix} \cosh\theta_y & -1 \\ -1 & \cosh\theta_y \end{vmatrix} \tag{11.25}$$

The considered structure of the TDHom FDE of (Fig. 11.8b) is a five-terminal that can be described by a certain *Y*-matrix with respect to the ideal conductor of the FinE. Hence, the following algorithm summarizes the procedure of constructing the *Y*-matrix.

First, a 4×4 zero square matrix is made. Then it is filled with values of the *y*-parameters calculated from the expressions (11.24)–(11.25). The rest of the entries are filled as shown in Fig. 11.9.

Insert the *y*-parameters of the rest of the two ODHom *RC*-EDP according to the described principle. Since $y_{11} = y_{22}$, $y_{21} = y_{12}$. Then the resulting *Y*-matrix of the TDHom FDE will be as follows:

	1	2	3	4
1	y_{11}^x	y_{12}^x	0	0
2	y_{21}^x	y_{22}^x	0	0
3	0	0	0	0
4	0	0	0	0

(*a*)

	1	2	3	4
1	y_{11}^x	y_{12}^x	0	0
2	y_{21}^x	$y_{22}^x + y_{11}^y$	0	y_{12}^y
3	0	0	0	0
4	0	y_{21}^y	0	y_{22}^y

(*b*)

Fig. 11.9 The TDHom FDE *Y*-matrix formation principle illustration: (**a**) matrix $[Y]^x$ elements inserted between nodes 1 and 2; (**b**) matrix $[Y]^y$ elements inserted between nodes 2 and 4

$$[Y]\text{TDHom FDE} = \begin{bmatrix} y_{11}^x + y_{11}^y & y_{12}^x & y_{12}^y & 0 \\ y_{12}^x & y_{11}^x + y_{11}^y & 0 & y_{12}^y \\ y_{12}^y & 0 & y_{11}^x + y_{11}^y & y_{12}^x \\ 0 & y_{12}^y & y_{12}^x & y_{11}^x + y_{11}^y \end{bmatrix}. \quad (11.26)$$

It is obvious that (11.26) represents a symmetric matrix; therefore, it is sufficient to use just the upper or the lower triangular matrix (upper or lower echelon forms) when using the analysis and the synthesis algorithms.

The assembly algorithm of the *Y*-matrix of the TDHom FDE devices is used to assemble the TDHom *R*-*C*-0-layered EDP global *Y*-matrix. The first stage in this case is to create a $(N_1 - 1)(N_2 - 1)$ zero matrix. Then the TDHom FDE *Y*-matrix elements are inserted into this matrix according to the aforementioned principle.

All the forthcoming operations (considering the boundary conditions and the calculations of highly sparse matrix of *y*-parameters of the TDHom FDE) focus on programming the matrix operations but not FDEM. That is why such operations are not studied in this chapter.

Practice

1. Figure out a *Y*-matrix for the *i*-th ODHom FDE of the *R*-*C*-0-layered EDP that is heterogeneous in the dielectric layer-specific capacitance.
2. Use the *Y*-matrix of the *i*-th ODHom FDE of the *R*-*C*-*NR*-layered EDP given in practical assignment 1 of Chap. 5 to derive a *Y*-matrix of the *i*-th ODHom FDE of the *R*-*C*-*NR*-layered EDP that is heterogeneous in the resistive layer *R* square resistance.
3. Derive *Y*-matrices of the *i*-th ODHom FDE of the *R*-*C*-*NR*-layered EDP that are the constituents of the ODC *R*-*C*-*NR*-layered EDP with specified cutting variation law $b(x)$.
4. Write an algorithm to calculate the *y*-parameters of the ODC *R*-*C*-*NR*-layered EDP considered in the previous practical assignment.

Test Questions

1. Describe the concept of the variable separation method used to solve the differential equations.
2. Describe the concept of the finite difference method used to solve the differential equations.
3. Describe the concept of the finite element method used to solve the differential equations.
4. Describe the concept of the finite distributed element method used to solve the differential equations.
5. Show that FDEM is much more efficient as compared to the classical FEM.
6. Provide an FDEM-based algorithm to calculate the *y*-parameters of the ODHet *RC*-EDP with width heterogeneity.
7. Provide an FDEM-based algorithm to calculate the *y*-parameters of the ODHet *RC*-EDP with heterogeneous square resistance.
8. What are the advantages of the ODC *RC*-EDP over the conventional ODHet *RC*-EDP to your mind?
9. Provide an algorithm of creation of the TDHom *RC*-EDP global *Y*-matrix.

Chapter 12
Primary Parameter Calculation of the *RC*-EDP Using Generalized *RCG*-EDP Devices

The concept of a generalized simulation environment proposed in Chap. 8 implies a generalized analysis, design, and fabrication of objects that can be represented with multilayered *RCGL*-EDP devices to meet performance characteristics.

In this regard, there is a necessity to extend finite distributed element method (FDEM) to the new class of objects with distributed nature. This chapter considers the basic ideas of the proposed FDEM enhancement and its application to calculate the *Y*-matrices of the generalized heterogeneous structure of *RC*-EDP.

12.1 Generalized Finite Distributed Element Method

12.1.1 Preliminaries

Recall that the FDEM is an electrical equivalence of the classical FEM "hybrid" algorithm, and the FinE equivalent circuit elements are the ODHom *RC*-EDP with different layer structure (FDE). Let us consider such a problem statement for the generalized *RCG*-EDP (see Chap. 9).

First, the *RCG*-EDP surface is divided into equal-shaped elements (triangular, rectangular, trapezoidal, etc.) (see Fig. 12.1a). Every such element is a spatial object within the multilayered structure (FinE) as depicted in Fig. 12.1b. Numbers *1–12* here mark the vertices of the FinE resistive layers.

Every resistive layer of the FinE should be determined with numerous node potentials and equations that approximately describe the potential distribution over every layer. For that purpose, the coefficients of these equations are calculated by solving a system of algebraic equations of these coefficients. However, FEM is not applied to electrodynamical objects with numerous different conductivity layers.

This book proposes implementing this algorithm by means of FEM using another approach to approximate the potential distribution over the selected

A.Kh. Gil'mutdinov et al., *Fractal Elements and their Applications*, Analog Circuits and Signal Processing, DOI 10.1007/978-3-319-45249-4_12

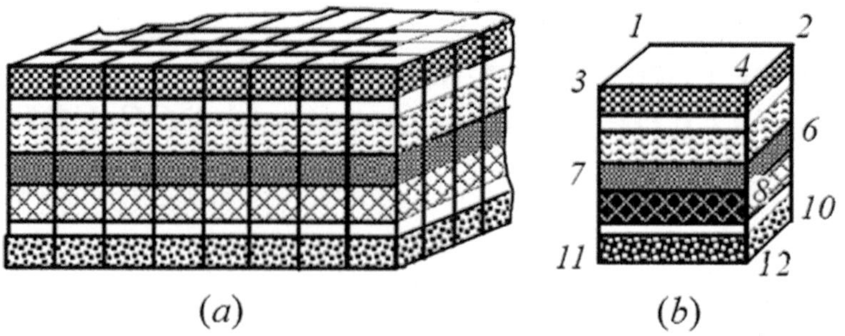

Fig. 12.1 A generalized *RCG*-EDP divided into rectangular finite elements: (**a**) a fragment of the *RCG*-EDP, (**b**) a finite element

fragment of the generalized *RCG*-EDP (see Fig. 12.1b). Let us use the terminology introduced in Chap. 11 and call the developed technique *the generalized finite distributed element method* (GFDEM).

12.1.2 The GFDEM Algorithm

Potential distribution over every FinE will not be approximated with algebraic polynomials. It will be simulated by means of a physical structure where node potentials cause electric currents and the potential distribution that reflects the character of the simulated medium.

It should be noted that a particular FinE can be represented as a two-port (four-terminal) circuit. The steady-state small-signal dependence between the currents and the node voltages is usually specified with multiterminal parameters (y- or z-parameters). Therefore, the next step is to develop an algorithm to create a FinE model that would make it possible to calculate the parameters of this multiterminal occupied by the area of the FinE of interest under the specified boundary conditions and the simulation parameters.

In case of the one-dimensional homogeneous generalized *RCG*-EDP devices, such model can be implemented by means of the $R1$-$G1$-$C1$-R-$C2$-$G2$-$R2$-layered one-dimensional homogeneous *RC*-EDP. Analytical expressions of the y-parameters of the latter have been found in Sect. 10.4. According to the terminology accepted earlier, such one-dimensional FinE will hereinafter be called the *one-dimensional homogeneous generalized finite distributed element* (ODHom GFDE).

In the general case, the FinE can be simulated with the ODHet GFDE that also should possess an analytical solution. Applying the ODHet FDE (here ODHet GFDE) will make it possible to reduce the number of FinE when approximating the potential distribution over the medium with heterogeneously distributed line parameters.

The graphical symbol of the ODHom GFDE is depicted in Fig. 12.2a and the general view of the conductivity matrix is shown in Fig. 12.2b.

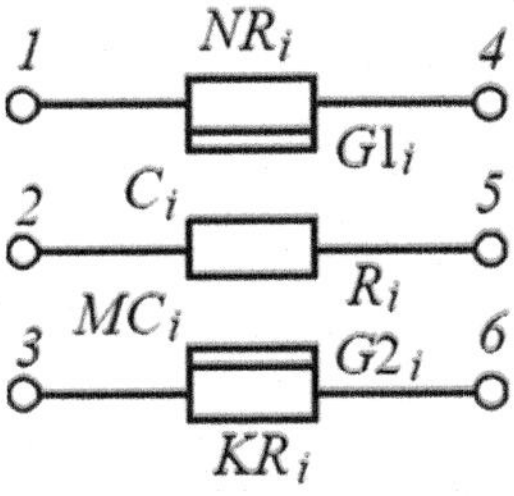

$$Y = \begin{bmatrix} y_{11} & y_{12} & y_{13} & y_{14} & y_{15} & y_{16} \\ y_{21} & y_{22} & y_{23} & y_{24} & y_{25} & y_{26} \\ y_{31} & y_{32} & y_{33} & y_{34} & y_{35} & y_{36} \\ y_{41} & y_{42} & y_{43} & y_{44} & y_{45} & y_{46} \\ y_{51} & y_{52} & y_{53} & y_{54} & y_{55} & y_{56} \\ y_{61} & y_{62} & y_{63} & y_{64} & y_{65} & y_{66} \end{bmatrix}$$

(*a*) (*b*)

Fig. 12.2 (**a**) The ODHom GFDE graphical symbol and (**b**) the corresponding *y*-parameter matrix general view

When the FinE is simulated with a GFDE-based model, the node potentials of all resistive layers can be calculated (for example, in order to study the potential distribution along the *RC*-EDP length) using the following matrix equation describing the equivalent circuit of the *RC*-EDP comprised of GFDE elements:

$$[\dot{U}] = [\dot{Y}]_{\Sigma}^{-1} [\dot{I}] \tag{12.1}$$

where $[\dot{U}]$ is the node voltage matrix, $[\dot{I}]$ is the matrix of currents through the branches of the equivalent circuit, and $[\dot{Y}]_{\Sigma}$ is the *y*-parameter global matrix comprised of separate GFDE conductivity matrices (12.1).

However, experience has proven that in order to analyze the *RC*-EDP performance characteristics, it is quite enough to use (12.1) to calculate the *y*-parameters of the *RC*-EDP with *n*-terminal circuit subject to specified boundary conditions at the external terminals, and apply the known methods of exception of the internal nodes of the electric circuit.

Just like in the case of the FDEM, the TDHom GFDE is built by uniting the four ODHom GFDE that represent the facets of the FinE (Fig. 12.3). The TDHom GFDE will be described with a 12th-order conductivity matrix comprised of the ODHom GFDE conductivity matrices (Fig. 12.4).

Notice that every FinE should be replaced with a TDHom GFDE in order to apply the GFDE method to determine the *Y*-matrix of the *RC*-EDP based on the generalized two-dimensional *RCG*-EDP (see Fig. 12.1). A fragment of the two-dimensional generalized *RCG*-EDP equivalent circuit, in addition to their currents and voltages of their external terminals, is shown in Fig. 12.5. Their *y*-parameters are calculated under the specified boundary conditions and parameters of the environment.

In order to enhance the GFDEM capabilities, the mathematical model of the GFDE should comprise not just analytical expressions of the conductivity matrix coefficients, but also the vector of the electro-physical parameters, Ψ. The components of this vector are the following:

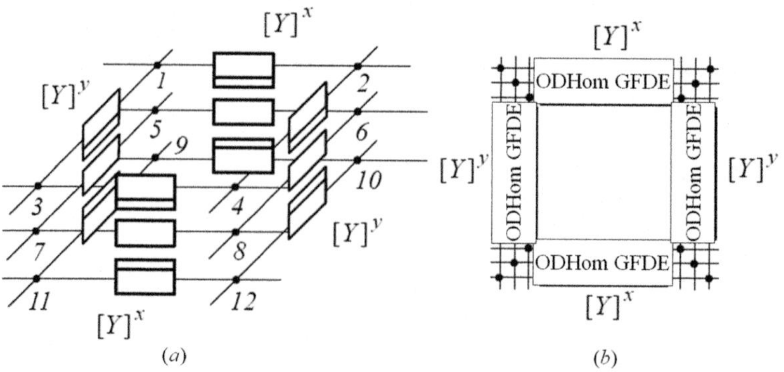

Fig. 12.3 The TDHom GFDE: (**a**) the equivalent circuit; (**b**) simplified graphical symbol

	1	2	3	4	5	...	8	9	10	11	12
1	$y_{11}^X + y_4^Y$	y_{14}^X	y_{14}^Y	0	$y_{12}^X + y_{45}^Y$	...	0	$y_{13}^X + y_{46}^Y$	y_{15}^X	y_{15}^Y	0
2	y_{14}^X	$y_{44}^X + y_{44}^Y$	0	y_{14}^Y	y_{42}^X	...	y_{24}^Y	y_{34}^X	$y_{46}^X + y_{46}^Y$	0	y_{34}^Y
3	y_{14}^Y	0	$y_{11}^X + y_{11}^Y$	y_{14}^X	y_{15}^Y	...	y_{15}^X	y_{16}^Y	0	$y_{13}^X + y_{13}^Y$	y_{16}^X
4	0	y_{14}^Y	y_{14}^X	$y_{44}^X + y_{11}^Y$	0	...	$y_{45}^X + y_{12}^Y$	0	y_{16}^Y	y_{34}^X	$y_{46}^X + y_{13}^Y$
5	$y_{12}^X + y_5^Y$	y_{24}^X	y_{15}^Y	0	$y_{22}^X + y_{55}^Y$	...	0	$y_{56}^Y + y_{23}^X$	y_{26}^X	y_{35}^Y	0
6	y_{15}^X	$y_{45}^Y + y_{45}^X$	0	y_{45}^Y	y_{25}^X	...	y_{52}^Y	y_{53}^X	$y_{56}^X + y_{56}^Y$	0	y_{35}^Y
7	y_{24}^Y	0	$y_{12}^X + y_{12}^Y$	y_{24}^X	y_{25}^Y	...	y_{25}^X	y_{26}^Y	0	$y_{23}^X + y_{23}^Y$	y_{26}^X
8	0	y_{24}^Y	y_{15}^X	$y_{45}^X + y_{12}^Y$	0	...	$y_{55}^X + y_{22}^Y$	0	y_{26}^Y	y_{35}^X	$y_{56}^X + y_{23}^Y$
9	$y_{13}^X + y_6^Y$	y_{34}^X	y_{16}^Y	0	$y_{23}^X + y_{56}^Y$	...	0	$y_{33}^X + y_{66}^Y$	y_{36}^X	y_{36}^Y	0
10	y_{16}^X	$y_{46}^X + y_{46}^Y$	y_{16}^Y 0	y_{16}^Y	y_{26}^X	...	y_{26}^Y	y_{36}^X	$y_{66}^X + y_{66}^Y$	0	y_{36}^Y
11	y_{15}^Y	0	$y_{13}^X + y_{13}^Y$	y_{34}^X	y_{35}^Y	...	y_{35}^X	y_{36}^Y	0	$y_{33}^Y + y_{33}^X$	y_{36}^X
12	0	y_{34}^Y	y_{16}^X	$y_{46}^X + y_{13}^Y$	0	...	$y_{56}^X + y_{23}^Y$	0	y_{36}^Y	y_{36}^X	$y_{66}^X + y_{33}^Y$

Fig. 12.4 The TDHom GFDE *Y*-matrix

$$\Psi_i = \left\{(\rho_{\square 1})_i, \left(g_{\text{spec1}}\right)_i, \left(C_{\text{spec1}}\right)_i, (\rho_{\square})_i, \left(C_{\text{spec2}}\right)_i, \left(g_{\text{spec2}}\right)_i, (\rho_{\square 2})_i\right\}, \qquad (12.2)$$

where $i = 1, \ldots, N$; N is the number of FinE elements; $(\rho_{\square 1})_i$, $(\rho_{\square})_i$, and $(\rho_{\square 2})_i$ are the square resistances of resistive layers $R1$, R, and $R2$ within the i-th GFDE, respectively; $(C_{\text{spec1}})_i$ and $(C_{\text{spec2}})_i$ are the specific capacitances of the dielectric layers $C1$ and $C2$ within the i-th GFDE, respectively; and $(g_{\text{spec1}})_i$ and $(g_{\text{spec2}})_i$ are the specific transversal conductivities of resistive layers $G1$ and $G2$ within the i-th GFDE, respectively.

In general, it is evident from (12.2) that the values of the specific parameters may vary for different GFDE under the influence of the synthesis algorithm using the desired performance characteristics of the ODHet *RC*-EDP. The formula also

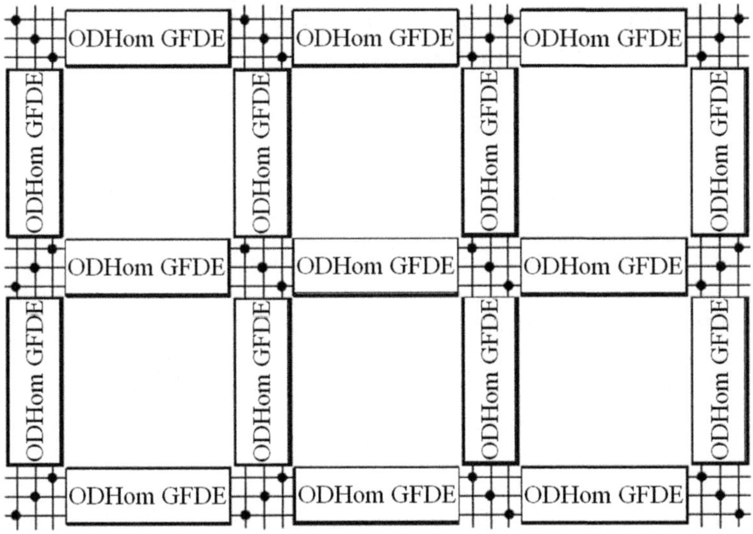

Fig. 12.5 A fragment of the two-dimensional generalized *RCG*-EDP equivalent circuit

shows that these values may vary when identifying object structure. These variations reflect the distinctive features of the simulated physical medium.

It may seem at first glance that GFDEM differs from FDEM only in the introduced concept of the GFDE. These differences, however, are much deeper. The authors suppose that the differences are the following:

1. The GFDE allows one to build various FDE with unique layer structures.
2. The analytical expressions of the GFDE *Y*-matrix help to calculate the *y*-parameters of all types of FDE without deriving these FDE *y*-parameter analytical expressions.
3. The heterogeneous *RC*-EDP with various combinations of structures can be synthesized and analyzed using the basis of the generalized design (i.e., the generalized *RCG*-EDP).
4. The heterogeneous structure of *RC*-EDP can be synthesized not just by varying the design and schematic parameters but also by applying infinite amounts of combinations of different structural options that can be put in different areas of the synthesized *RC*-EDP. This makes it possible to acquire the *RC*-EDP desired performance characteristics within a wide frequency range.
5. It is possible to identify the structure and the parameters of the physical objects (including the fractal ones) that consist of many layers and possess both structural and parametric heterogeneities.

12.2 Classification and Synthesis of FDE Using the GFDEM Elements

12.2.1 Classification of FDE Using the GFDEM Elements

It has been shown above that the two-dimensional FDE are made of one-dimensional ones. Therefore, the classification of these FDE forms derived from the GFDE will be performed only for the one-dimensional models. Figure 12.6 depicts the structure of the generalized *RCG*-EDP. This structure is divided into many FinEs. Notice that the picture also shows the GFDE graphical symbol and its simplified figure.

The GFDE can be used to generate the FDE layer structure, which either reflects the real local heterogeneity of the *RC* medium or is determined during the synthesis of the *RC*-EDP. Such conversion is possible by alternating the parameters of the layers during analysis or synthesis.

Figure 12.7 depicts the classification of the ODHom FDE with unique structure of their layers, which can be derived from the ODHom GFDE by changing the layer parameters. For simplicity, there are no signs "ODHom" in the picture.

The arrows between different types of FDE show the possible ways of transforming from one FDE into another. For example, the GFDE (model 2) can be converted into the 0-*C*1-*R*-*C*2-*R*′2-layered FDE (model 7) if resistive layers *R*1 and *G*1 are replaced with a conducting layer with $\rho_{\square} \approx 0$, i.e., $R'1 = 0$. The FDE (variant 7) can be transformed into the FDE (model 9) with fairly known layering 0-*C*1-*R*-*C*2-0 if resistive layers *R*2 and *G*2 are replaced with a conducting layer with $\rho_{\square} \approx 0$, i.e., if $R'2 = 0$ is provided. Finally, the FDE (option 9) can be converted into the FDE (model 18) with commonly known layering *R*-*C*2-0 just by neglecting the creation of the upper conducting layer.

This classification illustrates how the ODHom GFDE can be transformed and shows that it can be converted into the ODHom or ODHet (for the sake of simplicity, "ODHom" or "ODHet" notations prior to GFDE and FDE hereinafter

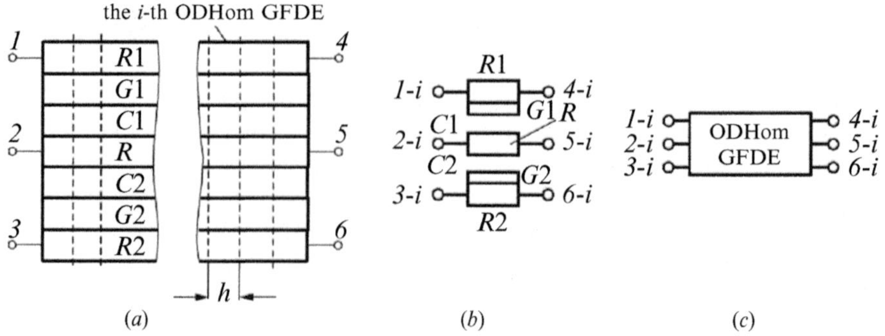

Fig. 12.6 One-dimensional *RCG*-EDP: (**a**) layer structure; (**b**) the *i*-th ODHom GFDE graphical symbol; (**c**) the *i*-th ODHom GFDE simplified graphical figure in the form of a six-terminal

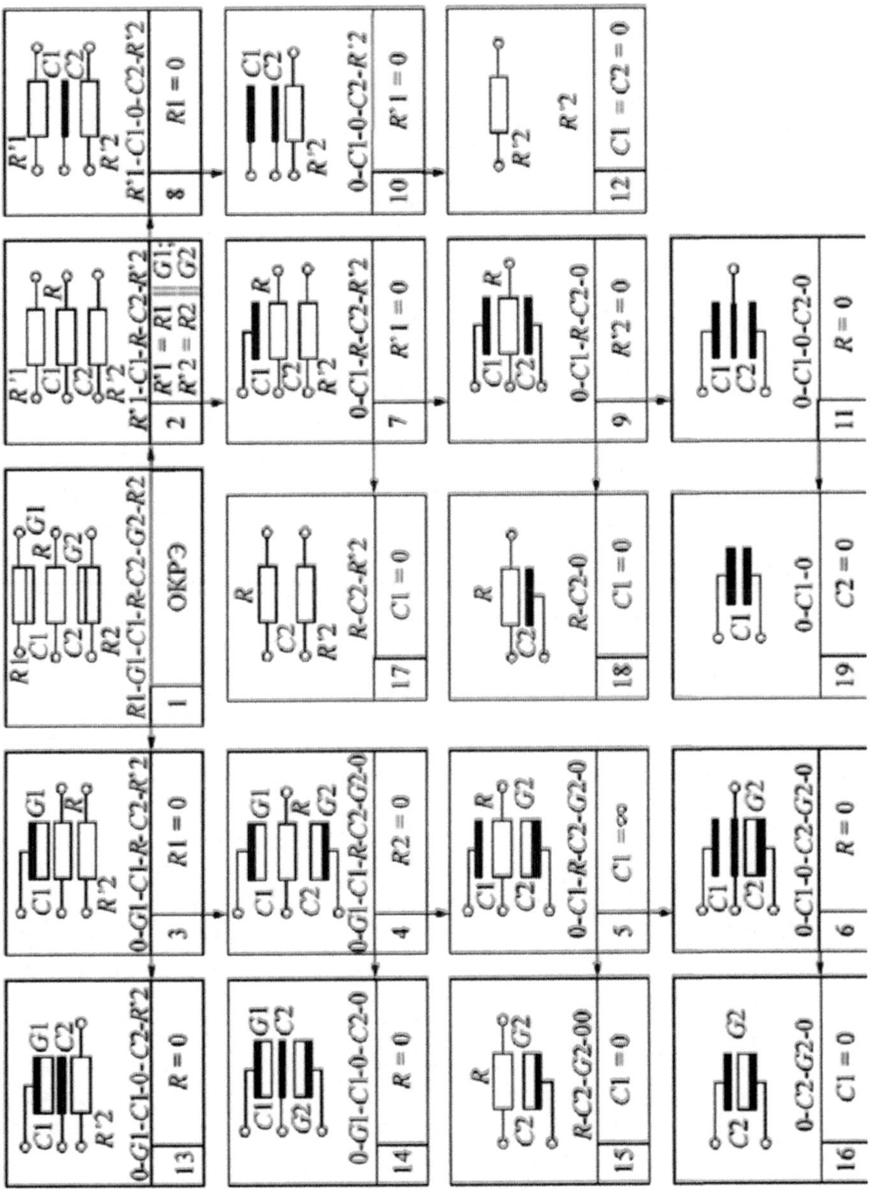

Fig. 12.7 ODHom FDE classification on the basis of the layer parameter values

are omitted) *RC*-EDP with known layer structures (variants 7, 9, 15, 17, 18, for instance) that already possess analytically derived *y*-parameters. The classification also shows that the ODHom GFDE can be converted into the ODHom or ODHet *RC*-EDP with a new structure that is never described before. Furthermore, the ODHom GFDE can be transformed into ordinary resistive *R* and capacitance *C* elements (models 11, 12, 19).

12.2.2 Generation of FDE Elements Based on Transformation of GFDEM

Conducting GFDEM to analyze and synthesize the *RC*-EDP of desired performance characteristics requires calculating the FDE *y*-parameters with any layering the GFDE can be transformed into.

However, the authors have analyzed various sources of information and have concluded that not all of the FDE that can be synthesized from the GFDE have analytical expressions of the *y*-parameters today. It is evident that the approach to calculate the ODHom *RC*-EDP *y*-parameter expressions studied in Chap. 10 can be used to determine such analytical expressions in this case.

Nevertheless, first, this way is quite laborious, and second, such analytical expressions have no practical significance if there are already GFDE *y*-parameter expressions. That is why the present section investigates the method that makes it possible to calculate the *y*-parameters for all kinds of FDE using the GFDE *y*-parameter expressions only.

In this case we do not need to additionally derive the *y*-parameter expressions and do not need to keep all expressions of the *y*-parameters for different FDE. This allows implementing one specific feature of GFDEM that differs from that of FDEM. Hence, it is possible to use the analytical expressions of the GFDE *y*-parameters to calculate the *y*-parameters of the FDE derived from this GFDE.

The previous statement has shown the possibility to transform the GFDE into 19 forms of FDE with different layering structures. That classification did not cover the structures of the FDE that mirror the FDE "base" of the middle resistive layer.

For example, the *R*-*C*2-*R*2- and the *R*1-*C*1-*R*-layered structures are the same at a first glance. However, when included into a compound *RC*-EDP (see Chap. 6), their influence on the element's electrical characteristics will sufficiently depend on *M*, *N*, and *K* parameters of a certain FDE, and on the connection circuit and their location in the general equivalent circuit of the *RC*-EDP. That is why these two structures should be considered when designing and analyzing *RC*-EDP devices. This increases the number of the FDEs produced from a single GFDE.

To automate the analysis and synthesis of the *RC*-EDP GFDEM requires figuring out an easily programmable and simple way to generate heterogeneous structure within the initially homogeneous seven-layered virtual *RC* medium. This should also provide determination of the FDE *y*-parameters simultaneously with the corresponding FinE structure variation.

It is a reminder that $R1$, R, and $R2$ resistive layers of the GFDE have line resistances Nr, r, and Kr, respectively, and the resistive layers $G1$ and $G2$ possess the corresponding line conductance g_1 and g_2, while the dielectric layers $C1$ and $C2$ have line capacitances c and Mc, respectively.

It is evident that if the line resistance of any resistive layer equals zero, then the corresponding resistive layer turns into an ideal conducting layer. If this line resistance equals infinity, the corresponding layer will not pass the electrical current and this layer is excluded from the structure. As we see in both cases, assigning the extreme values to the layer parameters produces the FDE structure with certain layering according to the classification given in Fig. 12.7.

The GFDE mathematical model, when used by the computer, can reflect these changes by assigning values close to zero (computer zero) or to infinity (computer infinity) to dimensionless coefficients N and/or K. Hence, the structure of the GFDE will change (i.e., it will turn into a FDE) along with changing the corresponding coefficients of the GFDE conductivity matrix. These coefficients reflect the specific features of this layer structure of the FDE. As for the dimension of the FDE conductivity matrix, it formally remains the same as the dimension of the initial conductivity matrix of the GFDE.

In the same manner, the GFDE layer structure can be changed by assigning extreme values to the line parameters of layers $G1$ and $G2$. If the parameters g_1 and/or g_2 are made zero (i.e., the conductivity along the direction perpendicular to the layer surface equals zero), then the transversal resistance and, correspondingly, the thickness of the $G1$ and/or $G2$ layers will be infinite. Under these conditions, the capacitance between layers $R1$ and R or R and $R2$ obviously loses its meaning. In this case, the layers $R1$ and $R2$ will turn into the resistive elements with lumped parameters.

If line parameters g_1 and/or g_2 are made infinite (i.e., the resistance of layers $G1$ and/or $G2$ along the direction perpendicular to the layer surface equals zero), then it will signify an absence of the corresponding layers $G1$ and/or $G2$ in the FDE.

There is no need to assign extreme values to the line parameters of $C1$ and $C2$ layers. In particular, if the parameter c or Mc equals zero, it means infinite thickness of the dielectric layers and the absence of capacitance coupling between layers $R1$ and R or R and $R2$, respectively. Such structure reproduces the case when line parameters g_1 and g_2 equal zero.

Hence, assigning extreme values to the parameters of layers $R1$, $R2$, $G1$, and $G2$ changes the structure formation of the GFDE, and thus the structure of the FDE. Assigning extreme values to parameters N, K, g_1, and g_2 makes it possible to change the multiterminal GFDE y-parameters in direct correspondence with the layer structure of the acquired FDE.

In order to construct the algorithms based on the performed reasoning with their further software application, the "zero" values of the GFDE parameters N, K, g_1, and g_2 will be coded with "0" digit and the "infinite" values with "1" digit. The unassigned parameters will be marked with "+" sign. In the course of analysis, the values of the latter parameters will correspond to the current values of the FDE layer line parameters. In the course of synthesis, their values can be chosen so as to optimize the *RC*-EDP performance.

Hence, all possible variants of FDE can be associated with a whole set of combinations of the four-digit *ternary code*. Here, the number of code digits equals the number of the varied parameters (N, K, g_1, and g_2) and the code base equals the number of the levels these parameters can possess ("0," "1," and "+").

As an example, Table 12.1 provides all cases of FDE layer structures at $r \neq 0$ and $r = \infty$. Another manifold of forms of the GFDE-based FDE layer structures appear at $r = 0$. One can see that some combinations of the coded variables produce the same variants of FDE structures. Obviously, using the logic algebra techniques is able to avoid such cases when the logic of determining the FDE structure is ambiguous.

Therefore, a mathematical model of the GFDE can be converted into mathematical models of FDE. This model also makes it possible to analyze the multilayered ODHet *RC*-EDP with various layers and their properties.

12.3 *Y*-Parameter Calculations of One-Dimensional Heterogeneous Structure of *RC*-EDP Devices Using GFDEM

12.3.1 Preliminaries

The universal design based on the form of a seven-layered structure along with GFDEM has promoted the appearance of a new class of the *RC*-EDP, i.e., the heterogeneous structure of *RC*-EDP. Some examples of the *RC*-EDP with heterogeneous layer structure have appeared in the *RC*-EDP design forms with incomplete layers overlapping (for example, see *RC*-EDP designs depicted in Figs. 4.12 and 4.13). However, the analysis and synthesis of the *RC*-EDP with such heterogeneities have not been supported with the corresponding mathematics.

For simplicity, let a one-dimensional *RC*-EDP be considered. Let us determine the concept of the one-dimensional heterogeneous-structure (ODSHet) *RC*-EDP. Let the initial one-dimensional virtual structure of the layers' materials with homogeneous distribution of specific parameters depicted in Fig. 12.6a be divided into n FinE elements with arbitrary length. The corresponding equivalent circuit of the initial virtual structure according to GFDEM will be as shown in Fig. 12.8.

When the structures of the layers within one or several FinE differ from the initial structure, the corresponding *RC*-EDP will be called *the one-dimensional heterogeneous structure*.

In this respect, in order to conduct GFDEM, the layer structure of every FinE should be derived from the initial seven-layered structure only by joining and removing the layers or by changing the layer materials' electro-physical performance characteristics. Then it will be possible to associate every FinE with an

Table 12.1 FDE layer structures that correspond to the GFDE layer parameter values ($r \neq 0$)

Item No.	g_1	g_2	N	K	FDE layer structure	Item No.	g_1	g_2	N	K	FDE layer structure
1	0	0	0	0	*R*	42	1	1	1	+	*R-C*2-*R*2
2	0	0	0	1	*R*	43	1	1	+	0	*R*1-*C*1-*R-C*2-0
3	0	0	0	+	*R*	44	1	1	+	1	*R*1-*C*1-*R*
4	0	0	1	0	*R*	45	1	1	+	+	*R*1-*C*1-*R-C*2-*R*2
5	0	0	1	1	*R*	46	1	+	0	0	0-*C*1-*R-C*2-*G*2-0
6	0	0	1	+	*R*	47	1	+	0	1	0-*C*1-*R*
7	0	0	+	0	*R*	48	1	+	0	+	0-*C*1-*R-C*2-*G*2-*R*2
8	0	0	+	1	*R*	49	1	+	1	0	*R-C*2-*G*2-0
9	0	0	+	+	*R*	50	1	+	1	1	*R*
10	0	1	0	0	*R-C*2-0	51	1	+	1	+	*R-C*2-*G*2-*R*2
11	0	1	0	1	*R*	52	1	+	+	0	*R*1-*C*1-*R-C*2-*G*2-0
12	0	1	0	+	*R-C*2-*R*2	53	1	+	+	1	*R*1-*C*1-*R*
13	0	1	1	0	*R-C*2-0	54	1	+	+	+	*R*1-*C*1-*R-C*2-*G*2-*R*2
14	0	1	1	1	*R*	55	+	0	0	0	0-*G*1-*C*1-*R*
15	0	1	1	+	*R-C*2-*R*2	56	+	0	0	1	0-*G*1-*C*1-*R*
16	0	1	+	0	*R-C*2-0	57	+	0	0	+	0-*G*1-*C*1-*R*
17	0	1	+	1	*R*	58	+	0	1	0	*R*
18	0	1	+	+	*R-C*2-*R*2	59	+	0	1	1	*R*
19	0	+	0	0	*R-C*2-*G*2-0	60	+	0	1	+	*R*
20	0	+	0	1	*R*	61	+	0	+	0	*R*1-*G*1-*C*1-*R*
21	0	+	0	+	*R-C*2-*G*2-*R*2	62	+	0	+	1	*R*1-*G*1-*C*1-*R*
22	0	+	1	0	*R-C*2-*G*2-0	63	+	0	+	+	*R*1-*G*1-*C*1-*R*
23	0	+	1	1	*R*	64	+	1	0	0	0-*G*1-*C*1-*R-C*2-0
24	0	+	1	+	*R-C*2-*G*2-*R*2	65	+	1	0	1	0-*G*1-*C*1-*R*
25	0	+	+	0	*R-C*2-*G*2-0	66	+	1	0	+	0-*G*1-*C*1-*R-C*2-*R*2
26	0	+	+	1	*R*	67	+	1	1	0	*R-C*2-*G*2-0
27	0	+	+	+	*R-C*2-*G*2-*R*2	68	+	1	1	1	*R*
28	1	0	0	0	0-*C*1-*R*	69	+	1	1	+	*R-C*2-*R*2
29	1	0	0	1	0-*C*1-*R*	70	+	1	+	0	*R*1-*G*1-*C*1-*R-C*2-0
30	1	0	0	+	0-*C*1-*R*	71	+	1	+	1	*R*1-*G*1-*C*1-*R*
31	1	0	1	0	*R*	72	+	1	+	+	*R*1-*G*1-*C*1-*R-C*2-*R*2
32	1	0	1	1	*R*	73	+	+	0	0	0-*G*1-*C*1-*R-C*2-*G*2-0
33	1	0	1	+	*R*	74	+	+	0	1	*R*1-*G*1-*C*1-*R*
34	1	0	+	0	*R*1-*C*1-*R*	75	+	+	0	+	0-*G*1-*C*1-*R-C*2-*G*2-*R*2
35	1	0	+	1	*R*1-*C*1-*R*	76	+	+	1	0	*R-C*2-*G*2-0
36	1	0	+	+	*R*1-*C*1-*R*	77	+	+	1	1	*R*
37	1	1	0	0	0-*C*1-*R-C*2-0	78	+	+	1	+	*R-C*2-*G*2-*R*2
38	1	1	0	1	0-*C*1-*R*	79	+	+	+	0	*R*1-*G*1-*C*1-*R-C*2-*G*2-0
39	1	1	0	+	0-*C*1-*R-C*2-*R*2	80	+	+	+	1	*R*1-*G*1-*C*1-*R*
40	1	1	1	0	*R-C*2-*G*2-0	81	+	+	+	+	*R*1-*G*1-*C*1-*R C*2-*G*2-*R*2
41	1	1	1	1	*R*						

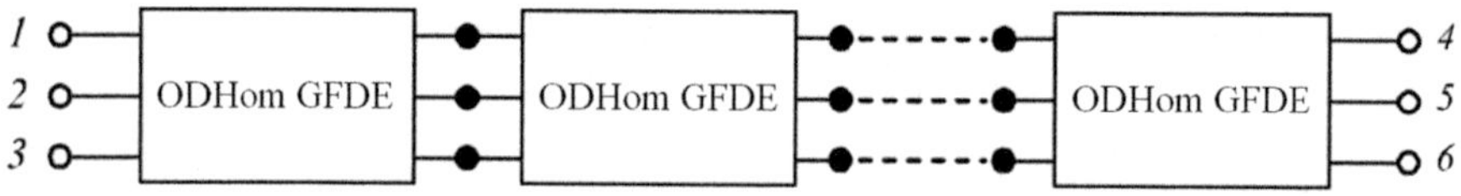

Fig. 12.8 The six-terminal *RC*-EDP equivalent circuit based on the virtual structure of the layer materials

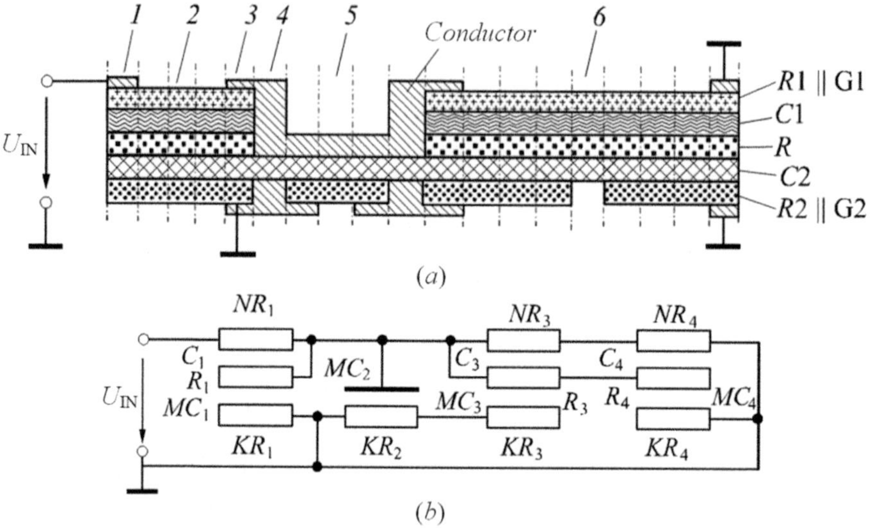

Fig. 12.9 ODSHet *RC*-EDP: (**a**) layer structure; (**b**) the equivalent circuit: (*1*) 0-*G*1-*C*1-*R*-*C*2-*R*2-layered FinE; (*2*) *R*1-*C*1-*R*-*C*2-*R*2-layered FinE; (*3*) 0-*G*1-*C*1-*R*-*C*2-*G*2-0-layered FinE; (*4*) 0-*C*2-0-layered FinE; (*5*) 0-*C*2-*G*2-*R*2-layered FinE; (*6*) *R*1-*G*1-*C*1-*R*-layered FinE

equivalent circuit in the form of an FDE made of the correspondingly transformed GFDE (see Fig. 12.7 and Table 12.1).

As an example, Fig. 12.9 depicts the ODSHet *RC*-EDP design comprising six different versions of FinE (layers *G*1 and *G*2 are not shown for clarity) and its equivalent circuit.

It is evident that the number of various combinations of heterogeneous structure can be great. Consequently, there will be a lot of the *RC*-EDP with unique performance that are made this way.

The equivalent circuit (see Fig. 12.9b) comprises only four FDE whereas the ODSHet *RC*-EDP structure contains much more finite elements. This difference is due to the following facts: firstly, several adjacent FDE with the same layer structure are united into one FDE, and secondly, some FDEs are the "technical" ones as long as proper operation of the ODSHet *RC*-EDP requires coupling of the adjacent "functional" FDE that differ in layer structures.

For example, the ODSHet *RC*-EDP with the structure given in Fig. 12.9a has areas with type *2* FDE. These areas neighbor the areas with type *3* FDE.

The implementation of the chosen circuit of electrical connection of these areas requires applying a conductor line that would couple resistive layers *R*1 and *R*2 and form a type *3* FDE. However, it cannot be done without overlapping *R*1 layer with this conductor line. As a result, the matching FDE appear. The latter are represented with *1*-, *3*-, *4*-, and 6-type FDE in Fig. 12.9.

It is obvious that the number of structures of the "technological" FDE will depend on the number of the adjacent structural heterogeneities, and the types of the "technological" FDE structures will depend on the interconnection circuits of the basic (functional) FDEs. This fact significantly complicates the process of analysis of the ODSHet *RC*-EDP.

That is why this book considers only the idealized ODSHet *RC*-EDP where FinE interconnection circuits are not limited with necessity to include the intermediate "technological" FDE connections (see Fig. 12.9b).

12.3.2 *ODSHet* RC-*EDP Mathematical Model*

It follows from the previous section that the availability of the expressions of the *Y*-matrix coefficients is not enough to calculate the *y*-parameters of the six-terminal ODSHet *RC*-EDP. Indeed, the analysis of this class of elements requires information about the extreme values of the parameters *N*, *K*, g_1, and g_2 for every FinE element, and about their current values if they have no assigned extreme values about the design parameters of the ODSHet *RC*-EDP areas (length), and about the interconnection circuit of these areas.

The areas of the ODSHet *RC*-EDP here mean the adjacent united FinEs with the same layer structures and layers' electro-physical properties. In other words, every such area can be considered as one FinE and can be simulated by one FDE—an ODHom *RC*-EDP with the corresponding layering.

The calculation of the ODSHet *RC*-EDP *y*-parameters is not the objective in itself and is needed only to analyze and synthesize the *RC*-EDP design structure. Therefore, the process of building the mathematical model that could be used further to solve these problems will be considered here.

To provide automated analysis and synthesis of the multilayered ODSHet *RC*-EDP, its mathematical model should consider:

1. All possible structural versions that have been implemented (in the course of analysis) and that are implemented (in the course of synthesis) at every ODSHet *RC*-EDP area (*the structural factors*)
2. Electro-physical properties of the layers' parameters of every area selected in the course of analysis of the ODSHet *RC*-EDP design structures (*the parametrical factors*)
3. Size of every such area (*the design factors*)
4. Connection of the layers between the adjacent areas and the forms of connecting the ODSHet *RC*-EDP with the external circuit (*the schematic factors*)

The aggregate of the listed factors that uniquely describe the ODSHet *RC*-EDP design structure can be represented in the form of manifold $\mathbf{\Psi}$ as follows:

$$\mathbf{\Psi} = \mathbf{P} \cup \mathbf{C} \cup \mathbf{S}, \tag{12.3}$$

where **P** is a set of parametric and design factors, **C** is a set of schematic factors, and **S** is a set of allowable FDE layer structures (a set of structural factors). An example of the GFDEM-based equivalent circuit of the ODSHet *RC*-EDP is given in Fig. 12.9b.

The manifold of parametric and design factors can be determined as the following union:

$$\mathbf{P} = \mathbf{N} \cup \mathbf{M} \cup \mathbf{K} \cup \mathbf{L}, \tag{12.4}$$

where the manifolds **N**, **M**, and **K** specify the parametric factors of the model, and the manifold **L** specifies the design factors. These manifolds can be determined in the following way:

$$\mathbf{N} = \{n_1, n_2, \ \ldots, n_i, \ \ldots, n_n\}, \ (n_1, \ \ldots, n_n) \in \mathbf{R}_N{}^+; \tag{12.5a}$$

$$\mathbf{M} = \{m_1, m_2, \ \ldots, m_i, \ \ldots, m_n\}, (m_1, \ \ldots, m_n) \in \mathbf{R}_M{}^+; \tag{12.5b}$$

$$\mathbf{K} = \{k_1, k_2, \ \ldots, k_i, \ \ldots, k_n\}, \ (k_1, \ \ldots, k_n) \in \mathbf{R}_K{}^+; \tag{12.5c}$$

$$\mathbf{L} = \{l_1, l_2, \ \ldots, l_i, \ \ldots, l_n\}, \ (l_1, \ \ldots, l_n) \in \mathbf{R}_L{}^+, \ \sum_{i=1}^{n} l_i = 1. \tag{12.5d}$$

Here $\mathbf{R}^+$ corresponds to the low indices for the sets of allowable positive real numbers for certain types of parameters. The allowable value ranges of parameters n_i, m_i, k_i, and l_i included into $\mathbf{R}^+$ depend on the *RC*-EDP production technology.

It is evident from Fig. 12.9b that $n_i = r_{1i}/r_i$, $m_i = c_{1i}/c_i$, $k_i = r_{2i}/r_i$, and $l_i = \Delta x_i/l$; $\sum_{i=1}^{n} l_i = 1$. Here Δx_i is the length of the i-th area, and l is the total length of the ODSHet *RC*-EDP resistive layer.

The manifold **C** of the schematic factors will be

$$\mathbf{C} = \mathbf{E} \cup \mathbf{A} \cup \mathbf{B}, \tag{12.6}$$

where **E** is the set of the allowable (from the implementation point of view) circuits of adjacent interconnected areas, **A** is the set of numbers of the grounded nodes of the adjacent areas of the interconnected circuits, and **B** is the set of the allowable forms of connection of the six-terminal of the ODSHet *RC*-EDP into the an external electrical circuit.

And finally, the manifold **S** is a set of the allowable code combinations that determine the structure of the FDE layers (see Table 12.1). The manifold of schematic factors requires additional explanation among the considered manifolds.

It is obvious that transferring the signal from the input to the output of a multiterminal circuit (in case of a two-terminal, it influences the following areas on the performance characteristics of the previous ones) requires at least connecting one resistive layer of the i-th area (i.e., one of the output terminals of the last FDE of the i-th area) with a resistive layer of the next $(i+1)$-th area (i.e., with one of the input terminals of the first FDE of the $(i+1)$-th area). As for the rest of the terminals in cases of synthesis problems, they can be "floating" (i.e., connected to nowhere) or be connected to the unused terminals of both the own and the adjacent area forming thus various connection circuits.

The synthesis problem should also take into consideration the fact that coupling areas should be provided when making the ODSHet *RC*-EDP design structure, i.e., in the course of continuous deposition of the layers. On this basis, the following *limitations* of the ODSHet *RC*-EDP mathematical model regarding the variants of the FDE terminal connection can be formulated:

1. Only terminals of the adjacent FDE can be connected (chain connection condition). Other coupling versions are just technically inefficient.
2. Short circuit of all three input and output terminals of one area at the same time is not allowed (no-short-circuit condition). This is not reasonable from the point of view of efficient use of the substrate area.
3. Simultaneous existence of three input (for the first FDE) or three output (for the n-th FDE) "floating" terminals is not allowed (circuit operation condition). Otherwise, there will be a spacing between the areas that will not let the signal to pass from the input to the output.

If the number of the ODSHet *RC*-EDP areas is specified, all possible versions of the adjacent FDE interconnection are analyzed. The corresponding manifold **E** is constituted only of those FDE that meet the listed above limitations.

In order to formalize the description of the manifold **E**, the allowable forms of connection circuits can be represented in the form of a connection matrix of dimension equals the number of the GFDE terminals. The elements of the matrix equal either "1" (when the corresponding terminals are connected) or "0" (when there is no such connection). As an example, Fig. 12.10 depicts a form of the connection circuit from manifold **E** with the corresponding connection matrix. Numbers 4 5 6 and 1 2 3 in the picture correspondingly designate the numbers of the output and the input poles of the adjacent six-pole FDE regardless the actual location of this connection in the ODSHet *RC*-EDP equivalent circuit.

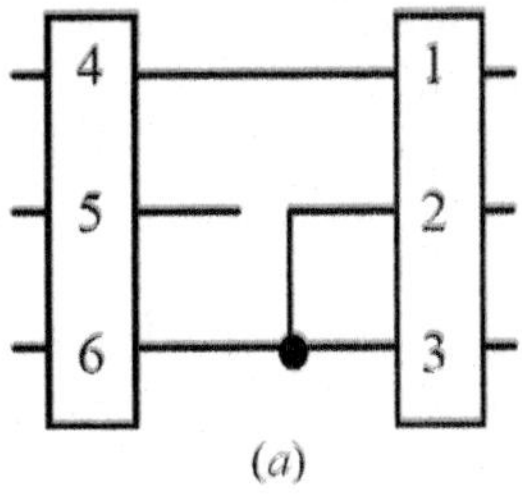

(*a*)

$$\begin{bmatrix} 1 & 0 & 0 & 1 & 0 & 0 \\ 0 & 1 & 1 & 0 & 0 & 1 \\ 0 & 1 & 1 & 0 & 0 & 1 \\ 1 & 0 & 0 & 1 & 0 & 0 \\ 0 & 0 & 0 & 0 & 1 & 0 \\ 0 & 1 & 1 & 0 & 0 & 1 \end{bmatrix}$$

(*b*)

Fig. 12.10 An example of the formalized description of two adjacent FDE connection circuit: **(a)** the connection circuit; **(b)** the connection matrix

It is convenient to store the connection matrix in the computer memory in the form of a one- or two-dimensional array. In the course of ODSHet *RC*-EDP synthesis, one or another connection matrix is supposed to be called from the memory of the synthesis algorithm. This requires numbering all possible connection circuits. When the number of circuits is substantial, it will be more convenient to use the table form of numbering. In this case, manifold **E** will comprise all combinations of two digits. The first digit is the number of the row; the second one is the number of the column of the table. Then, when performing synthesis, an arbitrary selection of any of these combinations will mean selection of a certain connection circuit.

In the course of synthesis, some nodes of the circuits from manifold **E** can be grounded; that is, it provides additional degrees of freedom for the design process. It can easily be seen that when removing the grounding of the internal node between resistive layers *KR*1 and *KR*2 of the ODSHet *RC*-EDP equivalent circuit depicted in Fig. 12.9b, then the two-port input impedance performance characteristics will change significantly.

The manifold **A** defines the grounding condition of the nodes. Since manifold **E** is represented in the matrix form, then it makes sense to do the same to manifold **A**. It can be done by describing the grounding condition of the *i*-th FDE with column matrix [$\mathbf{A}_i$] where the row numbers signify the *i*-th six-terminal FDE terminal numbers. Hence, in this case (the GFDE-form area) [$\mathbf{A}_i$] has six rows.

To describe the grounding condition it is convenient to apply binary code as follows: "1" for grounded, and "0" for not grounded. Grounding the *k*-th terminal of the *i*-th FDE is designated with "1" assigned to the *k*-th row of matrix [$\mathbf{A}_i$]. Otherwise, "0" is assigned.

When filling the matrix [$\mathbf{A}_i$], the following limitation should be taken into account: only terminals from layers *R*1 and *R*2, i.e., nodes 1, 3, 4, and 6 (see Fig. 12.10a), can be grounded if they, in their turn, have no hardline coupling with resistive layer *R*; that is, terminals 2 and 5 of the *i*-th FDE are not grounded and, consequently, "0" is always recorded into the second and the fifth rows of matrix [$\mathbf{A}_i$]. For example, assume that terminals 3 and 6 of the adjacent GFDE of the connection circuit depicted in Fig. 12.10a are connected to the common rail of the ODSHet *RC*-EDP equivalent circuit; that is, they are grounded. In this case, an element of manifold **A** that reflects this connection will be in the form of a column matrix where "1" will be recorded into the third and the sixth rows.

Since the manifold **B** comprises all allowable versions of connection of the six-terminal ODSHet *RC*-EDP into the external electric circuit, then the number of the manifold elements along with their description depends on whether the *RC*-EDP is a two-terminal of a four-terminal.

The tentative number of versions of connection circuits depending on the number of used *RC*-EDP external terminals can be estimated from the plots depicted in Fig. 5.2.

The authors proposed a GFDEM-based ODSHet *RC*-EDP mathematical model. It provides an analysis for almost all possible implementations based on the ODSHet *RC*-EDP fundamental design that may possess wide functional capabilities with substantial diversity and ranges of characteristics.

12.3.3 Y-Parameter Calculation Algorithm of Six-Terminal ODSHet RC-*EDP Devices*

The ODSHet *RC*-EDP mathematical model considered in the previous section is valid for arbitrary number of heterogeneous structure (SHet) areas ("functional" FDE). Apparently, the more SHet areas, the greater the number of model parameters that can be managed in the design phase. In addition, we can expect the ranges of the parameters that satisfy the desired characteristics using *RC*-EDP (such as the range of phase uniformity levels, fractal element frequency range, and filter MFR slope rate) to get wider.

However, the greater the number of SHet areas, the smaller their size and the greater the number of the transient areas ("technological" FDE). Despite the fact that the "technological" FDE will also have their own mathematical models, their reproduction accuracy will be poor under industrial manufacturing conditions of *RC*-EDP. The reason is their small size and the presence of transient parasitic resistances between the layers made during different fabrication stages and of parasitic capacitances appearing at inter-circuit connections. Hence, if the number of the SHet areas is substantial, then there is a real possibility that the synthesized performance characteristics will not practically be implemented.

Hence, the optimal choice of the number of SHet areas is difficult to formalize. It most likely can be done only in the course of real implementation of the ODSHet *RC*-EDP.

In order to illustrate application of GFDEM in case of ODSHet *RC*-EDP, let the number of the SHet areas be limited to four. It can be shown that the smaller number of the SHet areas does not allow implementing the operation range of the fractal element based on these areas that are wider than three frequency decades.

Despite the small number of chosen SHet areas, it is easy to calculate that the number of the unique combinations of 16 versions of FDE layer structures (16 of 19, when those with *R* and *C* structures are not considered) (see Table 12.1) taken each 4 at a time equals 1820 and of repeated combinations, i.e., as much as 3876! In addition, take into account the fact that a couple of adjacent six-terminal areas can be connected in numerous ways. Then an ODSHet *RC*-EDP comprised just of four areas can be converted into a substantial number of unique design structures only by means of structural and schematics factors.

In this case, without considering the "technological" FDE and with the aim to provide convenient construction of the analysis and synthesis algorithms of the ODSHet *RC*-EDP, its initial electrical equivalent circuit will be represented in the form of four unconnected six-terminal GFDE with the terminals numbered as depicted in Fig. 12.11.

Every GFDE in the initial circuit hereinafter will be simulated with the corresponding mathematical model with its own system of parameters. In the course of analysis, it will be defined by the user based on the known design-and-technology parameters of the ODSHet *RC*-EDP. The design procedure will be represented by the synthesis program. What is more, there are possible versions of implementation of the desired characteristics of the *RC*-EDP with lower number

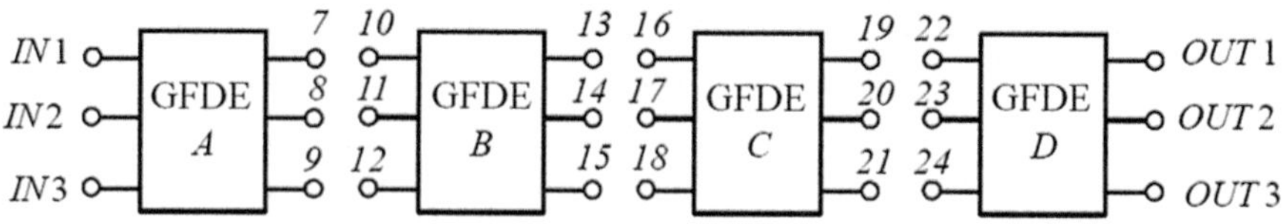

Fig. 12.11 The ODSHet *RC*-EDP initial equivalent circuit

of SHet areas (when it is possible to join areas with FDE that possess the same layer structure and to connect them in series).

The input and the output terminals marked differently from marking the internal terminals since the operations of these groups of terminals in the analysis and design are different.

Based on the initial circuit (see Fig. 12.11), let us constitute a 24×24 global *Y*-matrix out of the local matrices of the GFDE *y*-parameters. The principle of its construction (formation) is almost the same as the one that has been considered in Chap. 11. The only difference is that in the present case the numbers of the input and the output terminals are located as known in Fig. 12.12.

It allows not to interchange the numbers in order to further reduce the global *Y*-matrix dimension to 6×6, which is the dimension of the matrix that represents the ODSHet *RC*-EDP as a six-terminal and to derive its *y*-parameters.

The process of setting the parametric and design factors is defining the parameters of the manifold **P**, as given by (12.5a–12.5d). These parameters are chosen from the allowed range of values for the elements of every subset.

To make the construction algorithm more convenient, it makes sense to represent the information about parameters n_i, m_i, k_i, and l_i (manifold **P** of parametric and design parameters) of the ODSHet *RC*-EDP (according to expression (12.6)) given by the following matrix form:

$$\mathbf{P} = \begin{bmatrix} n_1 & n_2 & n_3 & n_4 \\ m_1 & m_2 & m_3 & m_4 \\ k_1 & k_2 & k_3 & k_4 \\ l_1 & l_2 & l_3 & l_4 \end{bmatrix}. \tag{12.7}$$

The process of setting the schematic factors is defining the manifold **C** from (12.6). In case of an ODSHet *RC*-EDP with four SHet areas, due to the limitations of the possible circuit connection of the adjacent GFDE, the allowable connection circuits depicted in Fig. 12.13 that determine manifold **E** can be acquired.

Apparently, the number of real (technically realizable) versions of the adjacent FDE connection circuits should be defined more accurately due to the capabilities of a certain manufacturing technology of the ODSHet *RC*-EDP.

Every connection circuit depicted in Fig. 12.13 will possess the corresponding connection matrix as formed in Fig. 12.10. If the connection circuits are defined (in the analysis phase), then the connection matrices of these circuits are used to form a global connection matrix according to the principle applied to build the global *Y*-matrix.

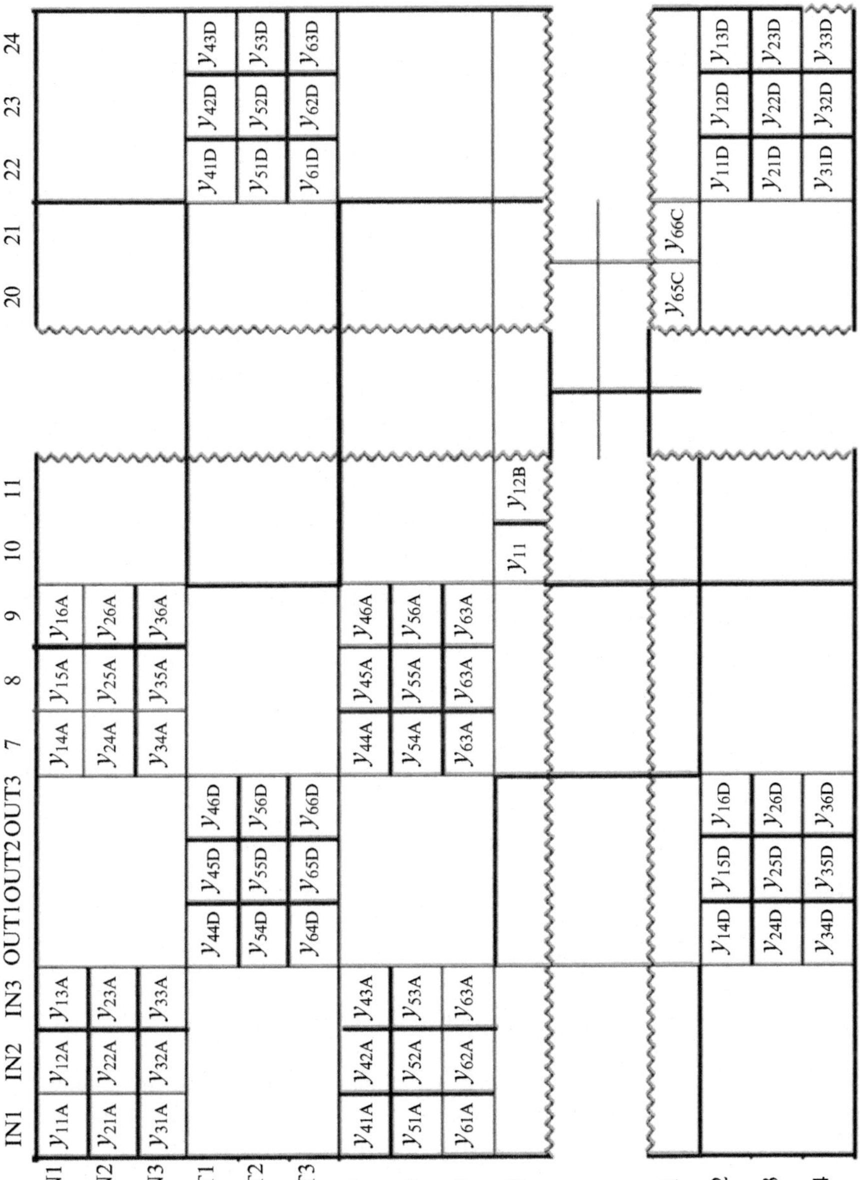

Fig. 12.12 Illustration of filling the ODSHet *RC*-EDP global *Y*-matrix that corresponds to the equivalent circuit depicted in Fig. 12.11

Fig. 12.13 The adjacent GFDE connection circuits' allowable versions according to manifold **E**

In the course of synthesis, the connection circuits are chosen arbitrarily from the circuit list. The constructed global connection matrix is further used to transform the global *Y*-matrix as follows: if the terminals are connected, the rows and columns with the corresponding numbers of terminals are united.

When the internal nodes of the ODSHet *RC*-EDP equivalent circuit are grounded (manifold **A**), the rows and the columns, where their numbers correspond to the numbers of the grounded nodes, are excluded from the global *Y*-matrix.

In order to calculate the *y*-parameters of the ODSHet *RC*-EDP indefinite matrix, the transformed global *Y*-matrix is divided into block matrixes as shown in Fig. 12.14 and its order is reduced to the external terminal number order (6×6) according to the block matrix expression as follows:

$$\mathbf{Y} = \mathbf{T} - \mathbf{U}\mathbf{W}^{-1}\mathbf{V}.$$

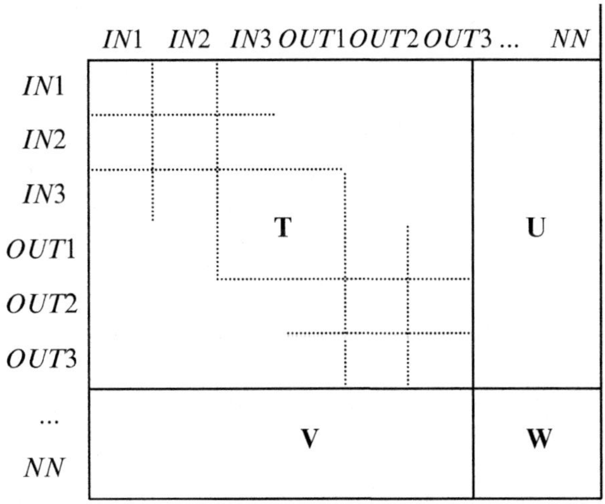

Fig. 12.14 Representation of the global Y-matrix in the form of a block matrix

Practice

1. Depict the ODSHet RC-EDP equivalent circuits whose parameters are defined in the form of the following manifolds (specific parameters of layers R *and* C are known):

(a) $\mathbf{S} = \{\text{“001+”}, \text{“000+”}\}; \begin{bmatrix} n & 0 \\ m & m \\ k & k \\ 0.4 & 0.6 \end{bmatrix}; \mathbf{E} = \{2, 3\}; A = \begin{bmatrix} 0 \\ 0 \\ 0 \\ 0 \\ 0 \\ 0 \end{bmatrix}.$

(b) $\mathbf{S} = \{\text{“000+”}, \text{“00+0”}, \text{“00++”}\};\ \mathbf{P} = \begin{bmatrix} 0 & n & n \\ m & 0 & m \\ k & k & k \\ 0.2 & 0.5 & 0.3 \end{bmatrix};\ \mathbf{E} = \{(1, 3),$

$\mathbf{A} = \begin{bmatrix} 0 & 0 \\ 0 & 0 \\ 1 & 1 \\ 0 & 0 \\ 0 & 0 \\ 1 & 1 \end{bmatrix}, (1, 1)\}.$

2. Depict some forms of “technical” FDE that correspond to these ODSHet RC-EDP.
3. Construct the global Y-matrices for these ODSHet RC-EDP.

Test Questions

1. Draw the algorithm of GFDEM.
2. Represent the concept of the electro-physical parameter vector and physical interpretation of its constituents.
3. Formulate the differences between GFDEM and FDEM.
4. Represent the principle of classification of the GFDE-based FDE.
5. What is the physical fundamental of building the FDE with a certain layer structure based on the GFDE?
6. What will be the relation between the structure of the FDE and the values of its *y*-parameters implemented by the automated analysis and synthesis system?
7. Represent the concept of one-dimensional structure-heterogeneous *RC*-EDP.
8. What are the “functional” and the “technical” FDE?
9. What is the limitation of the number of the ODSHet *RC*-EDP “functional” FDE?
10. Represent the concept of the manifold of the ODSHet *RC*-EDP model parametric and design factors.
11. Represent the concept of the manifold of the ODSHet *RC*-EDP model schematic factors.
12. Represent the limitations to the versions of connection of the ODSHet *RC*-EDP model FDE poles. Clarify the physical meaning of these limitations.
13. Draw the algorithm for calculating the six-terminal ODSHet *RC*- EDP *y*-parameters.
14. Describe the representation forms of the parametric, design, and schematic factors encapsulated in the ODSHet *RC*-EDP analysis and synthesis program.

References

1. Nigmatullin R.Sh., Belavin V.A. “Electrolyte fractional differential and integration two-pole element” Trudy (Trans.) of Kazan Aviation Institute (KAI), Issue 82, Radiotechnics and electronics, 1964. P. 58 – 65; In Russian. (Нигматуллин Р.Ш., Белавин В.А. “Электролитический дробно дифференцирующий и интегрирующий двухполюсник,” Труды КАИ, вып. 82, Радиотехника и электроника, 1964. С. 58 – 65.)
2. Oldham K. B. and Spanier J. The Fractional Calculus. – N. Y.: Academic Press, 1974.
3. Mandelbrot B.B. Les Objects Fractals: Forme, Hasard et Dimension. – Paris: Flammarion, 1975. – 187p.
4. Mandelbrot B.B. Fractals: Form, Chance and Dimension. – San-Francisco: Freeman, 1977. – 365 p.
5. Mandelbrot B.B. Fractal Geometry of Nature.– N. Y.: Freeman, 1982. – 468p.
6. Zeldovich Ya. B., Sokolov D.D. Fractals, similarity, intermediate dimension //Uspekhi Phys Sciences. 1985. – V. 146, N.3. – P.493 – 506; In Russian. (Зельдович Я.Б., Соколов Д.Д. Фракталы, подобие, промежуточная размерность // УФН. – 1985. – Т. 146, №3. – С.493 – 506).
7. Bagley R.L., Torvik P. On the appearance of the fractional derivative in the behavior of real materials // J. Appl. Mech. V. 51. 1984. P. 294 – 298.
8. Oustaloup A. Systemes Asservis Lineaires d’Ordre Fractionnaire: Theorie et Pratique. Editions Masson. – Paris. 1983. – 272 p
9. Stoynov Z.B., Graphov B.M., Savova-Stoynova B., Elkin V.V. Electrochemical Impedance. M. Nauka, 1991. – 336p.; In Russian. (Стойнов З.Б., Графов Б.М., Савова-Стойнова Б., Елкин В.В. Электрохимический импеданс. М.: Наука, 1991. – 336с.)
10. Gilmutdinov A.Kh. Mathematical model of two-dimensional homogeneous RC-elements with distributed parameters. // Vestnik of KSTU (Bulletin of the Kazan state techn. university named by A.N. Tupolev). 1997. – N.1, – P. 32–38; In Russian. (Гильмутдинов А.Х. Математическая модель двумерных однородных RC-элементов с распределенными параметрами // Вестник КГТУ им. А.Н. Туполева. – 1997. – №1, – С. 32–38.)
11. Potapov A.A. Fractal in Radiophysics and Radar: Toplogy of a Samples. – Moskow: Universitetskaya Kniga, 2005. – 848 p.; In Russian. Потапов А.А. Фракталы в радиофизике и радиолокации: Топология выборки. Изд. 2-е, перераб. и доп. – М.: Университетская книга, 2005. – 848 с.
12. Gil’mutdinov A.Kh. Resistive-capacitive elements with distributed parameters: Analysis, Sinthesis and Application. Kazan: Kazan state techn. university, 2005, 350p.; In Russian. (Гильмутдинов А.Х. Резистивно-емкостные элементы с распределенными

A.Kh. Gil’mutdinov et al., *Fractal Elements and their Applications*, Analog Circuits and Signal Processing, DOI 10.1007/978-3-319-45249-4 параметрами: анализ, синтез и применение. – Казань: Изд-во Казан. гос. техн. ун-та, 2005. – 350 с.)

13. Debnath L. Recent applications of fractional calculus to science and engineering / Intern. J. of Math. and Math. Science. v. 54 – 2003. – P. 3413 – 3442.
14. Uchaykin V.V. The method of the fractional order derivatives, "Artishok" Publ. house, Ulyanovsk. – 512 p.; In Russian. (Учайкин В.В. Метод дробных производных / В.В. Учайкин – Ульяновск: Изд-во «Артишок», 2008. – 512 с.)
15. New Trends in Nanotechnology and Fractional Calculus Applications / Editors: Baleanu D., Guvenc Ziya B., Machado J.A.T. – Springer Science+Business Media B.V. 2010. – 531 p.
16. Podlubny I. Fractional Differential Equations. Academic Press. – San Diego, 1999. – pp. 340.
17. Fractal in physics / Edited by L. Pjetronero, E. Tozatty. M.: Mir, 1988. – 672p.; In Russian. (Фракталы в физике / Пер. с англ. под ред. Я.Г. Синая и И.М. Халатникова. – М.: Мир, 1988. – 672 с.)
18. Rashid Shakirovich Nigmatullin and his scientific school / In: Fractal and fractional operators / Foreword Acad. Y.V. Gulyaev, Corresponding Member RAS S.A. Nikitov /Ed. by A.Kh. Gilmutdinov. – Kazan: Publishing house "Fan" the Academy of Sciences of RT, 2010 – 488 p.; In Russian. (Рашид Шакирович Нигматуллин и его научная школа / В книге «Фракталы и дробные операторы» / Предисловие акад. Ю.В. Гуляева, чл.-кор. РАН С.А. Никитова / Под общ. ред. А.Х. Гильмутдинова – Казань: Изд-во «Фэн» Академии наук РТ, 2010. – 488 с.)
19. Karamov F.A Superionic conductors. Heterostructures and elements of the functional electronics based on them. – M.: Nauka, 2002. – 237 p.; In Russian. (Карамов Ф.А. Суперионные проводники. Гетероструктуры и элементы функциональной электроники на их основе. – М.: Наука. – 2002. – 237 с.)
20. Ortigueira M.D. An introduction to the fractional continuous-time linear systems: The 21st century systems // IEEE Circuit and Systems Magazine, vol. 8, 2008, pp. 19–26.
21. Bohannan G. W. Analog Realization of a Fractional Control Element – Revisited, Oct. 27, 2002, http://mechatronics.ece.usu.edu/foc/cdc02tw/cdrom/aditional/FOC_Proposal_Bohannan.pdf
22. Haba T. Cisse, Ablart G., Camps T., "The Frequency Response of a Fractal Photolithographic Structure," *IEEE Transactions on Dielectrics and Electrical Insulation*, Vol 4, No. 3, June 1997. P. 321 – 327.
23. Haba T. Cisse, Ablart G., Camps T., Olivie F. "Influence of electrical parameters on the input impedance of a fractal structure realised on silicon," *Chaos, Solitons Fractals,* 24, 2005. P. 479 – 490.
24. Bohannan G.W., Hurst S.K., Spangler L., "Electrical Component with Fractional order Impedance," *United States Patent*, No.: US 2006/0267595 Al, Nov. 30, 2006.
25. Morrison R. RC Constant-Argument Driving-Point Admittances //, *IRE Transactions* on *Circuit Theory*, Vol. 6, No. 3, 1959 – Pp. 310 – 317.
26. Petraš I., Podlubny I., O'Leary P., Dorčak Ĺ., Vinagre B.M. Analogue Realization of Fractional Order Controllers. Fakulta BERG, TU Košice. 2002. – pp.86.
27. Charef A., "Analogue realisation of fractional-order integrator, differentiator and fractional PID controller," *IEE Proc. Control Theory Appl.*. V. 153. № 6. – pp. 714, 2006.
28. Charef A. and Idiou D., "Design of analog variable fractional order differentiator and integrator," *Nonlinear Dynamics*, , Vol. 69, Issue 4, pp 1577-1588, Sep. 2012.
29. Charef A. and Bensouici T., "Design of Variable Fractional-Order Integrators and Differentiators," *Signal, Image and Video Processing*, Vol. 6, Issue 4, pp 679-689 Nov. 2012,
30. Oustaloup A., Cois O., Lanusse P., Melehior P., Moreau X., Sabatier J. "The CRONE Approach: Theoretical Developments and Major Applications," Proc. 2nd IFAC Workshop on Fractional Differentiation and its Applications (Porto, Portugal, July 19-21, pp. 675 – 706, 2006.
31. Gil'mutdinov A.Kh., Potapov A.A., Ushakov P.A. Distributed resistive-capacitive elements with fractal dimension: the Past, the Present and the Future // Nelinejnyi Mir (Nonlinear

world). 2008. V. 6. N. 3. p.183 – 213.; In Russian. (Гильмутдинов А.Х., Потапов А.А., Ушаков П.А. Резистивно-емкостные элементы с распределенными параметрами и фрактальной размерностью: прошлое, настоящее и будущее // Нелинейный мир. 2008. Т. 6. № 3. С.183 – 213.)

32. Gil'mutdinov A.Kh., Potapov A.A., Ushakov P.A. // Proc. XIV Int. Scientific – Research Conf. "Radiolocation, Navigation, Communication" (Russia, Voronezh, 15 – 17 April 2008). Voronezh: NPF "Sakvoee", pp. 121 – 152, 2008.
33. Potapov A.A., Gil'mutdinov A.Kh., Ushakov P.A. "Systems Concept and Components of Fractal Radio Electronics: Part I. Development Stages and the State of the Art," *Journal of Communications Technology and Electronics*, Vol. 53, No. 9, pp. 977–1020, 2008.
34. Potapov A.A., Gil'mutdinov A.Kh., Ushakov P.A. "Systems Concept and Components of Fractal Radio Electronics: Part II. Synthesis Methods and Prospects for Application," *J. of Communications Technology and Electronics*, Vol. 53, No. 11, pp. 1271–1314, 2008.
35. Potapov, A. Gilmutdinov, P. Ushakov, Fractional-order radio-elements and radio-systems, Radiotechnika, 2009 – 200 p.; In Russian. (Потапов А.А., Гильмутдинов А.Х., Ушаков П. А. Фрактальные радиоэлементы и радиосистемы: Физический аспект. Монография / под ред. А.А. Потапова. – М.: Радиотехника, 2009. – 200 с.)
36. Fractal and fractional operators / Foreword Acad. Y. Gulyaev, Corresponding Member RAS S. A. Nikitov /Ed. by A.Kh. Gilmutdinov. – Kazan: Publishing house "Fan" the Academy of Sciences of RT, 2010 – 488 p.; In Russian. (Фракталы и дробные операторы / Предисловие акад. Ю.В. Гуляева, чл.-кор. РАН С.А. Никитова / Под общ. ред. А.Х. Гильмутдинова – Казань: Изд-во «Фэн» Академии наук РТ, 2010. – 488 с.)
37. R. El-Khazali, "On the biquadratic approximation of fractional-order Laplacian operators", *Analog Integrated Circuits and Signal Processing*, Vol. 82 (3), pp 503-517, 2015.
38. R. El-Khazali, "Fractional-order $PI^{\lambda}D^{\mu}$ controller design," *Computers & Mathematics with Applications,* vol. 66 (5), pp. 639-646, 2013.
39. W. Ahmad, R. El-Khazali, and A. Elwakil, "Fractional-Order Wein-Bridge Oscillator," *IET Electronic Letters,* Vol. 37, No. 18, pp. 1110-1112, 2001.
40. W.M. Ahmad and R. El-Khazali, "Fractional-order dynamical models of love," *Chaos, Solitons & Fractals* 33 (4), 1367-1375, 2007.
41. R. El-Khazali and N. Tawalbeh, "Realization of Fractional-Order Capacitors and Inductors," 5 th-IFAC Symposium on Fractional Diff. and its Applications, Nanjing, China, 19-21 May, 2012.
42. Elwakil A.S., "Fractional-order circuits and systems: An emerging interdisciplinary research area," *IEEE Circuit and Systems Magazine*, vol. 10, no. 4, 2010, pp. 40-50.
43. Gil'mutdinov A.Kh., Ushakov P.A. Fractal Elements: Tutorial / Foreword by Corresponding Member RAS S.A. Nikitov /Ed. by A.Kh. Gilmutdinov. Kazan: Kazan: Kazan State Techn. University, 2013. – 308 p.; In Russian. (Гильмутдинов А.Х., Ушаков П.А. Фрактальные элементы: Учебное пособие / Предисловие чл.-кор. РАН Никитова С.А. / Под ред. А. Х. Гильмутдинова. – Казань: Казан. гос. техн. ун-т, 2013. – 308с.)
44. Gil'mutdinov A.Kh., Porivaev N.V., Ushakov P.A. Active RC-filter with parametric RC-EDP for adaptive communication systems // Nelinejnyi Mir (Nonlinear world). 2011. T. 9. №11. P. 740 – 746.; In Russian (Гильмутдинов А.Х., Порываев Н.В., Ушаков П.А. Активный RC-фильтр на параметрическом *RC*-ЭРП для адаптивных систем связи // Нелинейный Мир. – 2011. – Т. 9. – №11. – С. 740 – 746.)
45. Gil'mutdinov A.Kh., Ushakov P.A., Gil'metdinov M.M. Fractal operators: Synthesis criteria and their implementation // Nelinejnyi Mir (Nonlinear world). 2008. V.6. Number 8. P. 452 – 463.; In Russian.. (Гильмутдинов А. Х., Ушаков П. А., Гильметдинов М. М. Дробные операторы: критерии синтеза и реализация // Нелинейный мир. 2008 – № 8. – С. 452 – 463.)
46. Gil'mutdinov A.Kh., Ushakov P.A. Modelling of fractal objects and processes by generalized finite distributed element method // Nelinejnyi Mir (Nonlinear world). 2008. V.6. N. 2, P. 114 – 120.; In Russian. (Гильмутдинов А.Х., Ушаков П.А. Моделирование фрактальных

процессов и объектов методом обобщенных конечных распределенных элементов // Нелинейный мир. – 2008. – Т.6. – №2. – С.114-120.

47. W. W. Happ, P. S. Castro, W. D. Fuller, "Synthesis of solid state distributed parameter functions," IRE International Convention Record, Part 6, pp. 262-265, 1962.
48. Gil'mutdinov A.Kh., Mokliakov V.A., Ushakov P.A. Outlook application of RC-elements with distributed parameters for analog signal processing, identification and management of fractal objects and processes // Vestnik of KSTU (Bulletin of the Kazan state techn. university named by A.N. Tupolev). № 3. 2007. P 24 – 29; In Russian. Гильмутдинов А.Х., Мокляков В.А., Ушаков П.А. Перспективы применения RC-элементов с распределенными параметрами для аналоговой обработки сигналов, идентификации и управления фрактальными объектами и процессами// Вестник КГТУ им. А.Н. Туполева. № 3. 2007. С. 24 – 29.
49. Gil'mutdinov A.Kh., Mokliakov V.A., Ushakov P.A. Resistive-capacitive elements with distributed parameters and fractal dimension: Design, analysis, synthesis and application // Nelinejnyi Mir (Nonlinear world). 2007. V.5. N. 10-11, P. 633 – 638.; In Russian. (Гильмутдинов А.Х., Мокляков В.А., Ушаков П.А. Распределенные резистивно-емкостные элементы с фрактальной размерностью: конструкции, анализ, синтез и применение // Нелинейный мир. 2007. Т.5. №10-11. С. 633 – 638).
50. Ushakov P.A., Maksimov K.O., Filippov A.V., "Research of fractal thick-film elements frequency responses," in Proc. 11th Int. Conf. Seminar on Micro/Nanotechnologies and Electron Devices, Novosibirsk, pp. 165-167, June 30 -July 4, 2010.